Cadmium and Health: A Toxicological and Epidemiological Appraisal

Volume II
Effects and Response

Editors

Lars Friberg, M.D.
Professor
Department of Environmental
Hygiene
Karolinska Institute
Stockholm, Sweden

Carl-Gustaf Elinder, M.D.
Associate Professor
Department of Occupational
Medicine
National Board of Occupational
Safety and Health
Solna, Sweden

Tord Kjellström, Ph.D. (Med. Dr.)
Senior Lecturer
Department of Community Health
and General Practice
Univesity of Auckland
Auckland, New Zealand

Gunnar F. Nordberg, M.D.
Professor
Department of Environmental
Medicine
Umeå University
Umeå, Sweden

CRC Press, Inc.
Boca Raton, Florida

Library of Congress Cataloging in Publication Data
Main entry under title:

Cadmium and health.

Includes bibliographies and index.
Contents: v. 1. Exposure, dose, and metabolism —
v. 2. Effects and response.
1. Cadmium—Toxicology. 2. Cadmium—Physiological
effect. 3. Poisons—Dose-response relationship.
I. Friberg, Lars. [DNLM: 1. Cadmium—metabolism.
2. Cadmium Poisoning. QV 209 C124]
RA1231.C3C28   1985   615.9'25662   85-5272
ISBN-0-8493-6690-9 (v.1)
ISBN-0-8493-6691-7 (v.2)

Direct all inquiries to CRC Press, Inc., 2000 Corporate Blvd., N.W., Boca Raton, Florida, 33431.

©1986 by CRC Press, Inc.

International Standard Book Number 0-8493-6690-9 (v.1)
International Standard Book Number 0-8493-6691-7 (v.2)

Library of Congress Card Number 85-5272
Printed in the United States

PREFACE

In 1971 the first edition of the monograph *Cadmium in the Environment* was published by CRC Press. A second revised and updated edition appeared in 1974. These works focused on information essential to the understanding of the potential toxic action of cadmium and the relationship between exposure and effects on human beings and animals.

Since the publication of the second edition of *Cadmium in the Environment,* a vast amount of new information has become available. Research has been stimulated by the great demand for more detailed and accurate information, which in turn is due to the concern about the health hazards for industrially exposed workers and for people living in cadmium-polluted areas. The incentive for research comes from international organizations, government departments, cadmium-using industries, the research community, public health officers, industrial physicians, and other decision-makers concerned with environmental and industrial toxicology.

The aim of *Cadmium and Health* is basically the same as that of the earlier work *Cadmium in the Environment,* i.e., to present a comprehensive and up-to-date treatise containing information relevant to an undestanding of the toxic action of cadmium and the relationship between exposures and effects. There are some passages in *Cadmium and Health* which are identical or similar to parts of *Cadmium in the Environment,* 2nd ed., but on the whole the present text is the result of complete rewriting.

In *Cadmium and Health* we have summarized many studies but do not treat all of them in detail. In order to facilitate access to some of the more significant information some studies are discussed and analyzed at length, including details of study design and results. We have also incorporated a number of references, occasionally in great detail, to studies which lack validity and for which the results and conclusions are highly questionable. The reason for this approach was to draw attention to the fact that these studies are often still cited as valid information.

The extensive data on the subject of cadmium have necessitated the division of the text into two volumes. Volume 1: *Exposure, Dose, and Metabolism,* treats analysis of cadmium, its uses and occurrence in the environment, cadmium and metallothionein, normal values in human tissues and fluids, metabolism, and presents a metabolic model for cadmium. Volume 2: *Effects and Response,* is primarily devoted to the toxicology of cadmium and includes effects on the respiratory system, kidneys, and bone as well as other toxic effects, including those from the hematopoietic and cadiovascular system, the liver, the reproductive organs, and the fetus. Carcinogenic and genetic effects are treated in a separate chapter as are the concepts of critical organ and critical concentration for cadmium. The last chapter in Volume 2 is devoted to some general conclusions and a general discussion of the toxicology of cadmium including diagnosis, treatment, and prevention. In an appendix we present a detailed discussion of the Itai-itai disease. Each chapter is followed by numerous references. In all, *Cadmium and Health* contains more than 1500 references.

As was the case with *Cadmium in the Environment* the preparation of *Cadmium and Health* is the result of teamwork on the part of the editors. All drafts of the chapters have been extensively discussed at numerous meetings and the conclusions drawn have general consensus among the editors. Nevertheless, the responsibility for the preparation of drafts of the different chapters has rested upon dividual editors, sometimes in collaboration with outside contributors. For this reason, names of authors are given for each separate chapter.

It is the editors' intention that these two closely related volumes should be read as a whole and not separately. Only then can one obtain a complete picture of the situation today. To maintain continuity the chapter numbers follow-on over both volumes and references are listed at the end of each chapter in which they are cited.

The authors are indebted to several colleagues in various countries, who critically reviewed the chapter manuscripts and provided both suggestions and corrections. However, sole responsibility for the statements which follow and for any errors which may remain rests with the authors. The names of those who reviewed the different chapters are given at the beginning of the book.

Parallel to the preparation of *Cadmium and Health* the Department of Environmental Hygiene of the Karolinski Institute, and the National Institute of Environmental Medicine in their capacity as a WHO Collaborating Centre for Environmental Health Effects have prepared a draft for a WHO Environmental Health Criteria Document on Cadmium. This draft was discussed at a WHO Working Group meeting in January 1984 with representatives from all WHO regions. To the extent that the discussion at this Working Group meeting motivated changes in *Cadmium and Health*, such changes have been introduced. Collaboration with the World Health Organization is gratefully acknowledged.

ACKNOWLEDGMENTS

The editors would like to thank Elisabeth Kessler who acted as administrative editor during the preparation of this book; Margit Dahlquist and Evi Werendel for valuable editorial assistance, for checking the references and for producing clean and correct manuscripts; Gunnel Gråbergs who drew most of the figures; Ilse Görke and Gun-Inger Loboda for compiling reference lists; and Diana Crowe for typing and checking some of the chapters.

Most of the original figures used in this volume have been slightly modified. In all cases reference is made to the original publications in the figure caption. The full sources can be found in the reference lists at the end of each chapter. The permission for the reproduction of this material is gratefully acknowledged.

THE EDITORS

Lars Friberg, M.D., is Professor and Chairman of the Department of Environmental Hygiene, Karolinska Institute and Director of the National Institute of Environmental Medicine, Stockholm, Sweden. These departments are a World Health Organization (WHO) Collaborating Centre for Environmental Health Effects.

Dr. Friberg graduated as M.D. from Karolinska Institute in 1945 and became Doctor of Medical Sciences in 1950. During 1967 he was Visiting Professor at the University of Cincinnati and during 1978 he worked at the Division of Environmental Health, WHO, Geneva.

Dr. Friberg is a member of the WHO Advisory Board on Occupational Health. In 1975 he served as Chairman of the WHO Task Group for an Environmental Health Criteria Document on Cadmium and in 1984 he served as Temporary Adviser to the WHO Working Group for the revision of the same document.

Dr. Friberg has been a member of the Board of the Permanent Commission and International Association on Occupational Health and is presently Chairman of its Scientific Committee on the Toxicology of Metals.

Since 1981 Dr. Friberg has been a member of the U.N. Joint Group of Experts on the Scientific Aspects of Marine Pollution (GESAMP) and also Chairman of its Working Group on Potentially Harmful Substances.

Dr. Friberg was awarded the prize of Jubilee by the Swedish Society of Medical Sciences for his work on cadmium. In May 1985 Dr. Friberg received the William P. Yant Award from the American Industrial Hygiene Association. He is Honorary Life Member of the New York Academy of Sciences.

Dr. Friberg has published about 200 scientific papers dealing primarily with the toxicology and epidemiology of metals and air pollution. He was Senior Editor of the monograph *Cadmium in the Environment,* 1st ed. 1971, 2nd ed. 1974, and one of the editors of the *Handbook on the Toxicology of Metals,* 1979.

Carl-Gustaf Elinder, M.D., is Associate Professor at the Department of Occupational Medicine at the National Board of Occupational Safety and Health, Solna, Sweden.

Dr. Elinder graduated as M.D. from the Karolinska Institute in 1978 and became Doctor of Medical Sciences at the Karolinska Institute in 1979.

Dr. Elinder has been involved in teaching environmental medicine and engaged in research in relation to metabolism and toxicity of metals, at the Department of Environmental Hygiene, at the Karolinska Institute since 1973. In 1983 he was appointed Associate Professor at the National Board of Occupational Safety and Health. On several occasions Dr. Elinder has acted as Temporary Adviser for the World Health Organization. Dr. Elinder is a member of an expert committee on hazards from environmental contaminants within the Swedish Society for the Conservation of Nature.

Dr. Elinder has published about 50 scientific papers, most of them dealing with metabolism and toxicity of metals, especially with regard to cadmium. He has also published review articles and criteria documents for several other metals.

Tord Kjellström, Ph.D. (Med. Dr.), M.Eng., is Senior Lecturer in occupational and environmental health at the Department of Community Health and General Practice, University of Auckland, New Zealand, since 1976.

Dr. Kjellström became Doctor of Medical Science at the Karolinska Institute, Stockholm, Sweden, in 1977 and has a Master's degree in Mechanical Engineering from the Royal Institute of Technology, Stockholm, since 1967.

Between 1970 and 1975 he was involved in environmental health research and teaching at the Karolinska Institute, Department of Environmental Hygiene. During 1978 he was a research student at Tokyo University. For several years he coordinated international cooperative cadmium research projects involving Sweden, Japan, and the U.S.

In 1975 he served as a member of the WHO Task Group for an Environmental Health Criteria Document on Cadmium and in 1984 he was appointed a member of the WHO Working Group for the revision of the same document. In 1985 he was appointed Medical Officer/Epidemiologist in the Division of Environmental Health, World Health Organization, Geneva.

Since 1979 Dr. Kjellström has been a member of the New Zealand Department of Health Advisory Committees on Occupational Threshold Limit Values. He was also Advisor to the New Zealand Federation of Labour on occupational health matters.

Dr. Kjellström has published about 100 scientific papers dealing mainly with cadmium toxicology and epidemiology. He was co-author of the monograph *Cadmium in the Environment,* 2nd ed., CRC Press, 1974.

Gunnar F. Nordberg, M.D., is Professor and Chairman of the Department of Environmental Medicine, Umeå University, Umeå, Sweden.

Dr. Nordberg graduated from the Karolinska Institute, Stockholm, Sweden, as Doctor of Medical Sciences in 1972 and as M.D. in 1977. He was Professor and Chairman of the Department of Environmental Medicine and Community Health, Odense University, Odense, Denmark in 1977 to 1979. Since 1979 he has been Professor and Chairman of the Department of Environmental Medicine at the University of Umeå.

During 1974 and 1975 Dr. Nordberg spent a year as a Visiting Professor at the Department of Pathology of the University of North Carolina, Chapel Hill, the National Environmental Research Center, and the National Institute of Environmental Health Sciences at Research Triangle Park, North Carolina. In 1978 to 1979 Dr. Nordberg worked as a consultant at the Division of Environmental Health, World Health Organization, Geneva.

Dr. Nordberg has also been engaged as a Temporary Adviser and Consultant for the World Health Organization, the CEC and NATO Scientific Committees, the Swedish National Environment Protection Board, and the National Board of Social Welfare.

In 1975 as well as 1984, Dr. Nordberg served as a member of the WHO Task Group for an Environmental Health Criteria Document on Cadmium representing the Permanent Commission and International Association on Occupational Health. He is also Secretary of the Scientific Committee on the Toxicology of Metals under the Permanent Commission and International Association on Occupational Health and in this capacity he has edited and co-edited a number of documents on metal toxicology (e.g., *Effects and Dose-Response Relationships of Toxic Metals,* Elsevier, Amsterdam, 1976 and *Reproductive and Developmental Toxicity of Metals,* Plenum Press, New York, 1983.

Dr. Nordberg has published over 100 scientific papers dealing mainly with the toxicology of metals. He was co-author of the monograph *Cadmium in the Environment,* 1st ed. 1971, 2nd ed. 1974, and co-editor of the *Handbook on the Toxicology of Metals,* 1979.

CONTRIBUTORS

Birger Lind
Chemical Engineer
National Institute of Environmental
 Medicine
Stockholm, Sweden

Monica Nordberg, Ph.D.
Associate Professor
Department of Environmental Hygiene
Karolinska Institute
Stockholm, Sweden

REVIEWERS

Volume I

Chapter	Name
1. Introduction	
2. Analyses	**Markus Stoeppler** Institute für Chemie der Kernforschungsanlage Jülich GmbH Jülich, West Germany
3. Uses and occurrence in the environment	**Arne Andersson** Department of Soil Sciences Swedish University of Agricultural Sciences Uppsala, Sweden
	Albert L. Page Department of Soil and Environmental Sciences University of California Riverside, California
	Markus Stoeppler Institut für Chemie der Kernforschungsanlage Jülich Gmbh Jülich, West Germany
4. Metallothionein	**M. George Cherian** Department of Pathology Health Sciences Centre University of Western Ontario London, Ontario, Canada
	Chiharu Tohyama National Institute for Environmental Studies Japan Environment Agency Tsukuba, Ibaraki, Japan

5. Norman values for cadmium in human tissues, blood, and urine in different countries

Reinier L. Zielhuis
Faculteit der Geneeskunde
Coronel Laboratorium
Universiteit van Amsterdam
Amsterdam, The Netherlands

6. Kinetics and metabolism

Thomas W. Clarkson
Division of Toxicology
University of Rochester
School of Medicine
Rochester, New York

Bruce A. Fowler
Laboratory of Pharmacology
National Institutes of Health
National Institute of Environmental
 Health Sciences
Research Triangle Park, North Carolina

7. Kinetic model of cadmium metabolism

Thomas W. Clarkson
Division of Toxicology
University of Rochester
School of Medicine
Rochester, New York

Bruce A. Fowler
Laboratory of Pharmacology
National Institutes of Health
National Institute of Environmental
 Health Sciences
Research Triangle Park, North Carolina

Volume II

8. Respiratory effects

Robert Lauwerys
Faculté de Médecine
Université Catholique de Louvain
Brussels, Belgium

David H. Wegman
Division of Environmental and Occupa-
 tional Health Sciences
School of Public Health Center for
 Health Sciences
University of California
Los Angeles, California

9. Renal effects

Ernest C. Foulkes
Institute of Environmental Health
Kettering Laboratory
Cincinnati, Ohio

Hiroshi Saito
Department of Hygiene and Preventive
  Medicine
Nagasaki University
School of Medicine
Nagasaki City, Japan

Jaroslav Vostal
Biomedical Science Department
General Motors Research Laboratories
Warren, Michigan

Sven Erik Larsson
Department of Orthopedic Surgery
University of Umeå
Umeå, Seden

Seiyo Sano
Department of Public Health
Faculty of Medicine
Kyoto University
Kyoto, Japan

Bruce A. Fowler
Laboratory of Pharmacology
National Institutes of Health
National Institute of Environmental
  Health Sciences
Research Triangle Park, North Carolina

George Kazantzis
TUC Centenary Institute of Occupational
  Health
London School of Hygiene and Tropical
  Medicine
London, England

Robert Lauwerys
Faculté de Médecine
Université Catholique de Louvain
Brussels, Belgium

TABLE OF CONTENTS

Volume I

Volume II

Chapter 8

RESPIRATORY EFFECTS

Carl-Gustaf Elinder

TABLE OF CONTENTS

I. INTRODUCTION

Inhalation of cadmium compounds can give rise to both acute and chronic effects in the respiratory system. The severe acute effects from inhalation of cadmium fumes, mainly cadmium oxide, are well established and have been known for a long time.[15] However, severe acute intoxication, sometimes with fatal outcome, still occurs.[64,76] Acute effects from inhalation are most likely to occur when workers not normally exposed to cadmium, solder or weld cadmium-containing materials.[51]

That long-term exposure to cadmium may cause chronic respiratory effects was recognized in the early 1950s[28-30] and has subsequently been confirmed in several studies.[1,48,59,60,86a] The existence of chronic respiratory effects as a result of cadmium exposure has, however, been questioned by others.[74,92]

In this review, special attention will be given to the effects of long-term exposure. As acute effects still constitute a hazard, such effects will also be discussed as well as the exposure levels associated with them.

II. ACUTE EFFECTS AND DOSE-RESPONSE RELATIONSHIPS

A. Animals

One of the first observations on the acute toxic effects of cadmium in the lung was published in 1932.[85] Since that time several reports, using more sophisticated methods for examining the lung, have been published, most of them verifying the deleterious effects on the lung.[6,36,75,91,99] Table 1 presents data on the acute toxicity of cadmium oxide fumes for various animal species. Exposure periods ranged from 10 to 30 min. The LD_{50} (cumulative mortality) up to 7 and 28 days, varied between 500 and 15,000 min × mg cadmium oxide per cubic meter, depending on the species of animal used. Harrison et al.[38] reported a 90% mortality among dogs exposed to 320 mg Cd per cubic meter in the form of cadmium chloride over a 30-min period (= 9600 min × mg/m^3). Snider et al.[91] mentioned that 17 out of 18 rats exposed to about 65 mg Cd per cubic meter in the form of chloride for 1 hr died within 3 days (= 3900 min × mg/m^3). Hadley et al.[36] exposed 61 rats to cadmium oxide at a concentration of 60 mg/m^3 for 30 min. Twenty-seven of the exposed rats died from acute pulmonary edema within 3 days of exposure (= 1800 min × mg/m^3). The average cadmium concentration in lungs from rats that had died was 26.9 mg/kg.

The acute toxicity of inhaled dust contained a mixture of cadmium and other compounds probably less than that of the pure oxide or chloride. Friberg[30] determined LD_{50} for seven groups of eight rabbits for cadmium-iron oxide dust (proportion Cd to Fe, 3:1; observation period, 14 days). On an average, about 95% of the particles were smaller than 5 μm and about 55% smaller than 1 μm (coniometer method). The rabbits were exposed for 4 hr and LD_{50}, calculated as the product of concentration and exposure time, was about 11,000 min × mg/m^3. Taking only the cadmium oxide part of the dust into consideration, the LD_{50} for cadmium oxide dust would be about 8000 min × mg/m^3, i.e., about three to four times the values for cadmium oxide fumes (for rabbits, 2500).

The lowest exposure levels of cadmium chloride and cadmium oxide sufficient to produce a significant increase in lung weight of animals after 15 to 120 min of exposure, indicating edema, were in the order of 5 to 10 mg/m^3 (Table 2). For example, Bus et al.,[16] 2 hr after a 1-hr exposure period to 6.5 mg Cd per cubic meter in the form of cadmium chloride, observed almost a doubling of the lung weight compared to that of controls.

Morphological changes in the lungs of rats that occurred 6 hr to 10 days after a

Table 1
LD$_{50}$ BY INHALATION OF ARCED CADMIUM OXIDE FUMES FOR VARIOUS SPECIES[7]

Species	LD$_{50}$ (min × mg/m³)	Remarks
Rats	500	160 animals used in groups of 10—25
Mice	Less than 700; probably about the same as for rats	Approximate only
Rabbits	2,500	Approximate only
Guinea pigs	3,500	Approximate only
Dogs	4,000	Approximate only
Monkeys	15,000	Approximate only

Note: Time exposed: 10 to 30 min.

Table 2
EXPERIMENTAL STUDIES ON ANIMALS WHERE A SIGNIFICANT INCREASE IN LUNG WEIGHT HAS BEEN OBSERVED SHORTLY AFTER EXPOSURE

Animal	Compound	Dose (mg Cd/m³)	Time of exposure (min)	Ref.
Rats	CdCl$_2$	6.5	120	40
	CdCl$_2$	6.5	120	24
	CdO	10	15	14
	CdO	6.5	60	16
Rabbits	CdO	6.4—22.4	15	33
Rats	CdO	1.5—8.6	30	11

single 2-hr exposure period to 10 mg CdCl$_2$ per cubic meter (6.5 mg Cd) have been described in detail by Strauss et al.[94] Initially, cytoplasmatic swelling and edema of type I cells were seen, occasionally with loss of the surface membrane. Two days later type II cells increased in number and 10 days later presumably replaced the damaged type I cells. Asvadi and Hayes[3] examined pulmonary lavage fluid in rats after using a similar exposure pattern. Their results showed a peak in the number of polymorpho-nuclear leukocytes 3 days after exposure, whereas macrophages dominated 4 days after exposure. Likewise Yamada et al.[107] noticed a dramatic increase in the number of al-veolar neurophilic leukocytes 6 to 48 hr after intrabronchial installation of 1 mg CdCl$_2$ into lungs of dogs.

In addition to morphological changes and increased lung weight, various types of biochemical effects have also been observed in animals shortly after exposure. Hayes et al.[40] found an increased total lipid content in the lung and increased activities of lactate dehydrogenase and glucose-6-phosphatase dehydrogenase the first 4 days after exposure. Similar findings have also been reported by Bus et al.[16] and Henderson et al.[42] Henderson et al.[42] also noticed increased activity of alkaline phosphatase after the exposure of Syrian hamsters to 16 mg CdCl$_2$ per cubic meter (10 mg Cd) for 30 min. Chichester et al.,[20] using rats, found a marked increase in the activity of two enzymes important for the production of collagen, lysyl-oxidase, and prolylhydroxylase. The changes occurred 4 days after a 2-hr exposure period using an aerosol containing 0.1% cadmium chloride (corresponding to 6.5 mg Cd per cubic meter). An increased content of collagen was also found and the authors suggest that these two enzymes are involved in the fibrinogenic effects seen in certain experiments following long-term cadmium exposure (see Section III.A).

It has also been shown that certain pulmonary enzymes are inhibited by cadmium. Boisset and Boudene[11] reported a significant decrease in benzo(a)pyrene hydroxylase and ethoxycoumarin deethylase in rabbits exposed to cadmium oxide fumes at concentrations exceeding 4.5 mg Cd per cubic meter for 30 min. In vitro studies on tracheal ring explant cultures have shown that cadmium chloride in concentrations exceeding 0.1 mM produces a pronounced drop in the normal ciliar activity as well as a decrease in the dehydrogenase activity and ATP content.[34]

There is a marked increase in susceptibility to bacterial infections in rats and mice shortly after exposure to cadmium oxides and cadmium chlorides.[14,35] Gardner et al.[35] exposed mice to cadmium chloride at a concentration of 0.08 to 1.6 mg $CdCl_2$ per cubic meter (0.05 to 1.0 mg Cd) for 120 min. They then exposed the mice to viable streptococci. Mortality increased from 15% in the lowest exposure group to 70% in the highest exposure group. The authors pointed out that enhanced mortality after bacterial infections occurred even at exposure levels in the order of 0.1 mg $CdCl_2$ per cubic meter, which is considerably lower than the levels which give rise to edema. In contrast to the increased susceptibility to bacterial infections, discussed above, Chaumard et al.[19] found a significantly lowered death rate when mice, after short-term exposure to cadmium oxide (9 mg Cd per cubic meter for 15 min), were challenged with influenza virus.

B. Humans

Acute inhalation of freshly generated cadmium fumes is a well-known hazard in industry, especially during the welding or cutting of materials containing cadmium. Usually there is only slight discomfort at the time of exposure and thus, lethal exposure is possible without prior warning. Typical cases of such poisonings have been described by Paterson,[75] Huck,[45] Reinl,[86] Lamy et al.,[56] Kleinfeld,[53] Blejer et al.,[10] Beton et al.,[9] Townshend,[101] and Tibbits and Milroy.[100] From a review prepared by MacFarland[65] it is apparent that more than 100 acute human intoxications with cadmium oxide fumes have taken place, and that at least 17 fatalities have occurred as a direct result of acute exposure. Initial symptoms include irritation and dryness of the nose and throat, cough, headache, dizziness, weakness, chills, fever, and chest pain. The initial symptoms are similar to metal fume fever, a benign condition which can be generated by exposure to, e.g., zinc fumes.[93] Severe pulmonary edema and/or chemical pneumonitis develop later, not infrequently leading to fatalities several days after exposure. The clinical signs and findings from autopsies of deceased subjects agree very well with what has been reported from findings in animals (Section II.A).

Taylor et al.[97] recently reported on another fatal intoxication which was considered to be caused by inhalation of cadmium subsequent to smelting lead. The authors stated (without any reference) that lead is usually contaminated with cadmium. The case was a 36-year-old man who initially suffered from vomiting and profuse watery diarrhea. About 48 hr after the exposure he developed pulmonary edema and died after 72 hr. The clinical features described[97] are not typical for acute cadmium poisoning and the levels of cadmium that were found in blood, urine, and tissues were not very excessively elevated. Therefore, in this particular case, it is not entirely clear whether cadmium was the sole cause of the severe intoxication.

Human data regarding dose-response relationships are scarce. Based on the amount of cadmium found in the lungs of two fatal human cases, described by Bulmer et al.[15] and Barrett et al.,[7] estimated the lethal dose. Estimation was based on the assumption that the percentage of retention of cadmium oxide fumes for man was the same as that for animals, i.e., 11%. The concentrations of cadmium found in the lungs of the two fatalities described by Bulmer et al.[15] were 17 and 18 mg/kg dry weight, corresponding to about 5 mg/kg wet weight. The figures were recalculated to give an estimated lethal dose of about 2500 min × mg/m³. An independent check on this estimated lethal dose was also made by the same authors.[6] They attempted to reproduce the actual exposure conditions and then measure the cadmium concentrations formed. They concluded that the lethal concentration of thermally generated cadmium oxide fumes for a man doing light work is not over 2900 min × mg/m³ and would possibly be as little as half of this value for fumes produced by arc welding. Beton et al.[9] applied the type of calculations used by Barrett et al.[7] to a fatal dose of cadmium poisoning. They reached the conclusion that with an observed concentration of cadmium oxide in the lungs of 2.5 mg/kg wet weight, the concentration multiplied by exposure time should have been around 2600 min × mg/m³ (fumes produced by arc welding). The exposure time in this case was 5 hr and the authors calculated the average exposure to have been around 8.6 mg/m³ for cadmium oxide fumes. For 8 hr of exposure a lethal concentration would then be around 5 mg/m³. Obviously these estimates, as pointed out by the authors, include a number of uncertainties. For example, the retention value of 11% is in all probability considerably lower than the retention value immediately after exposure. Also the amount received may well have been greater than that necessary to cause death.

In two more recent reports,[64,76] the cadmium concentrations in lungs from two fatal cases of cadmium oxide inhalation were 1.5 and 4.7 mg/kg wet weight, respectively. Since one of them is lower than the figures mentioned above, this could indicate that the minimum individual lethal dose might be lower. Therefore, 8-hr exposure to 5 mg/m³ should by no means be regarded as the lowest concentration that can give rise to a fatal poisoning. It is evident from animal experiments that exposure to lower concentrations can give rise to acute symptoms and a significant degree of lung damage. Even without applying a safety factor when extrapolating from animal data, it is reasonable to regard a cadmium oxide fume concentration of around 1 mg/m³ over an 8-hr period as immediately dangerous.

The question of whether nonfatal acute exposure can produce persistent effects in humans has been reported in one case only and, therefore, no general conclusions can yet be drawn. In 1968, Townshend[101] reported on one case of pulmonary edema caused by acute cadmium exposure after welding cadmium-silver alloy. The man was observed over a period of 17 years. In 1968 pulmonary function tests showed a gradual improvement during the first 6 months. The CO diffusing capacity was normal 4 years later, but the forced vital capacity was less than 80% of the predicted value. Seventeen years after the exposure this man developed a severe progressive pulmonary fibrosis.[102] The total lung capacity (TLC) at this point (1980) was only 54% of the predicted value and chest radiography showed patchy shadowing and widespread nodulations. Townshend, in his report,[102] concluded that the progressive pulmonary fibrosis most likely developed as a result of the cadmium inhalation 17 years earlier.

III. CHRONIC EFFECTS AND DOSE-RESPONSE RELATIONSHIPS

A. Animals

The bulk of experimental data available today shows that long-term exposure to cadmium at sublethal dose levels may give rise to microscopic signs of emphysema and chronic inflammatory changes in the respiratory system.

Friberg[30] exposed 25 rabbits for 3 hr/day, 20 days/month for 8 months to approximately 8 mg cadmium iron oxide dust per cubic meter (taken from the alkaline accumulator factory where the clinical studies were carried out; see Section III.B.1). This corresponded to approximately 5 mg Cd per cubic meter, or about 900 min × mg/m³ per exposure day. All rabbits showed signs of emphysema in addition to inflammatory changes. As the workers in the accumulator factory were also exposed to nickel-graphite dust, similar animal experiments were carried out with that dust. Emphysema and inflammatory changes were observed but to a much lesser extent, despite the fact that the actual dust concentrations were 10 to 20 times higher than those in the cadmium experiments. Vorobjeva[105] instilled cadmium oxide dust intratracheally in rats (2.5 mg/kg body weight), cadmium iron dust (3.5 mg/kg body weight) as well as nickel-graphite dust (20 mg/kg body weight). After 4 to 7 months the animals were killed. In the cadmium oxide and cadmium iron groups there were signs of interstitial pneumonia, sclerosis, and emphysema. Similar changes were not evident among the animals in the nickel-graphite group.

Prigge[82] exposed rats to much lower concentrations of cadmium oxide in air, 25 to 50 μg Cd per cubic meter, for 24 hr/day for a period of up to 90 days. Microscopic examination revealed the occurrence of emphysematous areas and cell proliferation of the bronchi and bronchioli in almost all exposed animals. Repeated exposure to cadmium chloride gives rise to essentially the same type of pulmonary effects. Snider et al.[91] observed granulation tissue and other morphological changes, mainly related to the respiratory bronchioli, and localized fibrosis, 10 days after 5 to 15 daily exposure periods to 10 mg $CdCl_2$ per cubic meter (6.5 mg Cd) for 1 hr. The authors mentioned that the lesions resembled human centrilobular emphysema. Daily intratracheal injections of cadmium acetate (1 mg/kg body weight) to squirrels resulted in emphysema after 4 to 6 weeks.[68,69]

Specific cadmium-binding proteins have been found in lungs from rabbits exposed to $CdCl_2$ aerosols at concentrations of 0.8 or 1.6 mg/m³ for 2 hr/day during a 5-day period.[80] After 1 hr exposure almost all cadmium was bound to high molecular weight proteins, but after 24 hr 85% was recovered in three different types of low molecular weight proteins. Two of these low molecular weight proteins were similar to metallothionein (Chapter 4) but a third type, which was a major cadmium-binding protein in rabbit lung, had ionic properties different from those of metallothionein. The significance of these low molecular weight cadmium-binding proteins in lung remains to be examined.

A detailed report on morphologic and biochemical effects of cadmium on the lung after long-term exposure to cadmium in the form of cadmium chloride in rabbits has been given by Johansson et al.[46,47] Eight rabbits were exposed to cadmium chloride at a concentration of 0.4 mg Cd per cubic meter 6 hr/day, 5 days/week for 4 to 6 weeks. Three days after cessation of exposure the animals were killed. Cadmium-exposed rabbits had significantly heavier lungs. Light microscopic examination revealed interstitial infiltration of white blood cells and accumulation of macrophages in the alveolar spaces. As in the acute exposures there was an increase in the number and size of type II cells. There was also an increase in the phospholipid content of the lung.[46] The number of alveolar macrophages lavaged from lungs of exposed rabbits were also markedly increased compared to controls.[47] The authors noted that pulmonary effects of cadmium chloride resembled those seen after similar exposure to nickel chloride ($NiCl_2$).

Princi and Geever[84] reported negative findings. They exposed dogs, via inhalation in special exposure chambers, to cadmium oxide dust (10 dogs, 6 hr/day, 5 days/week for 35 weeks) and cadmium sulfide dust (10 dogs, 6 hr/day, 5 days/week for about 30 weeks) without finding any respiratory changes at post-mortem compared with a control group. The concentration of cadmium in the air varied from 3 to 7 mg/m³ with an average concentration of 4 mg/m³. Of the particles, 98% were less than 3 μm in diameter. It should be mentioned that several of the animals had to be killed because of severe injuries received while fighting among themselves. Dogs that died of injuries from fighting had bronchopneumonia. It is obvious that this early study by Princi and Geever[84] had several methodologic drawbacks and that the findings presented should not be taken as an indication that cadmium does not cause lung lesions.

Peroral exposure to cadmium in rats can also give rise to effects in the lung. Miller et al.[67] gave rats water containing 17.2 mg Cd per liter for 6 to 41 weeks. Light and electron microscopic examination of the lung showed a thickened interstitium containing large bundles of collagen and elastic fibers. In rats fed a diet containing a marginal zinc content (10 mg/kg in diet) and given 17.2 mg Cd per liter in drinking water for 8 months, emphysematous changes were seen in the lung.[77] Compared to control rats fed the same diet, lungs obtained from cadmium-exposed rats exhibited increased alveolar spaces and dilated bronchioles. Another group of rats exposed to the same dose of cadmium in drinking water, but fed a diet containing more zinc, 40 mg/kg in diet, had less marked pathological changes in the lung indicating that zinc had a protective effect. Petering et al.[77] also refer to an internal report by Vinegar and Choudhury[104] who found increased static compliance in isolated lung from rats which had received 34.4 mg Cd per liter in drinking water for 520 days. In an additional unpublished study, quoted by Petering et al.,[77] it was found that the lowest dose of cadmium in drinking water that gave rise to pulmonary effects in the form of increased static lung compliance was 8.3 mg Cd per liter administered for 200 days.

It is likely that the effects of cadmium on the lung seen in rats exposed perorally are related, to some extent, to the influence of cadmium on other, essential metals such as zinc and copper. As mentioned earlier, additional zinc in the diet had a protective effect.[77] Furthermore, O'Dell et al.[73] found that severe copper deficiency in growing rats will give rise to pulmonary emphysema. It is known that excessive exposure to cadmium may precipitate a copper deficiency.[96] Indeed it has been shown that chicks fed a cadmium diet (100 mg/kg in diet) developed abnormalities in lung morphology that were very similar to those observed as a result of copper deficiency.[61]

Another mechanism which could be involved in the pathogenesis of lung lesions is a cadmium-induced decrease in the serum antitrypsin (SAT) activity.[22,23] Deficiency of this enzyme has been associated with pulmonary emphysema in humans.[26,55,63] After injections of cadmium chloride at doses in the order of 7.6 to 14 mg Cd per kilogram body weight in mice and rats, a decrease in SAT activity is seen. In one study it was possible to show that simultaneous exposure to cadmium and galactosamine resulted in a 50% reduction of plasma antitrypsin activity and pronounced morphologic changes in the lung.[23]

B. Humans

The first report on chronic respiratory effects of cadmium was published by Friberg.[28-30] He reported an increased quotient between residual volume and total lung capacity for workers exposed to cadmium dust in comparison to sawmill workers of similar ages. This residual quotient was estimated as the ratio between the residual volume (RV) and the total lung capacity (TLC) in percent (100% × RV/TLC). It is however, likely that Seiffert[87] as early as 1897 actually observed chronic cadmium poisoning and not plumbism, as he thought, among workers in a zinc smelter. Zinc ore often contains considerable amounts of cadmium and significant cadmium exposure may therefore occur in zinc smelters (Chapter 3). Seiffert reported emphysema among 83% and proteinuria among 82% of the workers examined. As discussed by Friberg,[29] there may well have been considerable exposure to cadmium as the symptoms described agree with those of chronic cadmium poisoning.

The Swedish results[30] were later confirmed by observations in Germany by Baader,[4,5] in the U.K. by, e.g., Lane and Campbell,[57] Bonnell,[12] Kazantzis,[48] Bonnell et al.,[13] and Kazantzis et al.,[49] in Belgium by Lauwerys et al.,[59] in the U.S. by Smith et al.,[90] in Poland by Kossman et al.,[54] in Australia by de Silva and Donnan[25] and in Japan by Sakurai et al.[86a]

There are, however, some studies in which the investigators reported no evidence of effects on the respiratory system from cadmium exposure, e.g., Princi,[83] Hardy and Skinner,[37] Suzuki et al.,[95] Tsuchiya,[103] L'Epée et al.,[62] Teculescu and Stanescu,[98] and most recently Stanescu et al.[92] Parkes,[74] in his textbook *Occupational Lung Disorders*, reviewed some of the studies discussed below. He relied heavily on the work by Stanescu et al.,[92] and was not convinced that cadmium exposure may produce chronic lung disease. We do not share this opinion; on the contrary, we feel that the bulk of data clearly shows that high cadmium exposure may cause chronic respiratory effects.

As the findings regarding chronic effects of the respiratory system may appear somewhat controversial and inconsistent, the main results from the studies will be presented in some detail in the following text. It should be emphasized that the diagnostic criteria used for detecting and diagnosing lung disease in cadmium workers have been variable and that different types of criteria were used by different investigators. Indeed, the clinical conditions which were regarded by the early investigators as emphysema would nowadays be classified as chronic nonspecific lung disease (CNSLD).[106] According to an expert group meeting,[106] the term chronic nonspecific lung disease can be used "to describe the group of conditions in which there is chronic sputum production and/or shortness of breath at rest and/or on exercise". Emphysema is one pathological condition which may be seen among patients suffering from CNSLD, but not always. It is difficult to diagnose emphysema correctly and it is often necessary to perform microscopic examinations of lung tissue specimens.

As will be seen in the text below, cadmium-induced chronic lung disease is better described by the term chronic nonspecific lung disease, especially when it is manifest in the form of chronic airflow obstruction ("obstructive syndrome") described by Fletcher et al.[27] and Peto.[78]

When evaluating occupational lung diseases, smoking habits should always be considered. The possibility that different smoking habits of exposed workers and controls might have invalidated comparison was not considered in the early studies. Smoking may not only be an important confounding factor if smoking habits are dissimilar in the exposed and control groups, but may also serve as an effect modifier. Cigarette tobacco contains cadmium and, furthermore, cigarettes are easily contaminated by cadmium-containing dust in the working environment[79] (Chapter 3, Section IV.B). Consequently, smokers always have an additional exposure to cadmium compared to nonsmokers in the same working environment and thus, potentially run a larger risk for negative health consequences.[39] The normal clearance of inhaled and deposited particles is impaired in smokers[18] and this may predispose smokers to respiratory lesions as a result of occupational exposure to cadmium.

The presentation below has been divided into two sections: studies showing strong evidence of respiratory effects and studies without strong evidence of respiratory effects. In the first section, examinations of workers predominantly exposed to cadmium dust or to cadmium oxide fumes are presented separately. The reports by Lauwerys and co-workers[59,60] which include data from workers with both types of exposure are presented in the section dealing with cadmium oxide fumes, as most of the workers in the study had long-term exposure to cadmium oxide fumes.

1. Studies Showing Strong Evidence of Respiratory Effects
a. Exposure to Cadmium Dust
    Friberg[28-30] examined 43 male workers exposed to cadmium oxide dust for 9 to 34

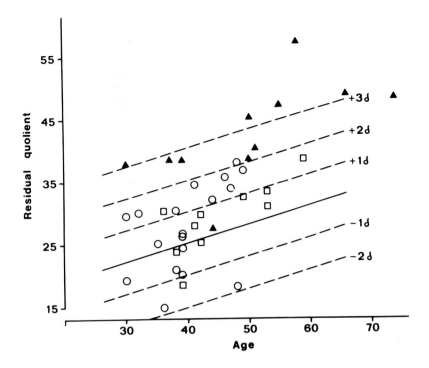

FIGURE 1.   Residual quotient (100% × RV/TLC) in cadmium workers (9 to 34 years of exposure) examined by Friberg.[30] Different symbols are used for workers having maximal physical working capacity: (▲) $\leq$ 600 kg/min; (○) = 900 kg/min; (□) $\geq$ 1200 kg/min. The lines in the figure denote the average ±1, 2, and 3 SD of a control group of 200 healthy men.

years in an alkaline accumulator factory in Sweden. Complaints of shortness of breath were common. Impaired lung function was defined and demonstrated as increased residual capacity in relation to total lung capacity. It was shown that the impairment of lung function was closely associated with poor physical working capacity evaluated by means of a standardized working test on a bicycle ergometer (Figure 1). Another, similarly exposed group of 15 workers, among whom lung function was normal, had been employed for only 1 to 4 years. All workers had been exposed to a mixture of cadmium iron oxide dust and nickel-graphite oxide dust. Quantitative data concerning the exposure were incomplete as air analyses were carried out on only one occasion and at only five locations (about 30 min at each location) in the working areas. The reported concentrations of cadmium in air varied between 3 and 15 mg/m³. Time-weighted 8-hr average values were not calculated but were probably lower.

Baader[4] in a similar German factory found that six out of eight workers exposed to cadmium for 8 to 19 years had emphysema as judged by the author from clinical and X-ray examinations. Complaints of coughing and shortness of breath were common. No quantitative analysis of the exposure was reported.

From the U.S.S.R., Vorobjeva[105] reported "diffuse pulmonary sclerosis" among female workers employed in the production of alkaline accumulators.

From another accumulator assembly plant, in Britain, Potts[81] reported that 6 men out of 70 examined workers with more than 10 years of exposure had bronchitis. In four of these cases the bronchitis was associated with emphysema. Diagnostic methods were not discussed. In a later report from the same factory, Adams et al.[1] found a forced expiratory volume (FEV_1) below the normal range in 5 male workers out of 27 examined. Furthermore, the group as a whole showed significantly ($p < 0.001$) lower than normal values. Their exposure duration was not obvious from the report but seems to have varied between 15 and 40 years. Beginning in 1957, estimations of cadmium in air were reported for this factory. Before 1957, regular analyses had not been carried out. In one area in the factory values varied between 0.5 and 5 mg Cd per cubic meter, in a second area between 0.1 and 0.5, and in a third area they were mostly around 0.1 mg Cd per cubic meter. It is not possible from the report to evaluate exposure before 1957. The extent to which respirators were worn was not stated.

Kazantzis et al.[49] reported obstructive lung changes in workers handling cadmium pigment. The men (n = 13) had been exposed to a variety of cadmium compounds of which cadmium sulfide-based pigments were by far the most abundant. In addition to cadmium sulfide, the compounds included cadmium selenosulfide, cadmium zinc sulfide, cadmium carbonate, cadmium hydroxide, and cadmium oxide dust and fumes. Out of six men engaged in the manufacture of cadmium pigments for 25 years or more, three had mild respiratory symptoms and showed slight, but definite, impairment of ventilatory function with a low $FEV_{1.0}$ (forced expiratory volume, 1 sec) percentage (56 to 61%) and a high time constant (1.08 to 1.38 sec). Chest X-rays were, however, normal. It was mentioned that a fourth man (not included in the study) had died at the age of 46 from respiratory insufficiency and right-sided heart failure due to emphysema. It was concluded that obstructive lung changes were caused by exposure to cadmium. Among the other seven workers, with less than 15 years of exposure, there was no evidence of pulmonary damage. There were no data on exposure levels in the report.

de Silva and Donnan[25] performed respiratory function tests on ten workers in an Australian plant manufacturing insoluble cadmium pigments; cadmium selenosulfide and cadmium sulfide. Concentration of respirable cadmium dust was found to range from 0.2 to 1.6 mg/m³ with a time-weighted average of 0.7 mg/m³. Exposure periods ranged from less than 1 to 19 years. Two of the workers, with 7 and 13 years of exposure, respectively, suffered from dyspnea and had respiratory function test results indicating severe airway obstruction. A third worker, exposed for 3 years, had signs of mild airway obstruction but no dyspnea. The remaining seven workers had normal lung function. Nine of the ten examined workers were smokers, including the two men with severe obstruction. However, the authors concluded that respiratory function was more severely affected than could be explained by smoking histories only.

b. Exposure to Cadmium Oxide Fumes

British researchers have published several reports on chronic cadmium poisoning in men exposed to cadmium oxide fumes in the casting of copper-cadmium alloys. Symptoms of chronic obstructive lung disease were prominent.

Lane and Campbell[57] reported two fatal cases of emphysema among 20 workers who had been manufacturing a cadmium-copper alloy. Both had been exposed to cadmium for less than 2 years. The general exposure in this smelter (called factory A below) was reported as 0.1 to 0.4 mg Cd per cubic meter. For short periods, however, the men had been exposed to higher concentrations of cadmium metal fumes (actual concentrations not given). The report includes detailed morphologic examination of lung specimens from one of the cases which revealed "extreme advanced emphysema". Similar findings were noted in three later fatalities among workers in the same factory, also exposed to high levels of cadmium fume.[89]

Bonnell[12] examined 100 exposed men and 104 controls from two factories, A and B, both producing cadmium-copper alloys. Among the exposed workers, 12 were diagnosed as having emphysema. Diagnosis was based on clinical examination, chest X-ray, and lung function tests. There was no mention of emphysema in the controls. In more recent terminology it would probably be more appropriate to use the term chronic obstructive lung disease or CNSLD. On a group basis significant differences were found between exposed workers and controls. In factory A the vital capacity and the maximum ventilatory capacity were similar in the two groups, but the mean swept fractions of 30, 50, and 70 respirations per minute were significantly lower in the exposed group indicating obstructive lung changes. A significant difference was also found between the groups for the mean time constant of the expiratory fast vital capacity curve. Also in factory B the mean time constant for the expiratory fast vital capacity curve differed significantly between the exposed group and the control group. The other test results were similar in the two groups. The results of these lung function tests are discussed in more detail by Kazantzis.[48]

Further studies were made on 37 of the exposed men from factory B.[17] Thirteen men with more than 10 years of exposure showed a significant increase in the mean value of the residual air expressed as a percentage of the total lung volume (same as "residual quotient" used by Friberg[30]) (mean: 43.9%) compared with a control group (n = 36, mean: 34.6%) and workers exposed for less than 10 years (n = 24, mean: 36.6%).

The exposure conditions in factory A and B at the time of Bonnell's, Kazantzis's, and Buxton's medical investigations have been evaluated by King[50] using stationary samplers. The cadmium concentrations were considerably lower than those in the accumulator industries referred to earlier in Section III.B.1.a. In one working area of factory A, King found average values for 8-hr working shifts over a 5-day period of 13 to 89 μg Cd per cubic meter, in one area of factory B (two different positions) over a 9-day period values of 4 to 132 μg/m^3, and in another working area (factory B, four different positions) over an 8-day period values of 1 to 270 μg/m^3. About 90% by weight of the particles had a size less than 0.5 μm. The mean exposure for the majority of the workers must have been considerably less than 0.1 mg Cd per cubic meter. Unfortunately, the study did not include any data on earlier exposure conditions. It was, however, pointed out that working conditions in the industry had improved in recent years.

In the above studies smoking was not considered and/or control groups were not examined. Lauwerys et al.[59,60] overcame several of the shortcomings of the early studies. Workers from different factories were examined, including an electronic workshop, nickel-cadmium storage battery factory, and two cadmium-producing plants. In each factory a control group was selected to match the exposed group according to sex, age, weight, height, smoking habits, and socioeconomic status.

A total of 115 male workers were divided into two groups according to duration of exposure. The first group was composed of 90 workers exposed to cadmium dust and fumes for less than 20 years, on average 7.5 years. The concentration of respirable dust ranged from 1 to 58 μg/m^3 at the time of examination. The second group consisted of 25 workers, all smokers, exposed to cadmium for more than 20 years, on average 27.5 years. The men in the second group were mainly from the two cadmium-producing plants and can thus be considered to be exposed predominantly to cadmium oxide fumes. The concentration of respirable cadmium was about 20 μg/m^3, and stated not to exceed 30 μg/m^3.[59] Additional data on workroom levels of cadmium in the same factories,[60] however, indicate that exposure levels were sometimes higher, respirable cadmium dust ranging from 3 to 67 μg/m^3 and the total dust ranging from 50 to 350 μg/m^3. As pointed out by the investigators, the major uncertainty in the study was the evaluation of the past average concentration of airborne cadmium, since the cadmium concentration in air was measured only at the time of the study. It was pointed out, however, that no important modifications had been made at the plants, therefore, the measured levels should be quite representative of past exposure. In both groups of male workers there was a significant reduction in forced vital capacity (FVC), forced expiratory volume in one second (FEV$_{1.0}$), and peak expiratory flow rate (PEFR) compared

Table 3
PULMONARY FUNCTION TESTS IN
Cd-EXPOSED (>20 YEARS) AND
CONTROL WORKERS

	Cd-exposed	Controls
n	25	25
FVC, 1	3.51[a]	3.97
FEV 1.0, 1	2.48[a]	2.81
T 0.50, sec	0.39	0.39
PEFR, 1/sec	7.22[a]	7.97
MEFR 50% VC, 1/sec	3.16	3.57
MEFR 75% VC, 1/sec	0.99	1.03

[a]   Significantly different from control value.

From Lauwerys, R., Roels, H., Buchet, J.-P., Bernard, A., and Stanescu, D., *Environ. Health Perspect.*, 28, 137—145, 1979. With permission.

to controls. Table 3 presents results from pulmonary function tests of the 25 most exposed workers compared to 25 matched controls. Other pulmonary function test results, such as maximal expiratory flow rate, were also lower in exposed workers though not significantly.

In addition one group of 31 women, exposed for a mean period of 4.4 years to an average total and respirable (aerodynamic diameter < 5 μm) airborne cadmium dust concentration of 31 and 1.4 μg Cd per cubic meter, respectively, were examined. No statistically significant effects on the respiratory system were observed.

Smith et al.[90] examined workers in a cadmium production plant in the U.S. A high exposure group was composed of 17 workers exposed to "cadmium concentrations commonly greater than 0.2 mg/m³", mainly in the form of respiratory cadmium oxide fumes, for more than 6 years (mean: 26.4 years). A low exposure group, matched with the high exposure group for age and cigarette smoking status, was selected from the same plant. A control group, composed of maintenance employees at a university, was also examined. The high exposure group had significantly reduced mean forced vital capacity (FVC), 90% of age and height predicted values compared to the internal and external control group, which had mean FVC of 100 and 97% of predicted values, respectively. The $FEV_{1.0}$, maximal mid-expiratory flow (MMEF), and total expiratory time were also reduced in the high exposure group, however, not significantly. Chest X-ray examination revealed mild to moderate pulmonary fibrosis in 5 of the 17 highly exposed workers. No such findings were observed in the controls. The authors[90] suggest that cadmium fume exposure may give rise to mild fibrotic reactions in the lung. Other substances, which might cause fibrotic changes in the lung, were not mentioned as occurring in the plant.

The lung function of seven highly cadmium-exposed workers who manufactured alloy-solders and nine, to a much lesser degree exposed, cadmium alloy workers in Japan was examined by Sakurai et al.[86a] The lung function data from the exposed workers were compared to a reference group of 122 men and adjusted for age and height. Smoking habits in the exposed groups and in the control group were similar. The alloy-solder workers had been exposed to cadmium in their breathing zone at concentrations ranging from 0.1 to 8.4 mg/m³ and were considered to have experienced an average cadmium concentration in workroom air of about 1 mg/m³ for about 5 years. That the men had been highly exposed to cadmium was confirmed by the observation of high levels of cadmium in blood and urine, and evidence of kidney damage in the form of total proteinuria and β_2-microglobulinuria. No data on atmospheric exposure were available from the other group of exposed workers, but the data on cadmium in blood and urine indicate that the exposure had been considerably less than for the men producing solders.

Forced vital capacity (FVC), forced expiratory volume in 1 sec (FEV$_{1.0}$), peak expiratory flow (PEF), maximum expiratory flow at 75, 50, and 25% of the FVC and FEV$_{1.0}$/FVC ratio were significantly reduced in the alloy-solder workers compared to the reference group. For most of the examined lung function tests there were also significant differences between the seven highly exposed workers and the nine men who had been considerably less exposed. When the low exposure group (n = 9) was compared to the reference group there was a significant drop in FVC and FEV$_{1.0}$ only.

Kossman et al.[54] reported on respiratory function of 42 cadmium-exposed workers in a nonferrous metal plant in Poland. Exposure periods ranged from 1 to 33 years with an average of 13 years. Exposure levels were not reported. Emphysema was diagnosed from chest X-ray examination in three of the workers. Reduced FEV$_{1.0}$ and reduced peak expiratory flow (PEF) were reported in 23.8 and 42.8% of the examined workers, respectively. Other signs of obstructive lung disease such as reduced mid-expiratory flow rate (V$_{50\%}$) and abnormal carbon dioxide diffusion curves were seen in 54.7 and 43.9% of the men. Smoking habits were, however, not considered and no control group was examined.

2. Studies without Strong Evidence of Respiratory Effects

As mentioned earlier, there are also studies in which respiratory effects have not been observed. These studies are presented below in chronological order. Some of the studies have methodological drawbacks and rather than providing evidence that cadmium does not cause pulmonary damage should, to our mind be regarded as "inconclusive".

Princi[83] examined 20 workers in a cadmium smelter. The average period of employment was 8 years (range 0.5 to 22). The men were exposed to cadmium in the form of dust or fumes of cadmium oxide and/or cadmium sulfide. Air measurements were carried out on 3 different occasions at 11 different working areas. Cadmium levels in air ranged from 0.04 to 1.44 mg/m³ (as cadmium). Some of the working areas had considerably higher concentrations of dust. Thus, in two places, where the men were probably exposed to cadmium sulfide, values of 19 and 31 mg Cd per cubic meter were measured. In another working area where the workers were probably exposed to cadmium oxide, a content of 17 mg Cd per cubic meter was measured. Since many of the men alternated between different working areas, an accurate estimation of individual exposure was not possible. It is stated in the report[83] that respirators were provided but worn infrequently. Clinical investigations revealed no subjective symptoms which could be related to exposure to cadmium. Pulmonary X-ray examination showed no "pneumonitis or fibrosis". Lung function tests were, however, not carried out, and no control group was examined.

Hardy and Skinner[37] examined five workers exposed to cadmium, probably in the form of cadmium oxide and dust, for periods from 4 to 8 years in the production of cadmium-faced bearings. The average exposure to cadmium was approximately 0.1 mg/m³. Chest X-rays were normal. Nevertheless, the workers complained of unspecific symptoms, including respiratory symptoms on "dump days". No lung function tests were carried out.

Suzuki et al.[95] examined workers exposed to cadmium stearate dust and lead in a vinylchloride film plant. The study covered 27 male workers in 1963 and 19 male workers (including 17 men examined in 1963), as well as 24 controls, in 1964. The average

age of the exposed group examined in 1964 was low, 23 years, SD 5.5, slightly lower than the age of the control group which was 27 years, SD 4.8. Accordingly, the average exposure period was quite short, 3.3 years, SD 1.9. In 1963 cadmium concentrations in air were measured for four different production operations, and ranged from 0.03 to 0.69 mg Cd per cubic meter. Upon examination in 1964 no increased occurrence of respiratory symptoms was found among the exposed workers in comparison to controls. Furthermore, lung function tests showed no difference between the two groups.

Tsuchiya[103] examined 13 workers (age range: 19 to 32 years) and 13 controls. The workers had been exposed to cadmium fumes while smelting alloys of silver and cadmium for periods ranging from 0.7 to 12 years. The cadmium concentration in air was reported as time-weighted averages (electrostatic precipitator) measured at nose level of the workers for 5 days, with values varying from 0.07 to 0.24 mg/m^3. The report does not present any detailed results with regard to respiratory effects, but the author states that no persistent subjective symptoms could be detected and that chest X-ray examinations revealed no abnormalities. No lung function tests were reported.

L'Epée et al.[62] examined 22 workers in an alkaline accumulator industry, 14 of whom had been employed for less than 5 years and 8 for more than 5 years. Unspecific respiratory symptoms were found among five workers. No pulmonary studies were made. The report included no quantitative data on cadmium exposure.

Teculescu and Stanescu[98] examined 11 refinery workers in Romania who were exposed for 7 to 11 years to cadmium oxide. Concentrations of cadmium in air were measured at the time of the study, and varied between 1.2 and 2.7 mg CdO per cubic meter. Eight of the exposed workers were smokers, one a former smoker, and two nonsmokers. Clinical examinations, pulmonary function tests including RV/TLC ratio $FEV_{1.0}$, and chest X-ray examinations according to the authors revealed no major abnormalities. However, when scrutinizing the data it appears that nine of the examined workers had $FEV_{1.0}$ values which were less than the predicted. In addition, ten of the examined men had a vital capacity (VC) which was less than predicted. These findings may indicate that these workers in fact had developed slight obstructive and restrictive lung changes. No control group was examined.

Stanescu et al.,[92] now in Belgium, examined 18 of the most exposed workers formerly examined by Lauwerys et al.[59] Workers who had been exposed to cadmium oxides and fumes for more than 22 years (average 32 years) were compared to 20 nonexposed workers. The control group was selected to match the exposed group with regard to age, weight, height, smoking habits, and socioeconomic status. The match regarding smoking was not perfect, since smokers in the control group smoked considerably more than those in the exposed group, 22.2 pack years among 13 smokers in the exposed group compared to 34.5 pack years among 12 smokers in the control group. The proportion of workers with grade I dyspnea without other respiratory symptoms was significantly higher among exposed workers. Of several lung function tests, including lung volume, specific airway resistance, maximal expiratory flow rates (MEFR), slope of the N_2 alveolar plateau, closing volume, elastic recoil, diffusing capacity, and closing capacity, only the closing capacity was significantly different (increased) in the exposed group compared to controls. The residual volume, the closing volume, and the slope of the alveolar N_2 plateau were slightly increased among the exposed workers, but the difference was not significant. The ratio of forced expiratory volume in 1 sec and the vital capacity ($FEV_{1.0}/VC$) was somewhat lower in the exposed group (92.6 ± 13.9% of predicted) compared to controls (97.7 ± 11.9% of predicted), but this difference was again not statistically ($p > 0.05$) significant. Chest X-ray examination gave no indication of any marked emphysema in exposed workers or controls.

3. Prognosis

Friberg[30] showed that cadmium workers with obstructive lung disease had poor physical working capacity. He also reported one fatal case in which the cause of death was pronounced emphysema, as well as another case in which pronounced pulmonary symptoms were present and hypertrophy of the right ventricular chamber was found at post-mortem. Baader[4] reported one fatal case due to severe emphysema. There are several case reports from the U.K. of fatalities due to cadmium-induced emphysema.[12,49,57,88,89]

The poor prognosis of severe chronic cadmium poisoning is further supported by data from two 5-year follow-up studies[13,32] of two previously examined occupational groups.[12,30] Exposed groups showed greater deterioration in respiratory function tests with increase in age than did the control group in the British study.[13] The results in individual cases showed deterioration in the men classified as having emphysema at the time of the original survey despite the fact that the majority of men with chronic cadmium poisoning had not been exposed to cadmium since the original examination. In the Swedish follow-up study,[32] subjective symptoms increased in several cases. Though a general tendency toward poor results in lung function tests remained, further impairment had not taken place. Most of the workers had not been exposed to significant concentrations of cadmium during the period of the follow-up study.

Mortality studies of the cadmium workers examined by Friberg in 1950 were made by Friberg and Kjellström[31] and Kjellström et al.[52] For the 43 men with the longest exposure periods a significant over-mortality was found during the first years following the examinations in 1948 and 1949. The observed cumulative mortality among the cohort was 6 up to 1952 compared with an expected mortality of 1.75, $p < 0.01$. As the cohort was observed for longer periods, the ratio between observed cumulative and expected mortality decreased. It was found that kidney and respiratory diseases were the main causes of over-mortality among the cadmium-exposed men.[52] Between 1949 and 1975, the cumulative mortality in respiratory diseases among men who worked at the battery plant for more than 5 years between 1913 and 1947 was 4 compared with an expected number of 0.4. Exposure to cadmium oxide dust was at levels in the order of 1 mg/m³.[30]

Holden[44] examined the mortality among 347 men exposed to cadmium fumes in two cadmium-copper factories (factory A and B in Section III.B.1) in the U.K., where several investigations concerning health effects of cadmium had been carried out earlier.[12,13,48,50] The cumulative mortality among the exposed men up to 1978 was compared to the expected mortality estimated from death rates in England and Wales. A highly significant over-mortality in diseases of the respiratory system was found; 36 fatalities compared to an expected number of 20.3 ($p < 0.01$).

In another cohort study on causes of death in workers in 17 cadmium-using industries (i.e., cadmium alloys, pigments, and stabilizers) from the U.K., Armstrong and Kazantzis[2] found a significant increase in deaths caused by respiratory disease, mainly chronic bronchitis. Among 199 men from different industries considered even to have been exposed to high concentrations of cadmium, there were 12 deaths in respiratory disease compared to an expected figure of 2.9.

4. Other Studies which have Related Cadmium Exposure to Effects on the Respiratory System

There is an association between hereditary alpha-1 antitrypsin deficiency and pulmonary emphysema.[26,55,58,63] Chowdhury and Louria[21] studied the effect of cadmium on this enzyme activity in vitro by adding 5 to 15 μg/mℓ to human plasma and incubating the mixture at 37°C for 1 hr. A marked decrease in enzyme activity was shown. Results from animal experiments have also been reported (see Section III.A). The in vitro studies on human plasma were, however, not confirmed by Bernard et al.[8] who also examined alpha-1 antitrypsin in vivo and were able to measure the activity of this enzyme in plasma samples obtained from cadmium-exposed workers and controls. On the other hand, Marek et al.[66] found slightly reduced alpha-1 antitrypsin levels in serum obtained from cadmium-exposed workers as compared to nonoccupationally exposed controls. However, the differences between average alpha-1 antitrypsin levels were small, about 10%, and may well be related to factors other than cadmium.

In autopsy studies it has been shown that persons with a diagnosis of emphysema and/or chronic bronchitis have increased levels of cadmium in lung, liver, and kidney.[43,70-72] Originally, it was suggested that cadmium might play a role in the development of these diseases. It has subsequently been shown that cigarette smoking in itself gives rise to cadmium exposure and subsequent accumulation in the body (Chapter 5). In view of the well-known association between smoking and chronic bronchitis, as well as emphysema,[41] it is most likely that the major causative factor for these diseases is smoking, and that cadmium retention is a confounding factor.

IV. CONCLUSIONS

Brief inhalations of high concentrations of cadmium compounds can give rise to severe, often fatal, pulmonary changes (pulmonary edema and/or chemical pneumonitis). This has been shown in human beings in regard to cadmium oxide fumes, and in animals in regard to cadmium oxide fumes, cadmium chloride aerosols, and cadmium oxide dust. Long-term pulmonary effects from cadmium inhalation or instillation have repeatedly been observed in animals. The mechanism underlying acute cadmium toxicity in the lung is not known, but it may involve damage of alveolar type I cells.

For human beings, a fatal exposure to cadmium oxide fumes and cadmium chloride aerosols is not higher (probably lower), than about 2500 min × mg/m³ (corresponding to about 5 mg/m³ for 8 hr). An exposure to about 500 min × mg/m³ (corresponding to about 1 mg/m³ over 8 hr) is considered directly dangerous. Dose-response relationships for cadmium oxide dust are scarce and questionable. In rabbits, LD_{50} was about 8000 min × mg/m³, which would correspond to about 15 mg cadmium oxide dust per cubic meter for an 8-hr exposure. If a person survives acute cadmium poisoning due to inhalation of cadmium compounds, severe lung changes may develop several years later.

It is evident that occupational exposure to cadmium compounds at relatively high concentrations for longer periods of time can give rise to chronic pulmonary disorders, characterized by obstructive changes. A significant increase in deaths caused by respiratory disease has been observed in several independent follow-up studies on workers who had been exposed to high levels of cadmium in air. As a rule these changes in human beings have taken several years to develop. Changes have, however, sometimes been observed after only a few years' exposure.

Lung lesions have been observed after exposure to cadmium oxide fumes, as well as after exposure to cadmium oxide dust and cadmium pigment dust. Dose-response relationships are not well established mainly because time-weighted averages are not usually available, or are only available for short-time periods. There is reason to believe that cadmium oxide fumes are more dangerous than cadmium oxide dust. The lowest levels of cadmium in workroom air which have been associated with respiratory effects have been in the order of 70 μg Cd per cubic meter as total dust and 20 μg/m³ as respirable dust. That respiratory effects may develop also at these relatively low levels of exposure has, however, not been confirmed. It may well be that lung lesions among cadmium-exposed workers are the result of repeated exposure to high peak values of cadmium in air, and are not related to the long-term exposure to a comparably low average level. At comparably low exposure levels smoking is an important confounding and effect-modifying factor. This has not been well examined.

REFERENCES

1. Adams, R. G., Harrison, J. F., and Scott, P., The development of cadmium-induced proteinuria, impaired renal function, and osteomalacia in alkaline battery workers, *Q. J. Med.,* 38, 425—443, 1969.
2. Armstrong, B. G. and Kazantzis, G., The mortality of cadmium workers, *Lancet,* June, 1425—1427, 1983.
3. Asvadi, S. and Hayes, J. A., Acute lung injury induced by cadmium aerosol. II. Free airway cell response during injury and repair, *Am. J. Pathol.,* 90, 89—96, 1978.
4. Baader, E. W., Chronic cadmium poisoning, *Dtsch. Med. Wochenschr.,* 76, 484—487, 1951 (in German).
5. Baader, E. W., Chronic cadmium poisoning, *Ind. Med. Surg.,* 21, 427—430, 1952.
6. Barrett, H. M. and Card, B. Y., Studies on the toxicity of inhaled cadmium. II. The acute lethal dose of cadmium oxide for man, *J. Ind. Hyg. Toxicol.,* 29, 286—293, 1947.
7. Barrett, H. M., Irwin, D. A., and Semmons, E., Studies on the toxicity of inhaled cadmium. I. The acute toxicity of cadmium oxide by inhalation, *J. Ind. Hyg. Toxicol.,* 29, 279—285, 1947.
8. Bernard, A., Roels, H., Buchet, J. P., Lauwerys, R. R., and Masson, P., α_1-antitrypsin in cadmium toxicity: an evaluation of its suggested role, *Toxicology,* 9, 249—253, 1978.
9. Beton, D. C., Andrews, G. S., Davies, H. J., Howells, L., and Smith, G. F., Acute cadmium fume poisoning, five cases with one death from renal necrosis, *Br. J. Ind. Med.,* 23, 292—301, 1966.
10. Blejer, H. P., Caplan, P. E., and Alcocer, A. E., Acute cadmium fume poisoning in welders — a fatal and a nonfatal case in California, *Calif. Med.,* 105, 290—296, 1966.
11. Boisset, M. and Boudene, C., Effect of a single exposure to cadmium oxide fumes on rat lung microsomal enzymes, *Toxicol. Appl. Pharmacol.,* 57, 335—345, 1981.
12. Bonnell, J. A., Emphysema and proteinuria in men casting copper-cadmium alloys, *Br. J. Ind. Med.,* 12, 181—197, 1955.
13. Bonnell, J. A., Kazantzis, G., and King, E., A follow-up study of men exposed to cadmium oxide fume, *Br. J. Ind. Med.,* 16, 135—145, 1959.
14. Bouley, G., Dubreuil, A., Despauz, N., and Boudene, C., Toxic effects of cadmium microparticles on the respiratory system. An experimental study on rats and mice, *Scand. J. Work Environ. Health,* 3, 116—121, 1977.
15. Bulmer, F. M. R., Rothwell, H. E., and Frankish, E. R., Industrial cadmium poisoning, *Can. Public Health J.,* 29, 19—26, 1938.
16. Bus, J. S., Vinegar, A., and Brooks, S. M., Biochemical and physiologic changes in lungs of rats exposed to a cadmium chloride aerosol, *Am. Rev. Respir. Dis.,* 118, 573—580, 1978.
17. Buxton, R. St. J., Respiratory function in men casting cadmium alloys. II. The estimation of the total lung volume, its subdivisions and the mixing coefficient, *Br. J. Ind. Med.,* 13, 36—40, 1956.
18. Camner, P. and Philipson, K., Tracheobronchial clearance in smoking-discordant twins, *Arch. Environ. Health,* 25, 60—63, 1972.
19. Chaumard, C., Quero, A. M., Bouley, G., Girard, F., Boudene, Cl., and German, A., Influence of inhaled cadmium microparticles on mouse influenza pneumonia, *Environ. Res.,* 31, 428—439, 1983.
20. Chichester, C. O., Palmer, K. C., Hayes, J. A., and Kagan, H. M., Lung lysyl oxidase and prolyl hydroxylase: increases induced by cadmium chloride inhalation and the effect of b-aminopropionitrite in rats, *Am. Rev. Respir. Dis.,* 124, 709—713, 1981.
21. Chowdhury, P. and Louria, D. B., Influence of cadmium and other trace metals on human alpha$_1$-antitrypsin: an *in vitro* study, *Science,* 191, 480—481, 1976.
22. Chowdhury, P., Louria, D. B., Chang, L. W., and Rayford, Ph. L., Cadmium-induced pulmonary injury in mouse: a relationship with serum antitrypsin activity, *Bull. Environ. Contam. Toxicol.,* 28, 446—451, 1982.
23. Chowdhury, P., Chang, L. W., Bone, R. C., and Rayford, P. L., Potentiation of Cd-induced pulmonary injury in alfa$_1$-antitrypsin suppressed rats, *Environ. Res.,* 30, 313—321, 1983.
24. Dervan, P. A. and Hayes, J. A., Peribronchiolar fibrosis following acute experimental lung damage by cadmium aerosol, *J. Pathol.,* 128, 143—148, 1979.
25. de Silva, P. E. and Donnan, M. B., Chronic cadmium poisoning in a pigment manufacturing plant, *Br. J. Ind. Med.,* 38, 76—86, 1981.
26. Eriksson, S., Studies in α_1-antitrypsin deficiency, *Acta Med. Scand.,* 177(Suppl. 432), 1—85, 1965.
27. Fletcher, C., Peto, R., Tinker, C., and Speizer, F. E., *The Natural History of Chronic Bronchitis and Emphysema,* Oxford University Press, Oxford, 1976.
28. Friberg, L., Proteinuria and kidney injury among workmen exposed to cadmium and nickel dust, *J. Ind. Hyg. Toxicol.,* 30, 32—36, 1948a.
29. Friberg, L., Proteinuria and emphysema among workers exposed to cadmium and nickel dust in a storage battery plant, *Proc. Int. Congr. Ind. Med.,* 9, 641—644, 1948b.

30. Friberg, L., Health hazards in the manufacture of alkaline accumulators with special reference to chronic cadmium poisoning. Doctoral thesis, *Acta Med. Scand.,* 138(Suppl. 240), 1—124, 1950.

31. Friberg, L. and Kjellström, T., in *Cadmium in the Environment,* 2nd ed., Friberg, L., Piscator, M., Nordberg, G. F., and Kjellström, T., Eds., CRC Press, Boca Raton, Fla., 1974, 100.

32. Friberg, L. and Nyström, Å., Aspects on the prognosis in chronic cadmium poisoning, *Lakartid-ningen,* 49, 2629—2639, 1952 (in Swedish).

33. Fukuhara, M., Bouley, G., Godin, J., Girard, F., Boisset, M., and Boudene, C., Effects of short-term inhalation of cadmium oxides on rabbit pulmonary microsomal enzymes, *Biochem. Pharmacol.,* 30, 715—720, 1981.

34. Gabridge, M. G. and Meccoli, R. A., Cytotoxicity and ciliostasis in tracheal explants exposed to cadmium salts, *Environ. Health Perspect.,* 44, 189—196, 1982.

35. Gardner, D. E., Miller, F. J., Illing, J. W., and Kirtz, J. M., Alterations in bacterial defence mechanisms of the lung induced by inhalation of cadmium, *Bull. Eur. Physiopathol. Respir.,* 13, 157—174, 1977.

36. Hadley, J. G., Conklin, A. W., and Sanders, C. L., Systemic toxicity of inhaled cadmium oxide, *Toxicol. Lett.,* 4, 107—111, 1979.

37. Hardy, H. L. and Skinner, J. B., The possibility of chronic cadmium poisoning, *J. Ind. Hyg. Toxicol.,* 29, 321—324, 1947.

38. Harrison, H. E., Bunting, H., Ordway, N., and Albrink, W. S., The effects and treatment of inhalation of cadmium chloride aerosols in the dog, *J. Ind. Hyg. Toxicol.,* 29, 302—314, 1947.

39. Hassler, E., Exposure to Cadmium and Nickel in an Alkaline Battery Factory — as Evaluated from Measurements in Air and Biological Material, Doctoral thesis, Department of Environmental Hygiene, Karolinska Institute, Stockholm, 1983.

40. Hayes, J. A., Snider, G. L., and Palmer, K. C., The evolution of biochemical damage in the rat lung after acute cadmium exposure, *Am. Rev. Respir. Dis.,* 113, 121—130, 1976.

41. Health Consequences of Smoking, A report of the Surgeon General: 1971, U.S. Department of Health, Education, and Welfare, Washington, D.C., 1971.

42. Henderson, R. F., Rebar, A. H., Pickrell, J. A., and Newton, G. J., Early damage indicators in the lung. III. Biochemical and cytological response of the lung to inhaled metal salts, *Toxicol. Appl. Pharmacol.,* 50, 123—136, 1979.

43. Hirst, R. N., Jr., Perry, H. M., Jr., Cruz, M. G., and Pierce, J. A., Elevated cadmium concentration in emphysematous lungs, *Am. Rev. Respir. Dis.,* 108, 30—39, 1973.

44. Holden, H., A mortality study of workers exposed to cadmium fumes, in *Cadmium 79, Proc. 2nd Int. Cadmium Conf. Cannes,* Cadmium Association, London, 1980, 211—215.

45. Huck, F. F., Cadmium poisoning by inhalation, report of a case, *Occup. Med.,* 3, 411—414, 1947.

46. Johansson, A., Camner, P., Jarstrand, C., and Wiernik, A., Rabbit alveolar macrophages after inhalation of soluble cadmium, cobalt and copper: a comparison with the effects of soluble nickel, *Environ. Res.,* 31, 340—354, 1983.

47. Johansson, A., Curstedt, T., Robertson, B., and Camner, P., Lung morphology and phospholipids after experimental inhalation of soluble cadmium, copper and cobalt, *Environ. Res.,* 34, 295—309, 1984.

48. Kazantzis, G., Respiratory function in men casting cadmium alloys. I. Assessment of ventilatory function, *Br. J. Ind. Med.,* 13, 30—36, 1956.

49. Kazantzis, G., Flynn, F. V., Spowage, J. S., and Trott, D. G., Renal tubular malfunction and pulmonary emphysema in cadmium pigment workers, *Q. J. Med.,* 32, 165—192, 1963.

50. King, E., An environmental study of casting copper-cadmium alloys, *Br. J. Ind. Med.,* 12, 198—205, 1955.

51. King, E., Cadmium fume, *Lancet,* Sept. 20, 641, 1980.

52. Kjellström, T., Friberg, L., and Rahnster, B., Mortality and cancer morbidity among cadmium-exposed workers, *Environ. Health Perspect.,* 28, 199—204, 1979.

53. Kleinfeld, M., Acute pulmonary edema of chemical origin, *Arch. Environ. Health,* 10, 942—946, 1965.

54. Kossman, S., Pierzchala, W., Rusiecki, Z., Scieszka, J., Andrzejewski, J., and Tomaszczyk, S., Estimation of ventilation efficiency of lungs in workers of cadmium division of non-ferrous foundry, *Pneumonol. Pol.,* 9, 627—633, 1979 (in Polish with English summary).

55. Kueppers, F. and Black, L. F., α_1-antitrypsin and its deficiency, *Am. Rev. Respir. Dis.,* 110, 176—194, 1974.

56. Lamy, P., Heully, F., Pernot, C., Anthoine, D., Couillaut, S., and Thomas, G., Pneumopathy caused by cadmium fumes, *J. Fr. Med. Chir. Thorac.,* 17, 275—283, 1963 (in French).

57. Lane, R. E. and Campbell, A. C. P., Fatal emphysema in two men making a copper-cadmium alloy, *Br. J. Ind. Med.,* 11, 118—122, 1954.

58. Laurell, C.-B. and Eriksson, S., The electrophoretic α_1-globulin pattern of serum in α_1-antitrypsin deficiency, *Scand. J. Clin. Lab. Invest.,* 15, 132—140, 1963.

59. Lauwerys, R. R., Bucket, J.-P., Roels, H. A., Brouwers, J., and Stanescu, D., Epidemiological survey of workers exposed to cadmium, *Arch. Environ. Health,* 28, 145—148, 1974.
60. Lauwerys, R., Roels, H., Buchet, J.-P., Bernard, A., and Stanescu, D., Investigations on the lung and kidney functions in workers exposed to cadmium, *Environ. Health Perspect.,* 28, 137—145, 1979.
61. Lefevre, M., Heng, H., and Rucker, R. B., Dietary cadmium zinc and copper: effects on chick lung morphology and elastin cross-linking, *J. Nutr.,* 112, 1344—1352, 1982.
62. L'Epée, P., Lazarini, H., Franchome, J., N'Doky, Th., and Larrivet, C., Contribution to the study on cadmium intoxication, *Arch. Mal. Prof. Med. Trav. Secur. Soc.,* 29, 485—490, 1968 (in French).
63. Liebermann, J., Human α_1-antitrypsin and its deficiency, *Med. Clin. North Am.,* 57, 691—706, 1973.
64. Lucas, P. A., Jarivalla, A. G., Jones, J. H., Gough, J., and Vale, P. T., Fatal cadmium fume inhalation, *Lancet,* 26, 205, 1980.
65. MacFarland, H. N., Pulmonary effects of cadmium, in *Cadmium Toxicity,* Vol. 15, Mennear, J. H., Ed., Marcel Dekker, New York, 1979, 113.
66. Marek, K., Wocka-Marell, T., and Marks, J., Effect of occupational exposure to cadmium on the activity of α_1-antitrypsin in the serum, *Pol. Arch. Med. Wewn.,* 63, 53—56, 1980.
67. Miller, M. L., Murthy, L., and Sorenson, J. R. J., Fine structure of connective tissue after ingestion of cadmium, *Arch. Pathol.,* 98, 386—391, 1974.
68. Mithal, R. K., Pulmonary lesions in experimental chronic cadmium poisoning of squirrels: histopathological and enzymological studies, *Ind. Health,* 18, 139—145, 1980a.
69. Mithal, R. K., Histopathological studies with reference to chronic cadmium acetate intoxication in kidneys of squirrels, *Ind. Health,* 18, 147—152, 1980b.
70. Morgan, J. M., Tissue cadmium concentrations in man, *Arch. Intern. Med.,* 123, 405—408, 1969.
71. Morgan, J. M., Cadmium and zinc abnormalities in bronchogenic carcinoma, *Cancer,* 25, 1394—1398, 1970.
72. Morgan, J. M., Tissue cadmium and zinc content in emphysema and bronchogenic carcinoma, *J. Chronic Dis.,* 24, 107—110, 1971.
73. O'Dell, B. L., Kilburn, K. H., McKenzie, W. N., and Thurston, R. J., The lung of the copper-deficient rat. A model for developmental pulmonary emphysema, *Am. J. Pathol.,* 91, 413—423, 1978.
74. Parkes, W. R., *Occupational Lung Disorders,* 2nd ed., Butterworths, London, 1982, 457—463.
75. Paterson, J. C., Studies on the toxicity of inhaled cadmium. III. The pathology of cadmium smoke poisoning in man and in experimental animals, *J. Ind. Hyg. Toxicol.,* 29, 294—301, 1947.
76. Patwardhan, J. R. and Finckh, E. S., Fatal cadmium-fume pneumonitis, *Med. J. Aust.,* 19, 962—966, 1976.
77. Petering, H. G., Choudhury, H., and Stemmer, K. L., Some effects of oral ingestion of cadmium on zinc, copper, and iron metabolism, *Environ. Health Perspect.,* 28, 97—106, 1979.
78. Peto, R., Hypothesis on CNSLD being two major syndromes, in *Early Detection of Chronic Lung Diseases,* Report on a WHO Meeting, Vienna, May 31 to June 2, 1978, Euro Reports and Studies 24, Regional Office for Europe, World Health Organization, Copenhagen, 1980, 29—30.
79. Piscator, M., Kjellström, T., and Lind, B., Contamination of cigarettes and pipe tobacco by cadmium oxide dust, *Lancet,* 2, 587, 1976.
80. Post, C. T., Squibb, K. S., Fowler, B. A., Gardner, D. E., Illing, J., and Hook, G. E. R., Production of low molecular weight cadmium-binding protein in rabbit lung following exposure to cadmium chloride, *Biochem. Pharmacol.,* 31, 2969—2975, 1982.
81. Potts, C. L., Cadmium proteinuria — the health of battery workers exposed to cadmium oxide dust, *Ann. Occup. Hyg.,* 8, 55—61, 1965.
82. Prigge, E., Early signs of oral and inhalative cadmium uptake in rats, *Arch. Toxicol.,* 40, 231—247, 1978.
83. Princi, F., A study of industrial exposures to cadmium, *J. Ind. Hyg. Toxicol.,* 29, 315—324, 1947.
84. Princi, F. and Geever, E. F., Prolonged inhalation of cadmium, *Arch. Ind. Hyg. Occup. Med.,* 1, 651—661, 1950.
85. Prodan, L., Cadmium poisoning. II. Experimental cadmium poisoning, *J. Ind. Hyg. Toxicol.,* 14, 174—196, 1932.
86. Reinl, W., On a mass-intoxication through cadmium oxide vapor, *Arch. Toxikol.,* 19, 152—157, 1961 (in German).
86a. Sakurai, H., Omae, K., Toyama, T., Higashi, T., and Nakadate, T., Cross-sectional study of pulmonary function in cadmium alloy workers, *Scand. J. Work Environ. Health,* 8 (Suppl. 1), 122—130, 1982.
87. Seiffert, Diseases in zinc smelter workers and hygienic precautions taken, *Dtsch. Vierteljahrschr. Oeff. Gesundheitspflege,* 29, 419, 1897 (in German).
88. Smith, J. C., Kench, J. E., and Smith, J. P., Chemical and histological post-mortem studies on a workman exposed for many years to cadmium oxide fume, *Br. J. Ind. Med.,* 14, 246—249, 1957.

89. Smith, J. P., Smith, J. C., and McCall, A. J., Chronic poisoning from cadmium fume, *J. Pathol. Bacteriol.*, 80, 287—296, 1960.

90. Smith, T. J., Petty, T. L., Reading, J. C., and Lakshminarayan, S., Pulmonary effects of chronic exposure to airborne cadmium, *Am. Rev. Respir. Dis.*, 114, 161—169, 1976.

91. Snider, G. L., Hayes, J. A., Korthy, A. L., and Lewis, G. P., Centrilobular emphysema experimentally induced by cadmium chloride aerosol, *Am. Rev. Respir. Dis.*, 108, 40—48, 1973.

92. Stanescu, D., Veriter, C., Frans, A., Goncette, L., Roels, H., Lauwerys, R., and Brasseur, L., Effects on lung of chronic occupational exposure to cadmium, *Scand. J. Respir. Dis.*, 58, 289—303, 1977.

93. Stokinger, H., in *Industrial Hygiene and Toxicology*, 2nd ed., Fasset, D. W. and Frish, F., Eds., Interscience, New York, 1963, 1185.

94. Strauss, R. H., Palmer, K. C., and Hayes, J. A., Acute lung injury induced by cadmium aerosol. I. Evolution of alveolar cell damage, *Am. J. Pathol.*, 84, 561—568, 1976.

95. Suzuki, S., Suzuki, T., and Ashizawa, M., Proteinuria due to inhalation of cadmium stearate dust, *Ind. Health*, 3, 73—85, 1965.

96. Task Group on Metal Interaction, Factors influencing metabolism and toxicity of metals: a consensus report, *Environ. Health Perspect.*, 25, 3—41, 1978.

97. Taylor, A., Jackson, M. A., Patil, D., Burston, J., and Lee, H. A., Poisoning with cadmium fumes after smelting lead, *Br. Med. J.*, 288, 1270—1271, 1984.

98. Teculescu, D. B. and Stanescu, D. C., Pulmonary function in workers with chronic exposure to cadmium oxide fumes, *Int. Arch. Arbeitsmed.*, 26, 335—345, 1970.

99. Thurlbeck, W. M. and Foley, F. D., Experimental pulmonary emphysema. The effect of intratracheal injection of cadmium chloride solution in the guinea pig, *Am. J. Pathol.*, 42, 431—441, 1963.

100. Tibbits, P. A. and Milroy, W. C., Pulmonary edema induced by exposure to cadmium oxide fume: case report, presented at the American Occupational Health Conference, New Orleans, La., April 10 to 14, 1978; *Milit. Med.*, June 1980.

101. Townshend, R. H., A case of acute cadmium pneumonitis: lung function tests during a four-year follow-up, *Br. J. Ind. Med.*, 25, 68—71, 1968.

102. Townshend, R. H., Acute cadmium pneumonitis: a 17-year follow-up, *Br. J. Ind. Med.*, 39, 411—412, 1982.

103. Tsuchiya, K., Proteinuria of workers exposed to cadmium fume. The relationship to concentration in the working environment, *Arch. Environ. Health*, 14, 875—880, 1967.

104. Vinegar, A. and Choudhury, H., 1976, as cited in Reference 77.

105. Vorobjeva, R. S., On occupational lung disease in prolonged action of aerosol of cadmium oxide, *Arch. Patol.*, 8, 25—29, 1957 (in Russian).

106. WHO, *Early Detection of Chronic Lung Diseases*, Report on a WHO Meeting, Vienna, May 31 to June 2, 1978, Euro Reports and Studies 24, Regional Office for Europe, World Health Organization, Copenhagen, 1980.

107. Yamada, H., Damiano, V. V., Tsang, A.-I., Meranze, D. R., Glasgow, J., Abrams, W. R., and Weinbaum, G., Neutrophil degranulation in cadmium-chloride-induced acute lung inflammation, *Am. J. Pathol.*, 109, 145—156, 1982.

Chapter 9

RENAL EFFECTS

Tord Kjellström

TABLE OF CONTENTS

I. INTRODUCTION

It is well documented that exposure to cadmium will cause impairment of the kidneys. This effect is the critical effect[247] after long-term exposure, and we will therefore give a detailed account of its mechanism in this chapter. Investigations have covered workers exposed to different cadmium compounds in industry and groups of the general population who are exposed to cadmium-contaminated food, as well as animal experiments. Almost all of the effects referred to have resulted from prolonged exposure to cadmium. A brief mention of the acute effects will be made. The acute effects of metallothionein-bound cadmium may furnish an explanation of the mechanism of the chronic renal effects.

II. ASPECTS OF RENAL PHYSIOLOGY OF RELEVANCE TO CADMIUM TOXICITY

About one fourth of the total body arterial blood flow passes the kidneys, where the blood is filtered through the glomeruli producing about 180 ℓ of primary urine each day in an adult.[30] The blood flow through the kidney is approximated as renal plasma flow (RPF), usually measured as the clearance of P-amino hippurate (PAH) or diodrast.[133] These substances are actively secreted by the normal renal proximal tubule to almost 100% and their clearance is not affected by glomerular function. In cases of disease, the tubular excretion of PAH or diodrast may be reduced by damage to tubular transport mechanisms. The normal PAH clearance for young adults is about 650 mℓ/min[30] (= 936 ℓ renal plasma flow per day) and this decreases with age.

The glomerular function is usually measured as glomerular filtration rate (GFR). The GFR is the volume of primary urine produced per unit time (milliliter per minute). The most common methods for measuring GFR are by the "clearance" of inulin or endogenous creatinine; both substances being completely filtrable at the glomerulus, not being reabsorbed or significantly secreted, and not being destroyed, synthesized, or stored within the kidney.[133]

"Clearance" is calculated as the urinary concentration (U) of the substance times urinary flow rate (V; volume per minute) divided by the plasma concentration (P; milliliters per minute). Thus, clearance = UV/P and GFR = clearance for inulin or endogenous creatinine. The normal GFR for young men is on average 125 mℓ/min (range 95 to 150 mℓ/min) and it decreases with age. At age 70 it is about 50 to 70 mℓ/min.[30]

The "clearance" of substances other than inulin and creatinine does not directly reflect how they are filtered. A better measure of how the glomeruli handle substances is the sieving coefficient (SC).[204] It is measured as the ratio between the clearance of the substance (assuming that it is not reabsorbed in the tubules) and the GFR, expressed as a percentage. It can be interpreted as the proportion of the theoretical maximum transglomerular flow of the substance that in fact does pass the glomeruli.

P times GFR times SC/100 denotes the amount of a substance available in the primary urine per minute. If the substance is reabsorbed but not secreted in the tubule, the amount of the substance in urine (U × V) will be less than GFR × P × SC/100. The difference GFR × P × SC/100 − U × V represents tubular reabsorption.

The reabsorptive function of the tubule is often measured as the tubular reabsorption of phosphorus (TRP).[141] It should be pointed out that if the substance can be introduced into the urine from the urinary tract (below the kidney), this contribution to the urinary level should be excluded in the calculation.

In studies on the health effects of cadmium, particular emphasis has been put on glomerular and tubular handling of plasma proteins. Many of these proteins are filtered through the glomeruli and reabsorbed in the tubuli.[14] Table 1 summarizes some of the relevant human data based mainly on studies by Mogensen and Sølling[138] using lysine injections to almost completely block tubular reabsorption. Lysine is more effective than arginine[139] for this purpose, and it has been speculated that site selectivity and competition for tubular reabsorption may be the mechanisms for this difference.[241] The data by Peterson et al.[192] regarding urinary proteins in normal and tubular proteinuria agree with those of Mogensen and Sølling[138] and we have included rounded figures in Table 1, taking into account that the lysine injections probably did not achieve a complete block of tubular reabsorption. We have also considered that urinary β_2-microglobulin in severe tubular damage can become as high as 200 mg/g creatinine.[171,172,226]

The sieving coefficient is very large (70%) for the low molecular weight proteins, such as β_2-microglobulin, and small (0.007%) for the higher molecular weight proteins, such as albumin. The proteins are almost totally reabsorbed in the tubules; the low molecular weight proteins are reabsorbed to the largest extent (Table 1).

Some of the proteins found in urine originate from posttubular sources: the urinary tract and the bladder.[8] These may primarily be unidentified mucoproteins, but they may also be specific plasma proteins. No data on posttubular excretion of albumin or β_2-microglobulin are available. However, even very small volumes (e.g., 0.1 to 0.2 mℓ plasma per 24 hr) may contribute significant amounts of albumin (5 to 10 mg/24 hr) in comparison with the normal amounts of albumin in the urine (about 10 mg/24 hr). As this source of urinary albumin has not been fully studied, the estimate of albumin reabsorption shown in Table 1 is very approximate. In the same volume of plasma (0.1 to 0.2 mℓ) there would be only 0.15 to 0.3 µg β_2-microglobulin, which is insignificant compared to the normal urinary excretion of 75 µg/24 hr (Table 1).

In spite of the very high tubular reabsorption rate for both albumin and β_2-microglobulin, there is a great difference between the extent to which urinary protein excretion changes after a similar percentage loss of tubular reabsorption.

If we assume that the relative loss of tubular reabsorption is the same for albumin, β_2-microglobulin, and total protein we can calculate that for a 10% loss the tubular reabsorption of albumin would decrease from 98 to 88.2% and the excreted fraction would increase from 2 to 11.8% (5.9 times). For β_2-microglobulin the tubular reabsorption decreases from 99.964 to 89.968% and the excreted fraction increases from 0.036 to 10.032% (278.7 times). A mathematical relationship between the urinary excretion rates can be calculated. Albumin in urine (A) increases as a function of β_2-microglobulin (B) such that A = 9.8235 + 2.353 B. The equivalent function for total protein (T) is T = 59.7 + 4.012 B.

At the same loss of reabsorptive capacity (Figure 1), the excreted *absolute amount* of albumin in urine will always be much greater than the amount of β_2-microglobulin,

Table 1

PLASMA PROTEINS. GLOMERULAR FILTRATION AND TUBULAR REABSORPTION IN HUMANS. APPROXIMATE AVERAGE VALUES FOR AN ADULT MALE (REFERENCES IN PARENTHESES)

Protein	Plasma conc. (P) (mg/mℓ)	Sieving coefficient (SC) (%)	Filtered amount in primary urine (F)			Urinary excretion[a] (U × V)			Excreted fraction (EF) (%)	Tubular reabsorption (100-EF) (%)
			mg/min	mg/24 hr	μmol/24 hr	mg/min	mg/24 hr	μmol/24 hr		
Albumin (69,000 mol wt)	40 (50)	0.0070[b]	0.35	500 (138, 192)	7.2	0.007 (30)	10 (138, 192)	0.145	2.0	98.0
β_2-microglobulin (11,600 mol wt)	0.0015 (60)	70[b]	0.14	200 (138, 192)	17.2	0.00005	0.075 (59, 120, 138, 192)	0.0065	0.036	99.964
Total protein	75 (50)	0.0067[b]	0.63	900 (138, 192)		0.042	60 (50, 192)		6.7	93.3

Note: U = urine concentration of protein (mg/mℓ); V = urine volume (mℓ/min); F = GFR × P × SC/100; EF = 100 × U × V/F.

[a] There may be a contribution of albumin and mucoproteins from posttubular sources. Any such contribution would decrease the EF value below the one given here.

[b] Based on a glomerular filtration rate (GFR) of 125 mℓ/min.

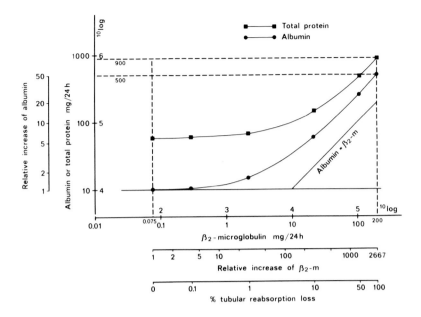

FIGURE 1.   Comparison of urinary albumin and urinary β_2-microglobulin, as a function of the relative loss of tubular reabsorptive capacity.

whereas the *relative increase* of urinary excretion (as compared to the normal excretion) will always be greater for β_2-microglobulin than for albumin (Figure 1). If the percentage decrease of tubular reabsorptive capacity is the same for albumin and β_2-microglobulin, we would find the relationship between urinary albumin and β_2-microglobulin depicted in Figure 1.

In a case with glomerular dysfunction leading to an increased filtration of all proteins, the urinary excretion of albumin would increase much faster than the excretion of β_2-microglobulin. Assuming that there is no change in tubular reabsorption, the latter protein can only increase a maximum of 1.43 times as a consequence of increased filtration (increased SC from 70 to 100%) (Table 1), whereas albumin can increase 14,300 times (increased SC from 0.007 to 100%). In Figure 1 urinary albumin and β_2-microglobulin data from cases with glomerular damage would appear above the curve and data from cases with increased internal production of β_2-microglobulin but without renal function changes[227] would appear below the curve.

The mechanisms for renal filtration, transport, and metabolism of low molecular weight proteins have been reviewed by Maack et al.[134]

III. ACUTE EFFECTS AND DOSE-RESPONSE RELATIONSHIPS

A. Animals

1. After Administration of Cadmium Salts and Oxides

Very high acute inhalation exposure and dietary exposure will cause severe local acute effects in the lungs (see Chapter 8, Section II.A) and in the gastrointestinal tract (see Chapter 11, Section II). Very little cadmium will have an opportunity to be absorbed and cause systemic acute effects. However, Prodan in 1932[209] found systemic effects in five cats exposed to cadmium oxide fumes at several milligrams Cd per cubic meter for 24 hr. There was fatty degeneration in the renal tubular epithelium. No biochemical tests were carried out.

Single subcutaneous (s.c.) injections of cadmium chloride (10 mg Cd per kilogram body weight) to rats[62] resulted in renal tubular lesions. Two s.c. injections of cadmium chloride (9 mg Cd per kilogram body weight) produced renal lesions in rabbits.[63] These excessive doses are similar to the LD_{50} (Chapter 11, Section III). The animals are likely to die of liver damage and a number of other acute effects do occur (Chapter 11). Severe acute vascular changes were seen in rabbits[63] and the effects on the kidneys were secondary to these changes.

At these high acute doses the most prominent effect is usually testicular necrosis (Chapter 11, Section VI.A). Some authors reported testicular effects, but no renal damage in rats.[113,187-189] However, these investigations did not specifically look for renal functional damage. Effects on the kidneys are not characteristic of acute cadmium poisoning in animals.[178]

2. After Administration of Cadmium Bound to Metallothionein and Other Low Molecular Weight Ligands

Acute renal effects following injections of cadmium chloride bound to mercaptoethanol have been reported by Gieske and Foulkes.[78] They found a decreased PAH clearance and a decreased tubular reabsorption of all amino acids, indicating a tubular damage 3 days after a single intravenous (i.v.) injection of $CdCl_2$-mercaptoethanol mixture (1.1 mg Cd).

When cadmium is injected intravenously or subcutaneously (1.1 to 6 mg Cd per kilogram body weight) into mice in the form of a cadmium-metallothionein complex, a very high acute mortality (30 to 100%) with serious renal tubular damage can be induced.[181] At the same dose levels of cadmium in the form of cadmium chloride, there was no acute mortality. It was estimated that a dose of about five times higher was needed to produce the same effect with cadmium chloride as with cadmium-metallothionein. The LD_{50} for cadmium salts is about 10 times higher than for cadmium-metallothionein.[266] The latter type of exposure also led to a higher cadmium concentration in the kidney and a lower concentration in the liver than did the same dose of cadmium chloride.[181] Because of the severe damage to the kidneys, oliguria and decreased urinary creatinine in the mice that died, it was concluded that the deaths were due mainly to renal damage. The type of renal damage (swelling and necrosis of proximal tubuli) was similar to that seen after chronic cadmium exposure (Section IV.A.4).

Metallothionein serves as a transport protein for cadmium in plasma (Volume I, Chapter 4, Section VI) and cadmium induces production of metallothionein in liver, from where it is released to plasma, transported to the kidney, filtered through glomeruli, and reabsorbed in the tubuli. After parenteral exposure to cadmium chloride the liver initially stores most of the cadmium in the newly formed metallothionein, and thereby prevents the kidney from receiving the full dose (Volume I, Chapter 4, Section VI). When the exposure is in the form of cadmium-metallothionein, the cadmium "bypasses" the liver, the renal dose becomes greater, and the effects are more severe. These phenomena are discussed in more detail in Volume I, Chapters 4 and 6.

Acute exposure to cadmium bound to low molecular weight ligands has been used in several studies as a model system for assessing the mechanism of chronic cadmium toxicity. The acute renal tubular damage recorded by histological examination in these studies is similar to the renal damage after long-term cadmium exposure (Section IV.A.4), and it leads to tubular reabsorption defects (Section IV.A.3).

Foulkes and Gieske[65] studied the specificity of metal effects on renal amino acid transport in a peracute study (50 to 60 sec). The venous blood flow from rabbit kidneys was diverted into a vessel where samples could be taken at short intervals (seconds). By infusing labeled amino acids together with metal compounds as well as inulin, the inhibition of renal peritubular uptake of amino acids could be measured. The authors reported that glutamate and aspartate transport in the rabbit tubular cells was specifically inhibited (24 to 68% inhibition) by an intraarterial cadmium dose of 2.5 μmol Cd per kilogram body weight (0.28 mg/kg) given together with mercaptoethanol. Arginine transport was inhibited 13 to 37% while the inhibition of phenylalanine, alanine, serine, and lysine was lower or zero. In further studies of tubular reabsorption of amino acids, Gieske and Foulkes[78] showed that reabsorption of all amino acids at the luminal membrane is strongly depressed by cadmium-mercaptoethanol.

In similar acute experiments Foulkes[64] found that injection of 1.1 mg Cd per kilogram body weight in rabbits together with mercaptoethanol, inhibited aspartate reabsorption in renal tubules. It is hypothesized that this effect is due to effects on tubular transport mechanisms.

These mechanisms have been elucidated by Squibb and co-workers.[234] They exposed rats to single intravenous injections of cadmium-metallothionein (0.17 or 0.017 mg Cd per kilogram body weight). At the lower dose a much greater proportion of the dose was retained in the kidneys after 24 hr. The cadmium-metallothionein is located in the lysosomes of the renal proximal tubular cells and the absorption was concluded to take place via pinocytosis. This would be the same reabsorption mechanism as for other low molecular weight plasma proteins.[240]

After a single dose of cadmium-metallothionein (0.6 mg Cd per kilogram body weight) to rats there were marked changes in renal function during 8 hr postinjection.[235] There was a 35-fold increase in urinary protein excretion, characterized by an increase in low molecular weight proteins. An accumulation of dense lysosomal-type bodies in the renal proximal tubular cells could be seen using electron microscopy. There were also a great number of vesicles in the cells, and these findings were interpreted as the result of an inhibition of the normal fusion of pinocytotic vesicles with cellular lysosomes. The increase of vesicles 4 hr after exposure coincided with a dramatic decrease in RNA synthesis.[235]

Further studies using a similar protocol[236] showed the same ultrastructural picture after a single injection of cadmium-metallothionein. Cytochemical analysis showed acid phosphatase activity in the dense cellular bodies, which indicated that these organelles were lysosomes. The vesicles did not contain acid phosphatase, but the inner side of their membrane contained a fibrillar coat similar to that of the glycocalyx surface coat on the outer side of the brush border membrane. The authors concluded that the vesicles were derived from the outer membrane and therefore could be classified as pinocytotic vesicles.[236]

By 24 hr after cadmium-metallothionein injection most cells were completely vacuolized and some showed signs of cell death and were beginning to slough into the tubular lumen. By 3 days many proximal tubules were filled with cellular debris and many cells had sloughed completely into the lumen, leaving only the bare basement membrane.[236] Throughout a 5-day observation period there were no apparent ultrastructural changes in glomeruli or distal tubuli and none of the rats died.

After 1 hr, morphometric study showed an increase in the number of lysosomes, but a decrease in their average size. The vacuole compartment increased considerably in volume, but there were only minor changes in the nuclear and mitochondrial compartments in the first 8 hr.[236] Signs of the tubular damage were a sevenfold increase in the urinary excretion of ribonuclease and a change in the urinary protein electrophoresis patterns towards tubular proteinuria patterns.

As explained in Volume I, Chapter 4 and Chapter 6, Section V.C, the available data are in favor of an important role for cadmium-metallothionein in transporting cadmium to the kidney and into the kidney cells. This happens also after long-term expo-

Table 2

RENAL CADMIUM CONCENTRATION AND SUBCELLULAR
DISTRIBUTION AFTER INJECTION OF CADMIUM-METALLOTHIONEIN
IN RATS[236]

Treatment	Time after injection	Total µg Cd/kidney		Lysosomal/ mitochondrial fraction[a] µg Cd/kidney		Cytoplasmic Cd			
						Total µg Cd/kidney		Non-MT[b]	MT[b]
Control		N.D.		N.D.		N.D.			
	1 hr	5.7[a]	2.6[c]			2.4	0.1[c]	97.0[d]	3.0[d]
	4 hr	10.8	0.4			3.8	0.2	37.8	62.2
Cadmium-	8 hr	11.2	0.9	1.48	0.07[c]	3.9	0.1	15.8	84.2
metallothionein	1 day	7.2	1.1	0.10	0.02	6.4	0.4		
	3 days	5.4	1.0	0.03	0.01	4.0	0.5		
	5 days	3.8	0.6	0.03	0.01	3.6	0.5		

Note: N.D., nondetectable.

[a] Ultracentrifugation.
[b] Measured by G-75 Sephadex gel chromatography.
[c] Mean ± S.E.M., m = 3.
[d] Average of two values.

sure to cadmium salts or other forms of nonmetallothionein-bound cadmium. These acute exposure experiments may mimic the mechanism for renal damage in chronic poisoning. The data by Squibb et al.[236] indicate that when cadmium enters the cytoplasma it is no longer bound to metallothionein but within 8 hr most of the cadmium is bound to metallothionein again (Table 2). This supports the idea that "free" cadmium is released from the reabsorbed cadmium-metallothionein in the renal tubular cells (Chapter 4 and Chapter 6, Section V.C), and that new metallothionein produced in the tubular cells eventually binds the cadmium.

Neither zinc-metallothionein nor lysosome injections caused the effects reported above. Pretreatment with zinc-metallothionein in fact protected against the cadmium effects[236] and the cytoplasmic nonmetallothionein-bound cadmium was reduced to a quarter of what it was without zinc-metallothionein pretreatment.

The lack of effects of zinc-metallothionein shows that it is not the metallothionein itself that is toxic. A study of rats exposed to an injection of iodine-metallothionein also showed no renal damage, but showed that the absorbed metallothionein ends up in lysosomes, where it is metabolized to lower molecular weight peptides.[142]

In another study of rats exposed to a single i.v. injection of cadmium-metallothionein (0.8 mg Cd per kilogram body weight)[143] very similar pathological changes to those reported by Squibb et al.[235,236] were seen. Murakami and co-workers[143] followed up the rats for up to 7 days and found that the maximum damage to renal tubular cells occurred after 2 days. After 7 days the necrotic cell debris in the tubular lumen had cleared in most rats and regenerating cells had replaced the damaged cells.

B. Humans

Acute exposure to high concentrations of cadmium oxide fumes has mainly caused severe lung damage (Chapter 8, Section II.B). In two fatal cases described by Bulmer and co-workers in 1938[37] kidney effects in the form of "cloudy swelling" were found on microscopic examination. The authors ascribed this to a general toxemia. In 1966, Beton and co-workers[29] reported a fatal case (age 53) in which bilateral cortical necrosis of the kidneys was found at autopsy. They thought that this was mainly due to vascular changes associated with massive pulmonary changes.

The concentration of cadmium in the kidneys was given as 5.7 mg/kg wet weight, which is below the average "normal" level for people of the patient's age (Volume I, Chapter 5, Section V). In two other patients, who recovered after exposure, transient proteinuria was noted. Electrophoretic examination of the urine from one of these patients showed that the main component was albumin.

In 1971, Wisniewska-Knypl and co-workers[271] reported a case of acute cadmium poisoning. A 23-year-old man drank 5 g of cadmium iodide dissolved in water in order to commit suicide (about 25 mg Cd per kilogram body weight). The patient developed transient anuria during the first day and later died of liver damage. The authors stated that the laboratory findings showed pronounced signs of damage to liver and kidneys, hypoproteinemia with hypoalbuminemia, and metabolic acidosis.

IV. CHRONIC EFFECTS

The chronic effects on the kidney have attracted more attention among researchers over the years than have any other cadmium-induced health effect. The discovery of the renal effects by Friberg in the late 1940s was unexpected. When such effects had been noted even earlier, they had been linked to lead exposure.

Substantial evidence on the occurrence of the chronic renal effects and the characteristics of these effects is now available both from animal studies and studies of exposed humans. The mechanism underlying the kidney damage is becoming more and more clear, and the link to the acute effects of cadmium-metallothionein (Section III.A.2) is of particular interest.

A. Animals
1. Proteinuria

The earliest animal studies of cadmium effects on the kidney[209,270] did not measure proteinuria, glucosuria, or any other signs of functional changes. The first study of proteinuria was reported by Friberg in 1950.[67] He made a long-term experiment in which he exposed rabbits for 3 hr/day for 8 months to inhalation of cadmium iron oxide dust (about 8 mg Cd per cubic meter). After 4 months of exposure, moderate proteinuria could be demonstrated with the trichloracetic acid test. When the animals were killed after 7 to 9 months of exposure, histological examination did not disclose any structural changes. Cadmium concentrations in the kidneys were generally between 300 and 700 mg/kg wet weight. When a group of rabbits was given s.c. injections of cadmium sulfate, 0.65 mg Cd per kilogram body weight, 6 days/week, proteinuria was demonstrated after 2 months. Analyses of cadmium in the kidneys were not performed. Electrophoretic analysis of urine proteins showed that the main component was not the same as that found in rabbits with kidney damage due to uranium salts. Friberg[68] repeated the experiment on rabbits given s.c. injections of 0.65 mg Cd per kilogram body weight as radiolabeled cadmium sulfate. After 8 to 10 weeks an increase in urinary protein was seen and at the same time there was a dramatic increase in urinary cadmium excretion (Volume I, Chapter 6, Figure 20).

Axelsson and Piscator[10] gave rabbits daily s.c. injections of cadmium chloride, 0.25 mg Cd per kilogram body weight, 5 days/week. After 23 weeks there was considerable proteinuria and excretion of cadmium. In another group of rabbits[10] exposure conditions were similar to those stated above, but exposure was discontinued after 24 weeks, after which the animals were followed up for another 30 weeks. Urine investigations showed that protein excretion increased during exposure and was about two to three times higher than the control group at 24 weeks. The excretion reached a maximum about 1 month after exposure had ceased, and was followed by an apparent reduction in protein excretion. At the end of the observation time the protein excretion was the same as that at 24 weeks, but due to individual variation, the difference between exposed animals and controls was not statistically significant. Electrophoretic examination of urinary proteins showed a predominance of proteins with a mobility like α- and β-globulins. It was concluded that this corresponded to the tubular proteinuria found in human beings.[38] The renal cortex in the exposed animals contained large amounts of cadmium,[10] about 235 mg/kg wet weight, after 11 weeks' exposure, about 460 mg/kg after 17 weeks, and about 250 mg/kg at the end of exposure (see further data in Section IV.D).

In mice the normal urinary protein excretion is relatively large, a fact which makes interpretation of some studies difficult. Nordberg and Piscator[180] found a decrease in the excretion of total protein in male mice given s.c. injections of cadmium chloride (0.25 or 0.5 mg Cd per kilogram body weight, 5 days/week for 6 months). The main urinary protein in male mice is a low molecular weight protein, synthesized in the liver, and this synthesis is stimulated by testosterone. A possible cadmium influence on testosterone synthesis could explain a decrease in the synthesis rate of that protein, and in turn, a decrease in total urinary protein excretion. After 21 weeks of exposure, cadmium excretion increased in the mice given 0.5 mg/kg. At that time there were also changes in the urinary protein electrophoretic pattern, which earlier had been dominated by the above-mentioned testosterone-dependent protein. Other proteins appeared which had a pattern indicating renal damage. Further examinations[179] at the end of the experiment, i.e., after 24 weeks of exposure, showed average concentrations of 170 mg/kg of cadmium in whole kidney in both exposure groups.

Suzuki[243] exposed rats to daily s.c. injections of 0.5 mg Cd per kilogram body weight, 6 days/week, for 15 weeks. Each week total proteinuria and cadmium in urine and blood were analyzed and three to five animals were killed in order to analyze the cadmium concentration in tissues. After about 5 weeks (Figure 2) there was a steady increase in total proteinuria. The rats also developed an increased daily urinary volume (Figure 2) most likely due to a decreased ability to concentrate urine. Cadmium was analyzed with atomic absorption spectrophotometry directly in the acid solution after wet ashing of the tissues. This method is likely to cause errors in analysis of low-level specimens, but at the high concentrations reported in the experiment by Suzuki[243] the method may be acceptable. At the time when the proteinuria started to increase the whole kidney concentration was about 100 mg Cd per kilogram wet weight and the liver concentration was about 180 mg Cd per kilogram wet weight.

When urinary proteins started to increase, there was a simultaneous increase in urinary cadmium, which progressed rapidly at 8 to 11 weeks (Figure 2). The blood cadmium concentration leveled off during this period as did the liver and kidney cadmium concentrations.[243]

Cadmium-induced proteinuria in rats has also been shown after long-term oral exposure.[24] Fifty female young rats were given 200 mg Cd per liter as $CdCl_2$ in deionized drinking water for up to 11 months. Fifty control rats received no cadmium in water, but like the exposed rats they received food pellets with 0.13 mg Cd per kilogram. Total proteinuria was measured by a Tsuchiya-Biuret method[193] and showed a progressive increase in the average proteinuria from the seventh month onwards (Figure 3). At this time the average cadmium concentration in renal cortex was about 200 mg Cd per kilogram wet weight (Figure 4), and urinary excretion of cadmium increased drastically in some rats (Figure 4).

Thirteen rabbits were given s.c. injections of cadmium chloride, 0.5 mg Cd per kilo-

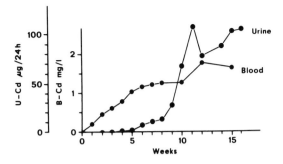

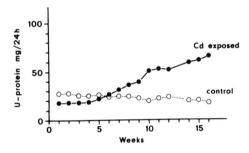

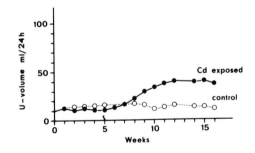

FIGURE 2.   Cadmium in urine, total proteinuria, urine volume, and blood of rats exposed to daily s.c. injections of 0.5 mg Cd per kilogram body weight.[243]

gram body weight a day for 42 weeks.[173] The average urinary total protein excretion (Tsuchiya-Buiret method) increased after about 10 weeks' exposure (Figure 5), at which time the renal cortex cadmium concentration was on average about 300 mg/kg wet weight.[173]

The cadmium-induced proteinuria in animals was first characterized by Friberg.[67] He pointed out the unusual electrophoretic mobility of the main components of proteinuria in cadmium-poisoned rabbits, but the proteins were not identified in detail. As reported earlier in this section, the cadmium-induced proteinuria is similar to the "tubular proteinuria" found in humans.[38] Low molecular weight proteins will be particularly increased in urine. Bernard and co-workers[24] used sodium dodecyl sulfate

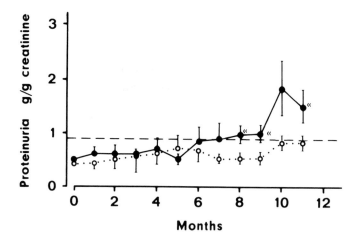

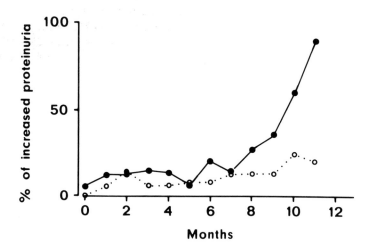

FIGURE 3.   Evolution of proteinuria and prevalence of increased proteinuria in rats given 200 mg Cd per liter in drinking water (●) and in control
rats (○). The horizontal dashed line represents the mean + 2 SD of the values
of proteinuria observed at the beginning of the experiment (0.9 g/g creatinine). In the upper graph each point is the mean of 15 animals. The bars
represent the SE *$p < 0.05$. (Redrawn from Bernard, A., Lauwerys, R., and
Gengoux, P., *Toxicology*, 20, 345—357, 1981a).

(SDS) polyacrylamide gel electrophoresis for detailed analysis of urinary proteins in
their cadmium-poisoned rats, but the results are made uncertain because of the method
used for concentrating the proteins which may have led to losses (concentration by
ultrafiltration with 31.75 mm Visking dialysis tubing to a creatinine concentration of
10 g/ℓ). The patterns on the gels were not very clear and some important bands were
not identified. Nomiyama and co-workers[173] used the same type of electrophoresis to
study the urinary proteins of the cadmium-poisoned rabbits. The method for concentrating the proteins was not mentioned, but the reproductions of the gels (Figure 6)
showed the low molecular weight proteins more clearly than those of Bernard and co-
workers.[24] There is an increase with exposure duration of both high and low molecular
weight proteins. This fits with the concept of a decrease in tubular reabsorption as a
cause of the proteinuria (Section IV.B.3). Similar patterns were seen in cadmium-exposed monkeys.[171]

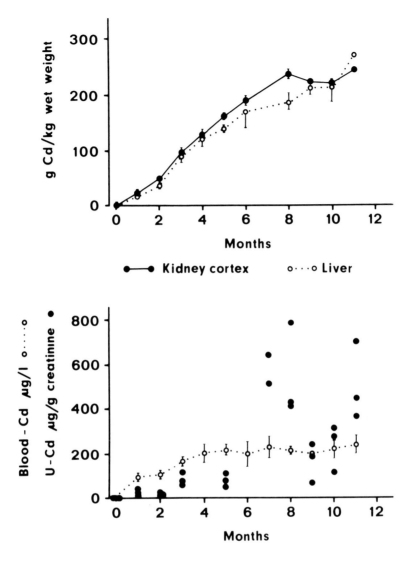

FIGURE 4. Evolution of cadmium concentration in kidney cortex, liver, urine, and blood in rats given 200 mg Cd per liter in drinking water. For kidney cortex, liver, and blood each point is the mean of three animals and the vertical bar represents its SE. For urine, the individual values have been indicated. (Redrawn from Bernard, A., Lauwerys, R., and Gengoux, P., *Toxicology,* 29, 345—357, 1981a).

In 1982, Nomiyama and co-workers[172] concluded that the SDS-electrophoresis pattern of rabbits with chronic cadmium poisoning was "hardly distinguishable" from the pattern of rabbits with acute renal damage caused by uranium or chromium. We believe that their data can be interpreted differently. Certainly, there is an increase in excretion of several proteins in urine both after cadmium exposure and after uranium and chromium exposure. However, the increase of albumin excretion appears more limited and the excretion of β_2-microglobulin is more prominent after exposure to cadmium. This difference may be in accordance with the difference between cadmium and uranium poisoning in rabbits reported by Friberg.[67] The effects of chromium and uranium reported by Nomiyama and co-workers[172] appear to be reversible, which the same study shows is not true for cadmium.

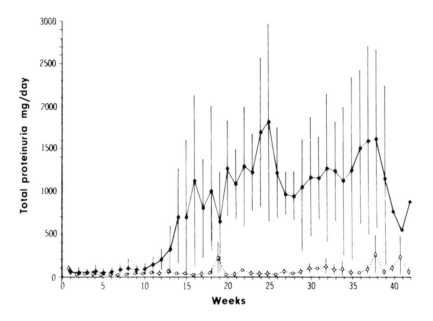

FIGURE 5.   Urinary excretion of protein during the period of cadmium administration to rabbits (s.c. injections of 0.5 mg Cd per kilogram body weight and day). Open circles and dots represent arithmetic means of the control group and the cadmium group, respectively. Vertical lines indicate one SD. (Modified from Nomiyama, K., Nomiyama, H., Yotoriyama, M., and Matsui, K., *Ind. Health,* 20, 11—18, 1982b).

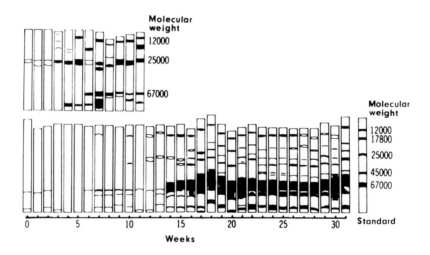

FIGURE 6.   Sodium dodecyl sulfate (SDS) polyacrylamide gel electrophoretic patterns of urine proteins of a rabbit exposed to s.c. injection of cadmium (0.5 mg Cd per kilogram body weight and day). Upper figures (silver staining) and lower figures (Amidoblack 10B staining). (Modified from Nomiyama, K., Nomiyama, H., Yotoriyama, M., and Matsui, K., *Ind. Health,* 20, 11—18, 1982b.)

Ten male rhesus monkeys were exposed for about 1 year to cadmium in food pellets at concentrations of 0, 3, 30, or 300 mg Cd per kilogram[170]. Proteinuria was found in one of two monkeys exposed to 300 mg Cd per kilogram food after about 30 weeks. The reported renal cortex cadmium concentrations in these two monkeys were 583 and

350 mg Cd per kilogram wet weight after 55 weeks' exposure. It was not reported which of the two monkeys had proteinuria, but it was stated that the 350 mg Cd per kilogram monkey had an increased urinary excretion of β_2-microglobulin after 30 weeks, so it is likely that this monkey had proteinuria. The ratio between cadmium concentration in renal cortex and in renal medulla varied between 3 and 11, which is higher than the generally reported values of around 2 to 4 for humans (Volume I, Chapter 6, Section IV.B.2). The kidneys in the rhesus monkeys used by Nomiyama and co-workers must be very different from human kidneys as the ratio of cortex weight to medulla weight in the monkeys was about 10,[171] whereas in humans the ratio has been estimated to be 2.4[244] and 3.0.[89]

In a later experiment with monkeys exposed to cadmium in food pellets for up to 3.5 years, at 0, 3, 10, 30, and 100 mg Cd per kilogram food, some monkeys in each group developed proteinuria measured by qualitative tests, but a clear quantitative increase was only seen in the 100 mg Cd per kilogram group.[171]

Nomiyama and co-workers[170,171] also reported β_2-microglobulin in urine values using single radial immunodiffusion with a detection limit of 3 mg/ℓ. This method could not detect β_2-microglobulin in urine of control monkeys and measured values were only recorded for the monkeys exposed to 100 mg/kg for 40 weeks. Nomiyama and co-workers[171] interpreted the data as negative for all groups except the 100 mg Cd per kilogram group. In view of the low sensitivity of the urinary β_2-microglobulin analysis, and the findings of qualitative proteinuria in several monkeys in the other exposed groups, a slight or moderate renal tubular effect in the other cadmium-exposed monkeys may well have escaped detection due to the methods used.

However, this extensive study of monkeys[171] has shown clearly the severe renal effects that occur as an effect of the high cadmium exposure in the 100 mg/kg group. Apart from the proteinuria, these monkeys had aminoaciduria, glucosuria, decreased creatinine clearance, increased urine volume, etc. The study has been continued with up to 5 years' exposure using the same methods.[5] Akahori and co-workers[5] conclude that the additional data indicate that cadmium effects on the kidneys were found only in the 100 mg/kg group. The same weakness with regard to methods persists, however.

There are other problems with this study. Thus, the concentrations of cadmium in urine in the exposed as well as the control group are unrealistically high, indicating either analytical problems or contamination. For example, the average urine cadmium in the control group was reported to be as high as 100 $\mu g/\ell$ on one occasion, and in the 100 mg Cd per kilogram group the values were 1000 $\mu g/\ell$ after only a few weeks of exposure. During the next 180 weeks of exposure there was no clear increase of urinary cadmium in the exposed groups and when severe renal damage occurred in the 100 mg Cd per kilogram group no increase of urinary cadmium was reported. This does not agree with other animal studies (Section IV.A.5).

Another extensive study of monkeys[115] was carried out primarily to investigate the bone effects of cadmium (Chapter 10, Section III.E). A partly similar protocol to the one used by Nomiyama and co-workers was used.[249] Four groups of young rhesus monkeys were exposed to cadmium chloride via diet (30 mg Cd per kilogram food) for 3.5 years. One of the four groups received normal diet as well as the cadmium (group 5); one received a low-calorie diet (group 6); one received a diet deficient in vitamin D (group 7); and one received a low-calorie and vitamin D-deficient diet (group 8). There was a control group without cadmium exposure equivalent to each of the four nutrition-specific groups.

The authors reported renal damage as judged by urinary β_2-microglobulin for group 8 only.[115] It appears from the presentation of the data that the detection limit of the radioimmunoassay method used is much lower than the earlier method used by Nomiyama and co-workers.[171] In all groups except for group 8, the β_2-microglobulin excretions were below 100 μg/day. The β_2-microglobulin increased after 3 years of cadmium exposure (1 year at 3 mg Cd per kilogram and 2 years at 30 mg Cd per kilogram) to 2000 μg/day in the two monkeys studied. It should be pointed out that these levels are still lower than the detection limit for the Nomiyama and co-workers study.[171]

The cadmium concentrations in urine were throughout considerably lower than the levels reported by Nomiyama and co-workers,[171] but increased slowly with body burden in a predictable way.[115] Three of the monkeys in group 8 died and were autopsied. After 3 years of the experiment, the renal cortex cadmium concentrations were 1016, 750, and 610 mg Cd per kilogram wet weight. Only the latter monkey was reported to have increased β_2-microglobulin in urine about 6 months before death.

One reason why kidney dysfunction was not seen in groups 5, 6, or 7 may have been that the cadmium uptake was lower than in group 8 where severe nutritional deficiency occurred (Volume I, Chapter 6, Section II.B.3). The study is still in progress and more information will be collected in the future. The authors themselves predict that kidney damage will occur in the future in groups 5, 6, and 7. It should be pointed out that even after 4 years, compared to their expected life span (about 30 years; Chapter 6, Table 7), the exposure period has not been as long as the one required to cause renal effects in some of the dietary cadmium experiments on rodents referred to above. The young age of the monkeys and the short exposure period compared to the life span may, however, be of particular importance for the causation of bone effects (Chapter 10, Section III.E).

2. Glucosuria and Aminoaciduria

Axelsson and Piscator[10] found glucosuria in some animals and a decreased average capacity of the kidneys to reabsorb glucose in the rabbits given daily s.c. injections of 0.25 mg Cd per kilogram body weight, 5 days/week for 23 weeks. As mentioned earlier (Section IV.A.1), the rabbits developed proteinuria. In another group of rabbits similarly exposed there was a correlation between glucose reabsorption and urinary protein excretion.

Nomiyama and co-workers[166] exposed rabbits to daily s.c. injections of $CdCl_2$ (1.5 mg Cd per kilogram body weight) for up to 45 days. Four of the seven rabbits died between days 15 and 36. The survivors had increased urinary glucose and amino acid excretion as well as proteinuria. Nomiyama and co-workers[167] reported that glucosuria and aminoaciduria can be induced in rabbits also by oral exposure to 300 mg Cd per kilogram in food pellets for 16 weeks (aminoaciduria) and 42 weeks (glucosuria). Total proteinuria was diagnosed (with the criteria used) after 38 weeks' exposure.

In a review of several of their studies on rabbits, Nomiyama and Nomiyama[165] reported on a group of rabbits exposed to s.c. injections of $CdCl_2$ (0.5 mg Cd per kilogram body weight), 6 days/week for 21 weeks. A slight increase of aminoaciduria accompanied by low molecular weight proteinuria occurred after 3 weeks when the cadmium concentration in renal cortex was about 200 mg/kg wet weight (Figure 7). It was not stated how the "increase" of amino acids in urine was assessed, but the relative increase in excretion cannot have been very great as, in rabbits with identical exposure conditions the increase was less than about tenfold for all individuals[172] and the increase of total protein excretion appeared greater (Figure 5).

Nomiyama and co-workers[170] reported glucosuria and aminoaciduria also in the monkeys exposed to 300 mg Cd per kilogram in food pellets for about 20 weeks (Section IV.A.1). Similar signs of renal damage occurred in monkeys exposed to 100 mg Cd per kilogram in food pellets for 54 to 60 weeks.[173] In one study[170] measurement of urinary concentration of specific amino acids was carried out, but as no quantitative reference values were given, it is difficult to interpret which increases were the greatest. As shown by Gieske and Foulkes[78] (Section III.A.2), cadmium may induce a pan-aminoaciduria due to inhibition of renal tubular reabsorption.

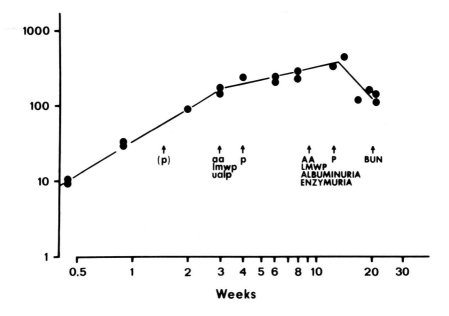

FIGURE 7.   Cadmium in the renal cortex from rabbits given s.c. injection of cadmium chloride over a period of 21 weeks. AA, LMWP, UALP, P, and BUN indicate aminoaciduria, low molecular weight proteinuria, alkaline phosphataseuria, proteinuria, and increased blood urea nitrogen, respectively. Small letters indicate slight changes and a small letter in parenthesis indicates a very slight change. (Modified from Nomiyama, K. and Nomiyama, H., *Biological Roles of Metallothionein*, Foulkes, E. C., Ed., Elsevier, New York, 1982, 47—67.)

3. Other Signs of Renal Dysfunction

The increased urinary excretion of different proteins, glucose, and amino acids caused by cadmium may be interpreted as indicators of decreased renal function. The tubular reabsorption of peptide enzymes and hormones is also affected and traditional tests of renal tubular function, such as tubular reabsorption of phosphorus (TRP), show a decrease of tubular function. Depending on exposure conditions and dose, glomerular function may also be affected.

Piscator and Larsson[201] gave cadmium as the chloride in drinking water (0 to 25 mg/ℓ) to 14 groups of female rats on normal and low calcium intake. The animals were killed after 2, 6, or 13 months. In rats on a calcium-deficient diet and drinking water with 7.5 mg Cd per liter, there was a highly significant increase in ribonuclease after 13 months, which may be a very early sign of renal tubular dysfunction (Section III.A.2). The mean cadmium level in renal cortex at that time was about 90 mg/kg wet weight. The total protein excretion was not significantly increased in this group of rats compared to control rats on a calcium-deficient diet but without exposure to cadmium.

Nomiyama and co-workers studied rabbits given repeated daily injections of nonradioactive $CdCl_2$ (1.5 mg Cd per kilogram body weight) for various periods of time up to 35 days, after which they measured excretion of PAH, inulin clearance, as well as clearance of ^{115m}Cd using single intraarterial infusions. The studies have been referred to in several publications, but were first described by Nomiyama in 1973.[160] The rabbits must have suffered from considerable renal dysfunction because both TmPAH and inulin clearance had decreased to half after 21 days with daily doses of 1.5 mg Cd per kilogram body weight.[161,162]

Increased urinary excretion of the enzymes alkaline phosphate, ASAT, ALAT, and LDH was found at the same time as proteinuria (after about 10 weeks) in rabbits exposed to cadmium subcutaneously (0.5 mg Cd per kilogram body weight for 21 weeks) (Figure 7). However, no change of creatinine clearance or TRP was found.[165] In monkeys exposed to 300 mg Cd per kilogram in food pellets[173] a decrease of TRP was seen after 28 weeks (about the same time that urinary β_2-microglobulin increased) and a decrease of creatinine clearance was seen after 47 weeks. Also at 100 mg Cd per kilogram in food[172] an effect on TRP and creatinine clearance was seen in monkeys. The day-to-day variation in TRP and creatinine clearance was quite large though, so it is possible that a single test of renal function using these methods would miss the effect.[171] Other signs of renal damage in the monkeys were increases of total protein and β_2-microglobulin in urine seen after 40 to 50 weeks and abnormal protein patterns in SDS electrophoresis (Section IV.A.1).

The biochemical significance of the cadmium accumulation in the kidney and the damage to the tubular reabsorption were discussed by Elinder.[53] He reviewed the data regarding cadmium effects on zinc-dependent enzymes in the kidney. Axelsson and Piscator[10] reported a marked decrease in renal cortex alkaline phosphatase activity of rabbits exposed to s.c. injections of $CdCl_2$ (0.25 mg Cd per kilogram, 5 days/week) for 24 weeks. However, no clear changes in enzyme activity with increasing cadmium dose were seen by Elinder[53] and Elinder and co-workers[54] who studied alkaline phosphatase and leucine amino peptidase activities in renal cortex of slaughtered horses with renal cortex cadmium concentrations up to 170 mg Cd per kilogram wet weight. The alkaline phosphatase activity in serum of rats exposed to cadmium orally was increased in some studies and decreased in other studies.[53] An effect of cadmium on zinc-dependent enzymes in vivo has not been unequivocally documented.

The fact that cadmium can induce both renal glomerular and tubular dysfunction was reported by Friberg as early as 1950. Further support for this combined effect has been given in the studies by Nomiyama and co-workers referred to above. In addition, Kawamura and co-workers[108] reported changes in inulin clearance and PAH clearance in rats given 50 mg Cd per liter in drinking water for 90 days, with or without a calcium-deficient diet. This cadmium exposure decreased the clearance of inulin to about half of the value of the control group and the clearance of PAH to a third of the control group values.

PAH clearance is often used as an indicator of effective renal plasma flow and the decrease was matched by the histological findings of ischemic changes in glomerular capillaries.[108] Histological sections of the tubules showed damage (Section IV.A.4). PAH clearance is not a good indicator of renal plasma flow when tubular damage has occurred. The decrease of inulin clearance shows an effect on the glomeruli as well. In the groups with significant functional change the average renal cadmium concentrations were 91 and 112 mg Cd per kilogram wet weight.[108] In the publication these concentrations are reported on a dry weight basis but this is a misprint (Itokawa, personal communication).

4. Histopathological Findings

In further experiments with rabbits similar to those by Friberg[67] mentioned earlier (Section III.A.1), Dalhamn and Friberg[47] found, by histological examination, that mainly tubular changes occurred, whereas the glomeruli were not affected. Bonnell and co-workers[33] gave rats of both sexes an intraperitoneal (i.p.) dose of cadmium chloride (0.75 mg Cd per kilogram body weight) three times weekly. After 5 to 6 months injections were discontinued for 1 to 2 months. Thereafter, some animals were given a reduced dose while others were given no further injections. The total time for the experiment was 1 year. Every month animals were killed and histological examinations performed. After 4 months of exposure, tubular damage was evident. From that time onwards despite continued exposure there was no further increase in kidney

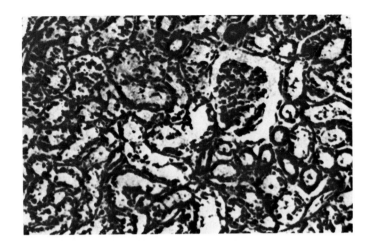

FIGURE 8.   Section of kidney in rabbit exposed to cadmium for 29 weeks (s.c. injections of 0.25 mg Cd per kilogram body weight, 5 days/week). There is marked damage to the tubular cells. They are vacuolated and most lack nuclei. The glomerulus is normal; hematoxylin and eosin staining, approximately × 220. (From Axelsson B., Dahlgren, S. E., and Piscator, M., *Arch. Environ. Health,* 17, 24—28, 1968. With permission.)

levels of cadmium; in fact, some decrease in kidney concentration was demonstrated.[33] After 4 months the renal concentration of cadmium was about 275 mg/kg wet weight and after 5 months about 225 mg/kg.

Histological examination[12] of the kidneys of the rabbits exposed by Axelsson and Piscator[10] (Section IV.A.1) revealed that already after 11 weeks mild alterations had occurred in the proximal tubules. At that time the cadmium concentration in the renal cortex was about 235 mg/kg wet weight (range: 194 to 315). After 23 weeks pronounced changes were found not only in the proximal tubules, but also in other parts of the nephron (Figure 8). The collecting tubules were not detectably affected. Though the glomeruli showed slight alterations in some rabbits, most of them were without demonstrable alterations.

Stowe and co-workers[239] gave ten rabbits cadmium in drinking water (160 mg/ℓ) for 6 months. The mean exposure was 15.5 mg Cd per kilogram body weight per day. Assuming that 1 to 2% had been absorbed, this dose would result in a mean daily absorbed dose of 0.16 to 0.32 mg/kg body weight. Cadmium was determined in whole kidney and histological examinations were carried out by light and electron microscopy. The mean renal concentration was 170 mg/kg wet weight (SD = 14). Extensive interstitial fibrosis was seen by light microscopy in the kidneys of exposed animals. Pronounced changes were seen in the proximal tubules, including pyknosis, karyorrhexis, and epithelial sloughing. Electron microscopy showed seemingly unaffected mitochondria. Collagen was deposited in the glomeruli. Amyloid deposits could not be demonstrated.

Murase and co-workers[144] studied the microscopic pathology of liver, kidney, and some other organs in dogs that had received 500 mℓ of drinking water, containing 100 mg Cd per liter, daily for 13 months. Cadmium in the cells was specifically stained using the "Okamoto" method. Two exposed dogs were compared with one control. There were pathological changes in 30% of the glomeruli and in about 25% of the proximal tubuli.[144] In the mesangium cells of the renal glomeruli and in the proximal tubular epithelium, fine cadmium-stained granules could be seen. These were diffusely spread throughout the cortex and in the same places pathological changes in the cells could be seen.

Among nine rats exposed to a low-calcium diet and 50 mg Cd per liter in drinking water for 90 days[108] there were some who developed severe renal function changes (Section IV.A.3). They had swelling and adhesion to Bowman's capsule of glomerular tufts and necrotic and hyalinized glomerular capillaries. They also had vacuolization, destruction, and desquamation of the tubular epithelial cells. At the end of the experiment the average renal cadmium concentration was about 91 to 112 mg Cd per kilogram wet weight (Section IV.A.3).

Scott and co-workers[220] gave 50 rats repeated s.c. injections of $CdCl_2$ (0.17 mg Cd per kilogram body weight) over a 6-month period. The maximum number of injections was five which is equivalent to a total dose of 0.85 mg Cd per kilogram body weight. No control group was included. All the surviving animals had testicular atrophy indicating acute damage of injected cadmium (Chapter 11, Section VI.A). This appears to conflict with the experience in other studies of cadmium-induced testicular damage (Chapter 11, Section VI.A.2), in which s.c. injection doses five to ten times higher were needed to induce damage. No cadmium analysis of the kidneys was carried out, but the concentrations must have been low as the total dose in, e.g., the study by Axelsson and Piscator[10] (Section IV.A.1), which was 29 mg Cd per kilogram body weight over a 6-month period (34 times greater dose) and those animals had a renal cortex cadmium concentration of about 250 mg Cd per kilogram wet weight after 6 months. However, Scott and co-workers[220] reported areas of desquamation and damage of tubular cells, and the tubules were filled with amorphous material. There was also hemorrhage, fibrosis, and cellular infiltration in the tubules. In the glomeruli there was an increased cellularity and an increase in the basement membranes. Electron microscopy revealed that the tubular epithelial microvilli had some blebbing and contained vesicles with particulate material. There was a general increase in dense bodies and the lining cells of the proximal convoluted tubules had a reduction in cell height and a loss of basal infolding of the plasma membrane.[220] Electron microscopy revealed a variety of findings in the glomeruli. Due to the lack of a control group of rats, and the surprisingly low dose both in terms of causation of testicular damage and in terms of causation of renal damage, these data by Scott and co-workers are difficult to interpret.

The same research group reported on a study of rats exposed to cadmium chloride in drinking water (50 mg Cd per liter) for up to 24 weeks.[7] Sixteen rats were exposed to cadmium and eight control rats were also included. There was a steady increase of kidney and liver cadmium concentrations with time. No quality control data on cadmium analysis were reported except a statement about satisfactory results on NBS bovine liver. After 24 weeks the average concentrations were 40 mg Cd per kilogram in kidney and 25 mg Cd per kilogram in liver. These levels are almost identical to those reached in rats after 37 weeks of exposure to 50 mg Cd per liter in drinking water in a study by Kawai and co-workers[107] (see below). The absorption rate in the study by Aughey and co-workers[7] may have been higher. Aughey and co-workers[7] report minor areas of tubular necrosis and increased cellular appearance of glomeruli already after 8 weeks of exposure, when the renal cadmium concentration was 10 to 15 mg Cd per kilogram wet weight. After 12 weeks the tubular injury was more extensive and after 24 weeks there were obvious areas of tubular necrosis, vascular changes with interstitial edema and glomerular fibrosis, and cell hypertrophy. The control rats showed no evidence of renal damage when examined with light microscopy.

An increase in the dense bodies of the proximal convoluted tubule cells was seen with electron microscopy already after 4 weeks of exposure.[7] By 12 weeks there was considerable tubular cell necrosis and also an increase in glomerular epithelial cell size and fusion of foot processes was common. It is noteworthy that Aughey and co-workers[7] make no reference at all to the earlier research by the same group on rats exposed to cadmium injections.[220]

Renal effects at these very low concentrations have not been reported previously. Kawai and co-workers[107] reported that three of five rats exposed to 50 mg Cd per liter in drinking water, one of five rats exposed to 10 mg Cd per liter, and three of five control rats developed "spontaneous nephropathy" as described by Snell.[232] Rats with higher cadmium exposures did not develop this type of nephropathy but instead developed cadmium nephropathy.[107]

Kawai and co-workers[107] described a series of studies of different animals using different cadmium doses. They divided the renal effects into "acute", "subacute", "subchronic", and "chronic". The more acute lesions had hydropic swelling and acidophilic necrosis as the main feature, whereas the more chronic lesions were characterized by typical tubular atrophy.

In one experiment[107] rats were exposed to s.c. injections of 0.5 mg Cd per kilogram body weight, 6 days/week for up to 25 weeks. After 7 weeks, when the renal cadmium concentration was 160 mg/kg, the kidneys showed slight diffuse interstitial edema and slight swelling of the tubular epithelium. As the exposure continued the damage progressed, until interstitial nephrosclerosis and collagen formation in the widened interstitium and tubular atrophy were seen after 22 weeks. There was no further increase in the renal cadmium concentration. In another experiment, groups of rats were exposed to 10 to 200 mg Cd per liter in drinking water[107] and in one rat, already at a renal cadmium level of 44.4 mg Cd per kilogram wet weight, there were early signs of tubular atrophy and interstitial edema. In the highest dose group the renal cadmium level after 18 months was 156 mg/kg and then a marked interstitial fibrosis could be seen. Electron microscopic observations disclosed marked thickening of the basement membrane around the proximal tubules and changes in the capillaries and arterioles.

Fowler and co-workers[66] also reported renal vascular changes in rats exposed to cadmium chloride in drinking water (0 to 200 mg Cd per liter) for up to 12 weeks. There were constrictions of smaller renal arteries, mild dilatation of larger arteries, and a diffuse dose-related scarring of peritubular capillaries.

The lesions reported in the experimental studies referred to above appear similar to those reported by Elinder and co-workers[54] for naturally exposed horses. These horses had renal cortex cadmium concentrations up to 186 mg/kg. The proximal tubules were dilated with flattened epithelium and often contained PAS-positive homogeneous cases. There was an interstitial reaction with accumulation of lymphocytes, plasma cells, and fibrous tissue. Glomerular changes of mesangial sclerosis and hyalinization were considered by the authors as secondary lesions resulting from tubular dysfunction. For each of the lesions there was a clear dose-response relationship.[54] The increasing prevalence of lesions started at renal cortex cadmium levels below 50 mg Cd per kilogram. At 75 mg Cd per kilogram the frequency of lesions was about 30%.

Recently, it has been reported that seabirds at St. Kilda, U.K., have very high cadmium and mercury exposures due to naturally high cadmium levels in their food and that this cadmium exposure is associated with renal damage in the birds.[148,149] The mean renal cadmium level in three "fulmar" birds was 228 mg Cd per kilogram dry weight (or about 57 mg Cd per kilogram wet weight). In three "Manx shearwater" birds it was 94.5 mg Cd per kilogram dry weight and in two "puffin" birds it was 114 mg Cd per kilogram dry weight. Renal damage in both glomeruli and tubuli as seen using light microscopy and electron microscopy is described in detail.[148] Patchy necrosis of the proximal tubular epithelium was observed, with some cells having pyknotic nuclei and vacuolized cytoplasm. There was only limited change in the glomeruli.

Three groups of starlings were exposed to cadmium in an experiment lasting 6 weeks.[148] One group received s.c. injections of cadmium chloride (2 mg Cd per kilogram body weight) three times per week giving a total dose of 36 mg Cd per kilogram body weight. Another group received 0.16 mg Cd per kilogram body weight up to 13 times, giving a maximum total dose of 2 mg Cd per kilogram body weight, and the last group was injected with saline. The high-dose birds reached renal cadmium concentrations above 160 mg Cd per kilogram dry weight (exact concentrations not given) and the low-dose birds reached 30 to 160 mg Cd per kilogram dry weight. The control starlings had an average renal cadmium concentration of 1.5 mg Cd per kilogram dry weight.

The kidney morphology was normal in all control birds, but in all the cadmium-exposed birds there were renal tubular changes similar to those seen in the wild seabirds from St. Kilda.[148] There was a dose-effect relationship as the severity of the tubular damage increased with increasing kidney cadmium concentration. The high-dose group had some glomerular damage as well as the tubular damage.

Nicholson and co-workers[149] compared the findings in the birds with the findings in a group of 12 mice exposed to cadmium via 10 to 13 s.c. injections. The renal cadmium levels in the mice (110 to 260 mg Cd per kilogram dry weight) were similar to those in seven experimentally exposed starlings (95 to 240 mg Cd per kilogram dry weight) and in nine wild seabirds from St. Kilda (80 to 480 mg Cd per kilogram dry weight). The pathological findings in kidneys were reported to be almost identical[149] even though the seabirds had less severe changes. Mercury injections caused similar type renal damage in the experimental animals[149] and it was pointed out that the seabirds had a combined exposure to both cadmium and mercury which may create more renal damage than each metal on its own.

An unusual their on an "immune complex glomerular nephritis" in rats exposed to either 100 or 200 mg Cd per liter in drinking water for up to 30 weeks[98] is notable, e.g., for the total absence of any literature references to other research on the renal effects of cadmium. It is stated that the "diffuse membranous glomerular nephropathy" was more pronounced in the low exposure group than in the high exposure group. The rats had hyaline droplets and vacuolations in the proximal convoluted tubules. The proximal tubular epithelial cells had granular irregular dense deposits of IgG. It was stated that the IgG was measured by immunofluorescence techniques, but no data were presented.[98] Such dense deposits were also seen in most glomeruli in the mesangial cells. There was irregular thickening of the glomerular basement membrane and the capillary walls were "diffusely and somewhat irregularly thickened". Joshi and co-workers[98] state that these findings are similar to those of clinical glomerulonephritis in man. The mechanism of damage is concluded to be antibody production against tissue damaged by cadmium. The antigen-antibody complex would be deposited in the glomeruli.[98] The study did not include data on the cadmium concentration in the renal cortex and no explanation was suggested as to why the lower dose animals had more severe effects than the higher dose animals.

Another study on rats concluded that cadmium may have induced autoantibody formation.[52] The rats were exposed for 13 months to 100 mg Cd per liter in drinking water (as $CdCl_2$). Tubular dysfunction was detected after 10 months and it is stated that the rats also developed glomerular proteinuria, but no data were presented. About 25% of the poisoned rats had IgG deposits in the glomerular mesangium area after 13 months and after 8 months the rats examined (not stated how many) had circulating antilaminin antibodies.[52] The authors stated that the pathogenic significance of these findings remains to be evaluated.

The monkeys studied by Nomiyama and co-workers[170] (Section IV.A.1) showed not only renal functional damage, but also histological changes. Some animals receiving 3 to 30 mg Cd per kilogram food pellet and all fed 300 mg Cd per kilogram had damage to the proximal tubular cells and changes in the glomeruli. The most common was swelling and disintegration of the brush border in the tubuli. This occurred to some extent also in the lower dose animals. The high dose animals all had atrophy and dilation of the tubules, proteinous fluid in the tubules and focal areas of degeneration to necrosis (Table 3). Interstitial round cell infiltration was seen in almost all exposed animals. Some animals also had thick basement membranes and a swelling of the lining of the Bowman's capsule in the glomeruli.[177] Among these cadmium-exposed monkeys, multiple definite lesions are first seen at a renal cortex cadmium level of 420 mg/kg and multiple slight lesions are seen at a level of 178 mg/kg.

5. Other Indicators of Renal Damage and Factors Affecting Renal Damage

It has been shown in several of the studies referred to above that the renal cortex cadmium concentration often decreases from the time damage becomes apparent, even if the exposure continues. The loss of cadmium can be explained by an increased excretion. Friberg[68] showed that rabbits given daily s.c. injections of radioactive cadmium sulfate had very low urinary excretion of cadmium and that after 2 months of exposure the cadmium excretion increased 50- to 100-fold (Volume I, Chapter 6, Figure 20). After 1 month of exposure the average renal concentration of cadmium was on an average 126 mg/kg wet weight, whereas after $1^1/_2$ months of exposure the concentration was 213 mg/kg. Friberg reported that the concentration in cortex was about 5 times higher than in medulla, which means that in cortex the concentrations would have been about 160 and 270 mg/kg, respectively (1.25 times the renal average), according to the assumption that the mass of cortex is 3 times greater than that of the medulla.[89] This calculated ratio renal cortex to renal average is the same as that estimated for human beings in Volume I, Chapter 6, Section IV.B.2.

In later experiments Friberg,[69] alternating between radioactive and nonradioactive cadmium, demonstrated that a large part of the excreted cadmium must have come from deposits in the kidneys. The findings have been confirmed by Suzuki[243] (Figure 2), Bernard and co-workers[24] (Figures 3 and 4), and Nomiyama and co-workers.[173] A rapid increase in the excretion of cadmium via urine is associated with the progressive renal tubular damage and a decrease of renal cadmium concentration.

In the studies by Nomiyama[160] in which rabbits were exposed to daily s.c. injections (1.5 mg Cd per kilogram body weight) of nonradioactive $CdCl_2$ (Section IV.A.3), there were signs of renal function changes. Nomiyama[160] studied the clearance of ^{115m}Cd using single intraarterial infusions. It increased with increasing renal cortex cadmium concentration. Already at 50 mg Cd per kilogram wet weight in renal cortex there was an apparent increase of cadmium clearance. Nomiyama and co-workers[169] have recalculated the cadmium clearance data to "tubular reabsorption rates" (TR) as TR (Cd) = 1 − (Cd clearance/inulin clearance).

A number of problems arise in the interpretation of these data. First, no allowance has been made for isotope exchange with cadmium in the kidney. Second, the major plasma fraction of ^{115m}Cd is not likely to be filtrable under normal conditions (Volume I, Chapter 6, Section III.A.3). Third, an increased passage of cadmium through kidney may just reflect a decreased availability of brush border binding sites[222] rather than decreased reabsorption into the cells.

Support for an increased excretion of cadmium as an effect of renal damage is given by Bernard and Lauwerys.[18] They gave rats cadmium ions in drinking water (200 mg Cd per liter) over a 1- or 4-month period. At that time intragastric injections of CCl_4 or s.c. injections of Na_2CrO_4 were given. The CCl_4 caused liver damage and a reduction in liver cadmium concentration. The Na_2CrO_4 caused kidney damage with a great increase (about ten times) in proteinuria and a decrease in kidney cadmium concentrations. When the kidney damage occurred there was a dramatic temporary increase in urinary excretion of cadmium. This cadmium is likely to have emerged from the dam-

Table 3
HISTOPATHOLOGICAL FINDINGS IN THE KIDNEYS[177]

Findings	21	30	2	20	32	13	36	29	14	27	15	11	31	25	28
Cadmium in diet (mg/kg)	0	0	3	3	3	10	100	100	10	100	100	30	10	30	30
Renal cortex cadmium concentration (mg/kg wet weight)	5	31	83	178	281	298	420	557	609	612	635	809	842	844	1033
Proximal convoluted tubules															
Dilatation of tubules	±	±		±	±		+	+	±		+			+	+
Proteinous fluid in the tubules	±	±	±	±	+		+	+	±	+	+	+	+	+	+
Swelling of the tubular epithelium		±	±	+						±	+				
Desquamation change									±	±	+			±	
Single cell degeneration					±		+	+	±	+			+	+	+
Focal area of degeneration to necrosis					±		+	+	±					+	+
Swollen eosinophilic large tubules							+	+	+		±			+	+
Mitosis in the tubular linings							+	+						+	+
Hemosiderosis	+							+						+	+

aged renal cells. The increase of urinary cadmium was a function of Na_2CrO_4 dose and of the original cadmium concentration in the kidney.

In addition, this study[18] showed that the acute liver damage also led to an increased urinary cadmium excretion. The increase was a function of the dose of CCl_4 and of the original cadmium concentration in liver. Seven days after the acute liver damage the cadmium concentration in kidneys had increased. This indicates that the liver damage releases cadmium to the blood, and that this cadmium is transported to the kidney. The increased plasma level of cadmium leads to an increased filtered amount of cadmium and an increased excretion.[18] This may explain part of the increased urinary cadmium after cadmium poisoning in the experiments by Suzuki,[243] Bernard and co-workers,[24] and Nomiyama and co-workers[173] (Section IV.A.1).

Glomerular damage in rats induced by puromycin aminonucleoside (AN) did not cause a great difference in renal cadmium accumulation when compared to control rats exposed to cadmium only.[272] The renal tubular effects of cadmium were delayed, however, and this was interpreted by the authors as a sign of an increased glomerular filtration of cadmium bound in such a way that it did not have nephrotoxic effects. Much of the cadmium in plasma is bound to high molecular weight proteins (Chapter 6, Section III.A.3).

The interaction between cadmium and gentamycin has also been studied.[263] Rats were exposed to cadmium in drinking water (300 mg Cd per liter) for 7 months. The average renal cadmium concentration at that stage was 168 mg Cd per kilogram wet weight. The rats did not have proteinuria, but showed a tendency for aminoaciduria.[263]

Daily s.c. injections of 10 mg gentamycin per kilogram body weight were given for up to 20 days. There was a tendency for an increased average proteinuria and aminoaciduria during treatment of the cadmium-exposed rats, but this was not statistically significant. In the control group and in a group exposed to lead, there was an increased aminoaciduria. The hypothesis that gentamycin interacts with cadmium by making the tubular damage worse was not proven.

There is considerable evidence (Chapter 4, Section V) that cadmium in the plasma is partly bound to metallothionein (MT) and that the cadmium-metallothionein complex constitutes the main part of the cadmium filtered through glomeruli and reabsorbed in tubuli (Chapter 4, Section VIII).

An in vitro study with rabbit kidneys[222] showed that cadmium-metallothionein complex binds to isolated renal proximal tubular brush border membranes. There were two parallel processes involved in this binding. It was inhibited by concurrent exposure of the brush border membranes to myoglobin and it was decreased in rabbits pretreated with 1 mmol $CdCl_2$ per liter in drinking water for 6 months. At the time of the binding test, the renal cortex cadmium concentration was 337 mg Cd per kilogram wet weight.

Nomiyama and Foulkes[164] collected urine from cadmium-poisoned rabbits (s.c. injections of 0.5 mg Cd per kilogram a day for 12 weeks; mean renal cortex cadmium was 262 mg/kg) with urethral catheters, while infusing inulin or cadmium-metallothionein (labeled with ^{115m}Cd) through a catheter in the thoracic aorta. Blood samples were collected via a femoral catheter. There was no significant uptake of cadmium-metallothionein by erythrocytes and after 20 min all plasma ^{115m}Cd remained in the low molecular weight fraction upon gel chromatography. The filtered load of cadmium-metallothionein was calculated as the product of its plasma concentration and inulin clearance. The excretion of cadmium was compared with the "filtered load" in control animals (Chapter 6, Figure 22) and in the poisoned animals. The range of filtered loads was 2.2 to 5.0 μg Cd per minute and the reabsorption was calculated as the difference between excretion and filtered load.

Taking the data at face value, Figure 22 in Chapter 6 indicates that there may be a tubular reabsorption threshold at the point where the excretion line reaches zero. This would be at about 1.2 μg Cd per minute. The extrapolation of the "reabsorption" line to zero was not accompanied by any confidence interval for the slope estimate.[164] In an earlier report with a different presentation of similar data the excretion "line" is not presented as a line but a curve, which appears to start close to the origin.[168] Further, as pointed out in Chapter 6, Section V.A.2, the plasma level of cadmium-metallothionein was more than 100 times higher than any levels ever measured in highly exposed humans. It appears unlikely that any person would get such high plasma concentrations that tubular reabsorption capacity is exceeded.

In Section III.A.2, the significance of acute exposure to cadmium-metallothionein as a model for renal handling was discussed. It is likely that the plasma cadmium-metallothionein determines the degree of accumulation of cadmium in the kidneys.

There is a great difference between the body distribution of cadmium after i.v. injections to mice as $CdCl_2$ or as cadmium-metallothionein[183] (see Chapter 4, Section III). Injected cadmium-metallothionein is sequestered almost totally by the kidney (Section III.A.2). The reabsorption of cadmium-metallothionein into renal tubular cells is normally high even though a significant amount is found in the urine of the bladder.[183] An increased urinary metallothionein may indicate tubular reabsorption damage.

Recently a radioimmunoassay method for metallothionein[251,261] has made it possible to study this protein and its association with cadmium in more detail (Sections IV.B.1.d. and IV.B.2.c).

Metallothionein was measured in urine, kidney cortex, and liver of rats exposed to s.c. injections of $CdCl_2$ (0.5 mg/kg body weight, 5 days/week for up to 15 weeks).[252] Similar to the findings of other proteins in urine (Section IV.A.1), there was a considerable increase when the kidney cortex level exceeded 200 mg Cd per kilogram wet weight (Figure 9).

Another aspect of the effect of cadmium on the metabolism of metallothionein is the Zn-Cu-Cd relationship in renal tissue. Nordberg and co-workers[184] measured the zinc, copper, and cadmium concentrations in kidney cortex of 20 horses. These horses had "normal" cadmium exposure from eating grass, etc. in a nonpolluted area of Sweden. The maximum cadmium concentration in renal cortex was 1.5 mmol Cd per kilogram wet weight (= 164 mg Cd per kilogram wet weight). There was an increasing zinc concentration but no increase in copper concentration when the cadmium concentration increased (Figure 10). Elinder[53] and Nordberg and co-workers[184] found that the molar ratio Zn/Cd in purified horse kidney metallothionein was continuously decreasing with the increasing cadmium concentration. The relationship between zinc and cadmium in the metallothionein appeared to be linear. Assuming that the zinc increase in kidney takes place mainly in metallothionein, it can be calculated that in renal tissue the increase of zinc concentration is best described by a second-order mathematical function ($Zn = 0.42 \times Cd_2 + 1.18 \times Cd + 0.36$). This curve was also included in Figure 10 and it fits well with the empirical data. Human data[94,95] also appear to fit best with a second-order mathematical function.

The fact that the zinc concentration in metallothionein decreases when the cadmium concentration increases may be seen as the first biochemical effect of cadmium on kidney cortex. Cadmium present in the kidney is mainly bound to metallothionein produced in the kidney (Chapter 4, and this chapter, Section III.A.2). This metallothionein will also bind zinc, but the affinity for cadmium is higher than for zinc (Chapter 4). A maximum of seven metal atoms can bind to each molecule of metallothionein (Chapter 4, Section IV).

Metallothionein thus acts as a sequestering agent for cadmium in plasma. Various chelating agents have been investigated as to their ability to increase urinary excretion or reduce cadmium damage to several organs (Chapter 6, Section VII.B). Ethylene-

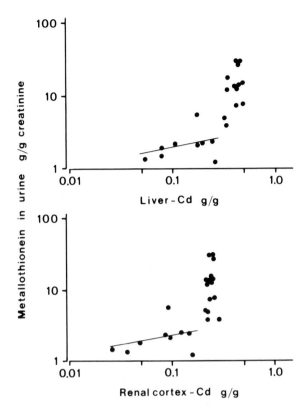

FIGURE 9.   The relationship between urinary metallothionein and tissue cadmium in cadmium-exposed rats. Least-squares lines predicting urinary MT concentrations from tissue cadmium levels are drawn. (Modified from Tohyama, C., Shaikh, Z. A., Ellis, K. J., and Cohn, S. H., *Toxicology,* 20, 181—191, 1981a.)

diamine-tetraacetic acid (EDTA) and diethylene-triamine-pentaacetic acid (DTPA) have been reported to decrease the renal uptake of cadmium,[61,70] whereas 2,3-dimer-capto-propanol (BAL) increased the uptake of cadmium in the kidneys.[150,248] The nephrotoxicity of cadmium was increased by combined BAL and cadmium exposure.[46]

Microscopy of the kidneys of cadmium-exposed rabbits[70] indicated that EDTA exposure after cadmium exposure increased the incidence of renal tubular changes (fatty infiltration and parenchymal degeneration).

The molar ratio cadmium to chelating agent is of importance for the type of interaction seen.[58] This may explain the variation in reported findings (Chapter 6, Section VII.B).

Most of the studies with combined exposure to cadmium and chelating agents have been short term and at high cadmium doses. Apart from the influence on metabolism, these ligands can also influence the acute liver toxicity of cadmium (Chapter 11, Section III.A.1). There is no firm evidence that chelating agents have any value in the treatment of acute or chronic cadmium poisoning.

B. Humans

1. Occupational Exposure

a. Proteinuria

In 1950 Friberg, having investigated a large group of workers exposed to cadmium

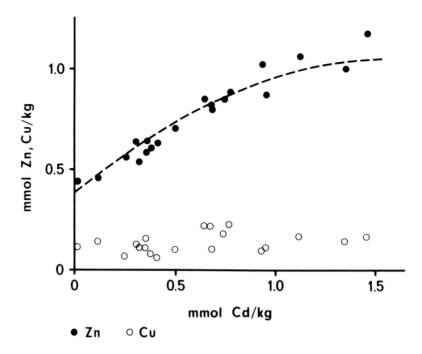

FIGURE 10.   Relationship between cadmium and zinc, and cadmium and copper in horse kidney cortex. Dotted line describes theoretical curve of zinc accumulation at increasing cadmium concentration in kidney. Zinc increase assumed to be related merely to formation of metallothionein, binding both cadmium and zinc. (From Nordberg, M., Elinder, C.-G., and Rahnster, B., *Environ. Res.*, 20, 341—350, 1979. With permission.)

oxide dust in an accumulator factory (for further details see Chapter 8), showed that prolonged exposure to cadmium gives rise to renal damage as well as lung damage. He studied two groups, one of 43 workers with a mean exposure time of 20 years (range: 9 to 34 years), and one of 15 workers with a mean exposure time of 2 years (range: 1 to 4 years). In the group with very long exposure, proteinuria could be demonstrated using the nitric acid test in 65% of the workers and using the trichloracetic acid test in 81%. In the group with shorter exposure, no positive reactions were obtained with any of these tests. The results of these precipitation tests for proteinuria differed from the results of "normal" albuminuria[67] and electrophoretic and ultracentrifugal analyses of the urinary proteins were carried out.[186] It was found that the amount of albumin was less than in other proteinuria cases and that a major fraction migrated as low molecular weight alpha-globulins.

The typical cadmium-induced proteinuria has later been characterized in detail (Section IV.B.3). Electrophoretic, immunoelectrophoretic, and gel filtration studies showed a high excretion of low molecular weight proteins.[197] Quantitative protein analysis showed an increase in the excretion of all plasma proteins, with particularly large increases of low molecular weight proteins.[23,197]

In 1950, Friberg[67] also described the common findings of proteinuria in workers in alkaline accumulator factories in Germany, England, and France. Further evidence for the high prevalence of proteinuria in cadmium workers has been provided by several investigators.[1,13,31,32,49,82,112,131,205,228,229,242,255] From these investigations it is seen that not only cadmium oxide dust, but also cadmium oxide fume, cadmium sulfide, and cadmium stearate may give rise to proteinuria, provided exposure has been prolonged.

The low molecular weight proteinuria in cadmium workers has been studied in detail (Section IV.B.3). β_2-microglobulin[15] has been given particular attention, as it was considered one of the best indicators of early cadmium-induced proteinuria.[122] Recent data show that measurement of some other low molecular weight proteins in urine may have certain advantages (Sections IV.B.4 and IV.B.5).

Some proteins in the urine originate from the renal tubular cells rather than from serum. One of them is carbonic anhydrase C (CAC), and a study of 29 battery factory workers with a mean urinary cadmium of 7.8 $\mu g/\ell$[246] showed that CAC was increased on average 207-fold (3086 $\mu g/g$ creatinine in cadmium-exposed group and 15 $\mu g/g$ creatinine in reference group), whereas β_2-microglobulin increased only 29 times (1429 vs. 49 $\mu g/g$ creatinine). This study did not show whether the increased CAC excretion started earlier than the increased β_2-microglobulin excretion, but the greater relative increase of CAC is of interest. This protein may be an even more sensitive indicator of renal tubular damage.

b. Other Signs of Renal Dysfunction

Tests of renal function[67] disclosed that in the high exposure group (n = 42), nine had decreased ability to concentrate urine (maximum specific gravity less than 1.019). Eight of these nine workers (89%) had persistent proteinuria. In 33 workers with a higher concentrating capacity (maximum specific gravity greater than 1.020), only 11 (33%) had persistent proteinuria. The glomerular filtration rate was affected in the high exposure group, as indicated by a considerable decrease in average inulin clearance (90 mℓ/min, n = 18) compared to a control group (130 mℓ/min, n = 61).[67] Of these 18 workers, 7 had inulin clearance below 80 mℓ/min and 2 workers had clearance below 50 mℓ/min.

Further evidence of disturbed renal function among cadmium-exposed workers was provided by the findings of glucosuria[3,32,112,195,231,242] and aminoaciduria.[3,42,79,112] Goyer and co-workers[79] could not find specific increases in the excretion of certain amino acids as reported by Clarkson and Kench.[42] Toyoshima and co-workers[254] found increased excretion of most amino acids in urine of cadmium workers, but particularly large increases occurred for citrullin and arginine. Bernard[17] found an increased urinary activity of β-galactosidase in workers with long exposure durations. This was interpreted by the author as a sign of cellular lesions in the epithelium of the urinary tract.

Kazantzis and co-workers[112] and Adams and co-workers[3] also found changes in renal handling of uric acid, calcium, and phosphorus. Cadmium-induced proteinuria can, however, appear alone without the other changes. Such changes are more common among the workers with the greatest cadmium dose, and the proteinuria can be considered a relatively early sign of cadmium intoxication.

Iwao and co-workers[93] compared the renal tubular reabsorption of phosphorus and β_2-microglobulin in ten cadmium workers and eight controls. TRP was 87.5% in the control group and 82.8% in the exposed group indicating a 38% increase of the fractional excretion of phosphate (increase from 12.5 to 17.2%). The control reabsorption of β_2-microglobulin was 99.964% and the fractional excretion 0.036%.[93] These empirical values are identical with the estimated "normal" values in Table 1. The equivalent excretion value for the exposed group was 0.42%, 11.7 times higher.[93]

Urinary metallothionein excretion is increased in cadmium workers[41] and there is a correlation between urinary cadmium and urinary metallothionein (Chapter 6, Section V.A.3). The relative increase of average β_2-microglobulin excretion was greater than the increase of average metallothionein excretion[41,182,214] in workers with cadmium effects on the kidney; in workers without such effects[252] there was a threefold increase of both urinary cadmium and metallothionein.

It should be pointed out that there is a considerable variation in the individual exposure and sensitivity to cadmium and that not all highly exposed workers develop proteinuria or other renal effects. Of the 43 workers with the highest exposure in Friberg's study,[67] some had kidney damage without lung damage and some had lung damage without kidney damage. There were five workers with no detected effects. The proteinuria can also vary in intensity and composition of proteins excreted (Section IV.C). Variations between workers can be due to the individual variations in dose inside the same factory (Chapter 3, Section III.B.2), variations in absorption, distribution, and metabolism of cadmium (Chapter 6), or variations in sensitivity of the critical organ (Section VI.B.2). When different studies are compared, differences in methodology may also influence the evaluations (Section IV.B.4).

c. Renal Stones

Friberg[67] noticed that renal stones were a common finding in cadmium-exposed workers. Ahlmark and co-workers[4] found in the same factory that 44% of a group of workers exposed to cadmium dust for more than 15 years had a history of renal stones. The stones were mainly composed of calcium phosphate.[9] An unusually high prevalence of kidney stones has also been found in British accumulator factory workers.[3] It is noteworthy that both studies found a higher prevalence of renal stones in nonproteinuric men, indicating that a more developed tubular dysfunction may mean less risk of developing kidney stones. This finding also indicates that changes in reabsorption of calcium might occur early. However, the methods to detect proteinuria employed by Adams and Axelsson were not sensitive enough to show minor increases in urinary protein excretion (Section IV.B.4).

Scott and co-workers[219] found that 22 of 27 coppersmiths exposed to cadmium had hypercalciuria. Five of these workers had developed renal stones. In another study[221] it was reported that the urinary calcium concentrations were mainly in the range of 6 to 15 mmol/ℓ, with a "normal" range of 3 to 6 mmol/ℓ. The workers had normal serum calcium values indicating that the cause of the hypercalciuria was renal functional change.

In a follow-up examination of 12 workers, Kazantzis[109] described recurrent renal colic with renal stone formation in two workers and nephrocalcinosis in a third worker. These three men all had more than 25 years' exposure to cadmium when they were first seen 7 years earlier.[112] A further follow-up of these workers[110,111] showed that none of the others had developed renal stones within 15 years of follow-up, in spite of three of them having renal tubular damage and hypercalciuria. The renal stone in one of the cases was composed mainly of hydroxyapatite together with a smaller amount of calcium oxalate dihydrate. Kazantzis[110] concluded that the stone formation was due to the high urinary calcium and that renal stones and osteomalacia were serious clinical effects which occurred in a proportion of workers with cadmium-induced renal damage. No reports of renal stone formation have emerged from the studies of cadmium-exposed workers in Japan (Tsuchiya, personal communication) or Belgium (Lauwerys, personal communication).

d. Histopathological Findings

Information about morphological changes in the kidneys is available from a total of 13 autopsies (7 workers exposed to cadmium oxide dust and 6 to cadmium oxide fume) and 10 biopsies (from workers exposed to cadmium oxide dust) (Table 4). In five of the autopsy reports (S.W.H., K.J., H.B., A.B., and case A1) the morphological changes in the kidneys were mainly confined to the proximal tubules, whereas the glomeruli were less affected. Similar but less pronounced changes were seen in three biopsies. In two autopsy cases (cases 1 and 2) there were only arteriosclerotic changes

Table 4

CONCENTRATIONS OF CADMIUM IN KIDNEY CORTEX (MG/KG WET WEIGHT), MORPHOLOGICAL CHANGES (AUTOPSY OR BIOPSY), PROTEINURIA, OTHER RENAL EFFECTS, AND EXPOSURE DATA FOR WORKERS EXPOSED TO CADMIUM OXIDE DUST OR FUMES

Worker	Cadmium in cortex[a] (mg/kg wet weight)	Cadmium in liver	Morphological changes[b]	Protein-uria[c]	Other renal effects	Age (years)	Years exposed	Exposure type and level (mg/m³)	Years since last exposure	Ref.
S.W.H.	68 (A)	88	++	+		46	28	D	1	112
K.J.	63 (A)	30	++	+		49	22	D	9	67, 71
K.N.	16 (A)	145	d	+		57	18	D	6	67, 71
H.B.27	(A)	23	++	+		60	26	D	3	67, 71
G.H.	100 (A)	60	N.M.	+	G	57	20	D	6	67, 71
A.B.	145 (A)	95	++	+		39	16	D	4	13, 88
Case A1	77 (A)	330	++T +G	+	C	54	32	F	2	31
				(urine Cd 30—40 µg/l)						
O.J.	52 (A)	N.M.	++	+		62	20	D	1	198
X.X.	N.M. (B)	N.M.	(+)(T)	+T	G, β_2 PSP, C	51	21	D, 0.14	0	83, 84
		(restrictive respiratory incompetency in spirometry)								
		(urine Cd 43—143 µg/l)								
G.J.	321 (B)	N.M.	(+)	+		44	11	D	12	11
G.K.	152 (B)	N.M.	(+)	+		46	15	D	0	11
L.F.D.	97 (A)	172	(+)	+	Am, G	71	36	D, 0.4	11	2
Case 1	N.M. (A)	N.M.	(+)Art	−		53	2	F	5	230
Case 2	150 (A)	160	(+)Art	+T		46	9	F	5	230
A.L.	220 (B)	N.M.	−	(+)		36	6	D	0	11
E.Y.	446 (B)	N.M.	−	−		39	7	D	0	11
E.H.	320 (B)	N.M.	−	+		40	15	D	10	11
N.U.	180 (B)	N.M.	−	−		44	12	D	0	11
H.N.	21 (B)	N.M.	−	+		45	13	D	0	11
K.N.	190 (B)	N.M.	−	−		50	15	D	2	11
Case 3	360 (A)	300	−	−	C	55	7	F	6	230

Table 4 (continued)
CONCENTRATIONS OF CADMIUM IN KIDNEY CORTEX (MG/KG WET WEIGHT),
MORPHOLOGICAL CHANGES (AUTOPSY OR BIOPSY), PROTEINURIA, OTHER RENAL EFFECTS,
AND EXPOSURE DATA FOR WORKERS EXPOSED TO CADMIUM OXIDE DUST OR FUMES

| | (mg/kg wet weight) | | | | | | | | Exposure type and level | Years since last | |
Worker	Cadmium in cortex[a]	Cadmium in liver	Morphological changes[b]	Protein- uria[c]	Other renal effects	Age (years)	Years exposed		level (mg/m³)	exposure	Ref.
Case 4	395 (A)	145	—	+T		57	8		F	19	230
Case 5	165 (A)	65	—	+		67	8		F	10	230

Note: A = autopsy; B = biopsy; G = glucosuria; β_2 = high β_2-microglobulin; PSP = low PSP test; C = low concentrating ability; Art = arteriosclerosis; Am = aminoaciduria; T = tubular changes or proteinuria; D = dust; F = fume.

[a]   Underlined figures are based on cadmium concentrations in whole kidney, assuming that the cadmium concentration in cortex is 1.25 times the average kidney concentration (Volume I, Chapter 6, Section IV.B.2).

[b]   — = no morphological changes; (+) = slight morphological changes; ++ = profound morphological changes.

[c]   — = negative results on repeated testing with trichloroacetic acid; (+) = varying results; + = positive results on repeated testing.

[d]   Results from histological examinations not reported, but at examination in 1946[67] this worker had the lowest kidney function tests of all (inulin clearance: 42 ml/min; concentration capacity: 1.016; NPN: 44 mg%).

in the kidneys; the other cases showed morphological changes, the exact nature of which was not specified. A detailed description of the changes has been given by Bonnell:[31] "The tubules were often grossly atrophied, containing hyaline casts. Other tubules were dilated containing brightly eosinophil granular casts mingled with leucocytes. The epithelium was flattened and the distinctive features of the proximal and distal parts of the nephron were lost. There were fatty-hyaline changes in the arterioles, the larger arteries showed moderate hypertrophy. Many of the glomeruli had undergone ischemic atrophy in association with fatty-hyaline degeneration of the smaller arterioles. Residual glomeruli frequently appeared unchanged."

The data in Table 4 indicate that cadmium levels in kidney cortex of some of the workers investigated were low, comparable "normal" levels (Volume I, Chapter 5, Section IV.B). The renal cortex levels were lower in cases with morphological changes than in cases without detectable morphological alterations, or with only minor changes. As exposure times were longer in the former cases and the air concentrations of cadmium conceivably higher (after 1950 improvements in working conditions were made in, e.g., the Swedish factory studied by Friberg in 1950; see Kjellström and coworkers[122]), the lower renal levels in these cases cannot be explained by less exposure.

There is no apparent relationship between renal cadmium concentration and the time that had elapsed between the end of exposure and autopsy or biopsy. The combination of heavy exposure to cadmium, severe renal damage, and relatively low levels of cadmium in the kidneys indicates instead a considerable excretion of cadmium (Section IV.A.5). This is likely to be associated with tubular cell necrosis and sloughing into the lumen, as seen in acute exposure experiments with cadmium-metallothionein (Section III.A.2). In workers with no impairment or only minor dysfunction, the renal losses must have been smaller and thus, the renal levels of cadmium were higher.

Animal experiments (Section IV.A) have shown that when renal signs and morphological changes become evident, the renal levels of cadmium cease to rise (Figure 4) and may decrease[172,243] in spite of continued exposure. These findings agree with the findings in human beings.

To summarize, pathological findings in severely cadmium-poisoned workers include damage to the proximal renal tubules, sometimes associated with glomerular damage. In several cases proteinuria was found without morphological changes, but there were no cases with the opposite picture. Apparently the functional changes in the kidney can occur before the microscopic structure of the kidney cells is severely damaged. On the other hand, the human data on pathological changes are limited and animal data (Section IV.A.4) show that in some studies morphological changes in the kidney tubules emerge before measurable proteinuria. When a person suffers severe renal damage through the toxic action of cadmium, his kidney concentration of cadmium will decrease considerably. Thus, he may have lower levels in the kidneys than a person with only slight or no renal dysfunction.

2. General Environment Exposure
a. Proteinuria and Glucosuria

In the cadmium-polluted areas of Japan, proteinuria, glucosuria, and other signs of renal dysfunction have been relatively common findings. The most extreme cases have been those with renal dysfunction and bone symptoms, often diagnosed as "Itai-itai disease" (see Chapter 10, Section IV and Appendix). Hagino[80] found proteinuria in 82% of 71 surviving cases of Itai-itai disease and 32% had glucosuria (analytical methods not stated). A detailed clinical study of 30 patients,[146] all of whom had severe cadmium-induced bone effects, showed that all patients had proteinuria (sulfosalicylic acid method) and some had glucosuria as well (Nylander method). These patients had received no treatment with vitamin D before the tests (Nakagawa, personal communi-

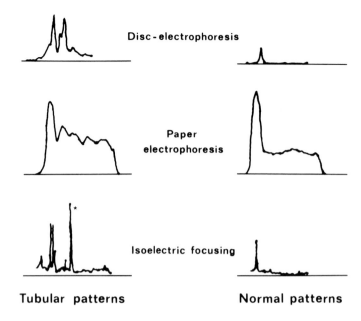

FIGURE 11.   Comparison of typical tubular and normal patterns of
three electrophoretic methods in analysis of the same urine samples. The
large peak to the left contains albumin and the star denotes the distinct
β_2-microglobulin peak in isoelectric focusing. (From Shiroishi, K., Kjellstrom, T., Piscator, M., Iwata, I., and Nishino, H., *Environ. Res.,* 13,
407—424, 1977. With permission.)

cation). Treatment with large doses of vitamin D to alleviate the bone effects has been
hypothesized as a cause of the renal damage seen in Itai-itai disease,[99] but the available
data (see Appendix) does not support this. A number of other studies have confirmed
the findings of proteinuria and glucosuria in the most severe cadmium-poisoned cases
(see for instance the reviews by Tsuchiya[257] and Nogawa[151]). The majority of affected
people may have only proteinuria, with or without glucosuria. The sensitivity of the
methods for proteinuria and glucosuria measurement strongly influences the findings
(Section IV.B.4). In one study of 138 exposed women (not Itai-itai cases),[121] 33 had
proteinuria according to a qualitative precipitation method (sulfosalicylic acid
method). Fourteen of them had glucosuria (enzymatic tape method) and only 8 of the
other 105 women had glucosuria without proteinuria (according to this method). However, 48 of the women had tubular patterns with disc electrophoresis and 63 had an
increased urinary excretion of β_2-microglobulin.[121]

The proteinuria of the people exposed in the general environment is characterized by
a high proportion of low molecular weight proteins,[86,153,216] (Figure 11). This is in
agreement with the findings in cadmium-exposed workers (Section IV.B.1). Quantitative measurements[226] have shown that when total urinary protein excretion was increased 7 to 11 times, the urinary excretion of β_2-microglobulin was increased 92 to 280
times in 10 Itai-itai patients and 32 people with renal dysfunction from the same area.

The health screening programs in the Japanese cadmium-exposed areas have included measurement of proteinuria and glucosuria. In several areas there was an increased prevalence of proteinuria and/or glucosuria as compared to reference areas
(Section V.B).

b. Other Signs of Renal Dysfunction

In patients with Itai-itai disease the renal tubular dysfunction was evident also in the form of decreased excretion of phenolsulfonphtalein (PSP).[146] The lower normal limit of excretion is 25% after 15 min, but in seven of the ten patients tested the excretion was 10 to 22.5% and for the other three patients it was 25%.[146] Many of the patients also had aminoaciduria, reduced ability to dilute and concentrate urine and to reabsorb phosphorus.[151] The thiosulfate and PAH clearance were low.[146]

Of 147 cadmium-exposed people in another Japanese area,[216] 33 had concurrent proteinuria and glucosuria, and 7 of them had generalized aminoaciduria. Of the 33 cases, 13 were studied in detail in a hospital. Measurement of fasting blood sugar and the glucose tolerance test showed that the glucosuria was of renal origin in all 13 cases. Seven had a decreased tubular reabsorption of phosphorus and eight had a reduced ability to concentrate urine. Saito and co-workers[216] also found some indications of glomerular damage. Most of the 13 hospitalized patients had decreased endogenous creatinine clearance at below 60 mℓ/min; in 2 cases the clearance was very low and the plasma creatinine concentration was elevated. The general picture was that of "multiple proximal tubular dysfunction".[216]

The bone disease may modify the findings. For instance, the excretion of arginine, citrulline, proline, and hydroxyproline are particularly increased in Itai-itai disease,[76] whereas among cadmium workers[254] the latter two amino acids are not so greatly increased (Section IV.B.1.b) The increased urinary proline and hydroxyproline excretion in Itai-itai disease are likely to be a sign of severe bone disease, which the cadmium workers studied by Toyoshima and co-workers[254] did not have. The mechanism underlying this may be an effect of cadmium on collagen metabolism (Chapter 10, Section III.C). Aminoaciduria with a particularly high proline excretion was also found in the five cases of bone disease outside the original Itai-itai disease area.[153]

In 1973, Watanabe and co-workers[265] reported an increased urinary copper excretion among people in a cadmium-exposed area. It was proposed by Watanabe and co-workers that high copper exposure in the area could be the cause of the proteinuria seen in the area. A recent study by Nogawa and co-workers[159] has shown that the increased copper excretion is most likely another indicator of renal tubular dysfunction caused by cadmium. From a polluted area, 69 women, including 40 Itai-itai disease patients or suspected patients, and 31 controls were studied for urinary cadmium, copper, zinc, and β_2-microglobulin and serum copper, zinc, and β_2-microglobulin. There was no difference in the serum copper or zinc levels between exposed (excluding Itai-itai cases) and controls,[159] but there was a tendency for a slight increase (5 to 10%) in serum β_2-microglobulin. The same comparison for urinary levels showed a 30 to 50% increase of copper in three age groups and a 100 to 6700% increase of β_2-microglobulin. The Itai-itai cases have greater increases of these urinary excretions. The increased cadmium excretion in urine in each of the three age groups of exposed women was about 300% and this did not correlate with the urinary copper excretion. It appears therefore that urinary copper may be seen as another indicator of renal dysfunction after cadmium exposure.

Some studies have also used urinary activity of N-acetyl-β-D-glucosaminidase (NAG) as an indicator of such dysfunction.[87,158] This enzyme is abundant in renal tubular cells and is released to urine if tubular damage occurs.[48,267] Honda and co-workers[87] showed an increased urinary NAG activity in women and men from a cadmium-polluted area. There was also a lesser increase with age. In the age group 60 to 69 years the NAG increase when compared to a control group was about 100% and the increase of β_2-microglobulin excretion was about 400 to 600%.[87] Itai-itai disease patients have somewhat greater increases of NAG activity and dramatically (100 times) greater increases of β_2-microglobulin excretion.

In a study of Itai-itai disease patients and other people from Japanese cadmium-polluted areas, Tohyama and co-workers[253] measured cadmium in urine as well as metallothionein and β_2-microglobulin. The inhabitants of a "nonpolluted area" had a higher than "normal" urinary cadmium at 5.7 μg/g creatinine (Table 5). In spite of the severe reabsorption losses as indicated by an increased excretion of β_2-microglobulin, up to more than 300 times (Table 5), the metallothionein excretion only increased about 5 times, or to a similar degree as the cadmium excretion. This agrees with the findings in cadmium workers (Section IV.B.1.b). Thus, in human beings at least, the urinary excretion and renal reabsoprtion mechanism does not seem to be the same for methallothionein and β_2-microglobulin. Urinary metallothionein may just reflect urinary cadmium excretion.

Thus, in terms of the relative increase of urinary excretion, β_2-microglobulin (and other low molecular weight proteins, e.g., RBP) (Section IV.B.2.a) are the most sensitive indicators of the cadmium-induced renal damage.

c. Histopathological Findings

No biopsy data about renal effects are available from people exposed to cadmium in the general environment, but there are both published and unpublished data on a total of 26 autopsy cases from Japanese cadmium-polluted areas. Some control cases of lower cadmium exposure also have been reported. A summary of data in several reports is given in Table 6. In cases where histological investigations of the kidney were carried out, damage to the tubules was seen often.

A detailed description of the findings in one case is given in Takebayashi[245] (case 1T): "The renal tubules were damaged remarkably, resulting in obstruction of the lumen. Many of the epithelial cells were degenerated or exfoliated. The thickening and wrinkling of the basement membrane was remarkable. Some dilated renal tubules were filled with casts. The interstitium showed infiltration by lymphocytes at some localities but little fibrosis. The glomeruli were relatively well maintained in number and size, and those with pertinent glomerular capillary loop were numerous. The arcuate arteries showed a moderate intimal thickening, mild arteriosclerosis, and no hyaline thickening. There was no inflammatory reaction in the renal pelvic mucosa and papilla. The collecting tubule was relatively well maintained. In some glomeruli, fibrin thrombi were formed in the capillary loops".

This description of the renal tubular damage is rather similar to that occurring in workers with cadmium poisoning (Section IV.B.1.d) even though natural arteriosclerotic and other age-related changes in the kidneys are likely to be more common in the Japanese cases, because of older age than was the case for the autopsied workers (Tables 4 and 6).

In agreement with the finding for the workers (Section IV.B.1.d), those cases with the most extreme damage to the kidneys seem to have lower renal cadmium levels than those with less damage (Table 6). This supports the concept of a considerable loss of cadmium from the kidney when damage is severe (Section IV.A.5). The liver values are very high (20 to 150 mg/kg wet weight) (Table 6). Details about 11 cases of Itai-itai disease were published by Kajikawa et al.[100] (see Appendix). Five of these cases were among those shown in Table 6. The additional cases also had histopathological signs of renal damage, but often not as specific findings of "tubular nephropathy" as in four of these five cases (see Appendix).

Data on cadmium concentrations in renal cortex of six highly exposed cases and three control cases, but with no accompanying morphological data, are also available (Table 7). These people lived their entire lives in a polluted area of Akita prefecture, a different polluted area than those represented in Table 6 and in the Appendix. In this polluted area the average daily intake of cadmium was reported to be 139 μg/day[128] and in a control area it was 40 μg/day. The average renal cortex level for the six cases in the polluted area was 146 mg Cd per kilogram wet weight and in the three control cases it was 45 mg Cd per kilogram wet weight, matching closely the daily intake data.

Table 5

URINALYSIS OF CADMIUM-EXPOSED FEMALE SUBJECTS

Group	No. of subjects	Age (years)	μg/g creatinine		
			Cadmium[a]	Metallothionein[a]	β_2-microglobulin[a]
I. "Itai-itai" disease patients	18	75.3[b] (60—84)	13.5[b] (9.1—18.8)	1880[b,c] (1580—2230)	91,200[b,c] (75,400—110,000)
II. Suspected patients	21	72.1 (61—84)	16.4[b] (13.4—19.9)	2000[b,c] (1630—2450)	102,000[b,c] (86,200—112,000)
III. Inhabitants of Cd-polluted area	15	74.5 (66—90)	13.8[b] (10.1—18.7)	880[b] (588—1320)	5,150[b] (1,590—16,700)
IV. Inhabitants of nonpolluted area	13	68.7 (60—83)	5.7 (4.3—7.5)	394[d] (292—533)	294 (84—1,030)

Note: Statistical expressions used are age — arithmetic mean and range; cadmium, metallothionein, and β_2-microglobulin — geometric mean and 95% confidence intervals.

[a] Because of the use of spot urine samples, urinary levels of cadmium, metallothionein, and β_2-microglobulin were corrected by specific gravity (not shown here) and by creatinine. Statistical treatment of the data corrected by specific gravity gave results similar to those obtained after correction by creatinine excretion.

[b] Significantly different from group IV ($p < 0.05$).

[c] Significantly different from group III ($p < 0.05$).

[d] The metallothionein level of one sample was below the detection limit (146 μg/l); for conservative statistical comparison, the MT level in this sample was assumed to be equivalent to the detection limit.

From Tohyama, C. and Shaikh, A. A., *Toxicology*, 20, 181—191, 1981. With permission.

Table 6

RENAL AUTOPSY DATA FROM PEOPLE EXPOSED IN THE GENERAL ENVIRONMENT

Patient	Cadmium in kidney cortex (mg/kg wet weight)	Cadmium++ in liver	Morphological changes[a]	Protein-uria[a]	Other renal effects[b]	Age/sex	Exposure area	Cause of death	Ref.
CITE-1	41	94	+(T)	+	Glu	79F	Fuchu	Stomach cancer	92
CITE-2	32	118	++(T)	++	Glu	71F	Fuchu	Bronchopneumonia	92
CITE-3	20	63	Pyelonephritis	+		60F	Fuchu	Endocarditis	92
CITE-4	89		++(T)	++	Glu	73F	Fuchu	Stomach cancer	92
CITE-5	12	132	++(T)	++	Glu	67F	Fuchu	Uremia	91
CITE-6	29	35	+				Annaka		162
CITE-7	134	32	+				Annaka		162
CITE-8	264	29					Annaka		162
Case 1N	53	75		++(T)	Glu	74F	Ikuno	Osteomalacia	153
Case 1M	52	76	+(T)	+(T)	Glu, Am	76F	Ikuno		137
Case 2M	80	92	+(T)	+(T)	Glu, Am	83M	Ikuno		137
Case 3M	146	44	+(T)	+(T)	Glu, Am	84M	Ikuno		137
Case 4M	31	71	+(T)	+(T)	Glu, Am	76F	Ikuno		137
Case 5M	87	21				69M	Ikuno		137
Case 1T	24	153	++(T)	+(T)	Glu, Am, TRP	75F	Tsushima		245

Note: Cases 1N and 1M are possibly the same person. + CITE is short for "Cadmium in the Environment".[74] These cases were mentioned in that book. ++ "Normal" values in Japan 2 to 5 mg/day/kg (Volume I, Chapter 5, Section V).

[a] (T) = tubular damage or tubular proteinuria.

[b] Glu = glucosuria; Am = aminoaciduria; TRP = tubular reabsorption of phosphorus decreased.

Table 7

RENAL CADMIUM AND ZINC CONCENTRATIONS IN AUTOPSY CASES
FROM KOSAKA TOWN

| | (mg/kg wet weight) | | | | | | |
| | Cadmium | Zinc | Protein- | Urine cadmium | | | |
Patient	in cortex	in cortex	uria	(μg/l)	Age	Sex	Cause of death
In exposed area							
4	155	64			45	M	Meningitis
7	114	60			68	F	Stroke
19	167	96			64	M	?
31	114	120			73	F	—
37	75	43	(+)(T)	7.3	68	F	Heart disease
36	248	168	(+)(T)	36.0	73	M	Cancer
In control area							
21	48	49			68	F	?
24	41	92			73	M	—
17	47	57			63	M	?

From Kojima, S., unpublished data.

Two exposed cases with renal cortex cadmium levels of 75 and 248 mg/kg wet weight were reported to have tubular proteinuria. No proteinuria data were seen in the other cases.

The autopsy data from the people exposed in the general environment showed a good correlation between findings of microscopic changes and proteinuria or other signs of renal effects. People with severe renal damage had low renal cortex cadmium levels in spite of liver levels showing high body burdens.

3. Characterization of the Cadmium-Induced Proteinuria and the Early Renal Effects

In 1958 Butler and Flynn[38] described the so-called tubular proteinuria which they found in cases with tubular dysfunction, such as in the Fanconi syndrome, galactosemia, etc. This proteinuria was characterized by a relatively small albumin fraction and dominance of proteins with the mobility of α-, β-, and γ-globulins in paper electrophoresis. By comparing urinary proteins from cadmium workers with urinary proteins from persons with other tubular disorders, Butler and Flynn,[39] Butler and co-workers,[40] Piscator,[193] and Kazantzis and co-workers[112] were able to show that the proteinuria in chronic cadmium poisoning was similar to the tubular proteinuria described by Butler and Flynn in 1958.[38] Typical electrophoretic patterns of urinary proteins in chronic cadmium poisoning are shown in Figure 11.

Further investigations by Piscator[195,196] showed that the proteins excreted by cadmium workers consisted mainly of low molecular weight proteins, chiefly derived from the serum. Among these were β_2-microglobulin, enzymes such as muramidase and ribonuclease, and immune globulin chains, the latter constituting nearly half of the low molecular weight fraction. It was not possible to show the existence of any proteins specific for cadmium poisoning.

As discussed in Section II, it is to be expected that there will be an increase in all serum proteins in the urine when the renal tubules are damaged. Damage to the glomerulus causing an increased filtration of proteins into the primary urine would, however, selectively increase the excretion of high molecular weight proteins, such as albumin (Table 1).

Electrophoretic studies[81] using isoelectric focusing[262] established quantitatively the

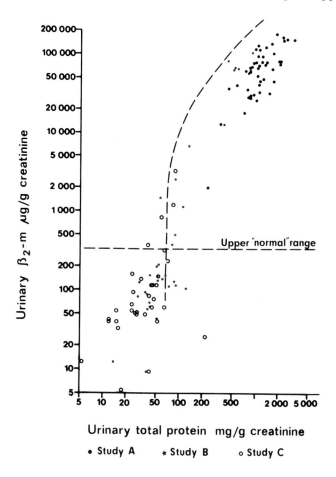

FIGURE 12.   Comparison of urinary concentrations of total protein
(Tsuchiya-Biuret method) and β_2-m (radioimmunoassay method).
Study A = certain and suspected Itai-itai disease patients.[226] Study B
= workers in a cadmium battery factory with 1 to 30 years' exposure.[199] Study C = workers in a cadmium battery factory with 1 to 14
years' exposure. The dashed line indicates the equal tubular reabsorption loss line, Figure 1, microglobulin, and total protein. (Modified
from Kjellstrom, T. and Piscator, M., *Quantitative Analysis of β_2-
Microglobulin in Urine as an Indicator of Renal Tubular Damage Induced by Cadmium,* Phadedoc No. 1, Diagnostic Communications,
Pharmacia Diagnostics AB, Uppsala, Sweden, 1977, 3—21.)

relative excretion of different proteins in 55 cadmium workers. The group of workers
with more than 25 years' exposure duration had a tenfold increase in urinary albumin
excretion compared to the group with less than 2 years' exposure and for β_2-microglo-
bulin the corresponding relative increase was about 100-fold. Hansen and co-workers[81]
found the highest absolute excretion of an individual protein zone for the zone containing albumin. In the highest cadmium dose group the absolute excretion of albumin
was about three times higher than the absolute excretion of β_2-microglobulin.

In another report analyzing the characteristics of the cadmium-induced proteinuria,[120] data on total protein and β_2-microglobulin in urine were compared. Figure 12
shows the combined findings of three studies and the "equal reabsorption reduction
curve" from Figure 1. It is seen that in the cases with a limited increase of β_2-microglo-
bulin over the upper "normal" range, the data closely follow the curve. In the cases

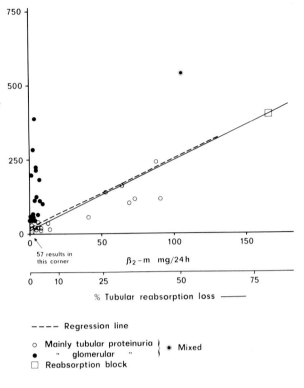

FIGURE 13.   Comparison of urinary β_2-microglobulin and albumin in individual cadmium-exposed workers (O,●), and the average of a group with almost complete tubular reabsorption block (□).[138]

with extremely high β_2-microglobulin excretions there is a higher total protein excretion than the values indicated by the curve. This may indicate a certain glomerular component of the proteinuria in cases with severe cadmium effects.

Human experimental studies, in which volunteers were given i.v. injections of amino acids to block renal tubular reabsorption,[138] supported the above findings. The "base line" urinary excretion of albumin (11 mg/24 hr) was about 100 times greater than the excretion of β_2-microglobulin (0.10 mg/24 hr). When tubular reabsorption of certain substances was blocked by high doses of lysine, the urinary excretion of albumin increased 40 times to 400 mg/24 hr, whereas the excretion of β_2-microglobulin increased 1600 times to 160 mg/24 hr. Still, the latter protein was excreted at less than half the level of albumin.[138]

The data from several recent studies of urine proteins in cadmium-exposed workers and from human experimental studies of urine proteins during i.v. infusions of amino acids that block the renal tubular reabsorption were pooled by Kjellström and Elinder.[119] Figure 13 shows the good agreement between the theoretical relationship of albumin to β_2-microglobulin and the observed relationship in many of the cases of cadmium poisoning and the cases of renal tubular reabsorption block. The findings in most cadmium poisoning cases appear therefore to be explainable mainly on the basis of a tubular reabsorption defect. In some cases an effect on glomerular filtration of proteins may exist concurrently with the tubular defect, but glomerular effects are not required to produce the characteristic "cadmium proteinuria". As mentioned in Section IV.B.1.b, glomerular damage does occur in some cadmium workers. In such cases the proteinuria will contain both tubular and glomerular components with a proportionately higher excretion of albumin than the values given by the curve in Figure 1.

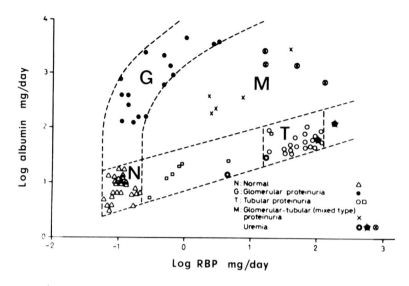

FIGURE 14.   Relationship between urinary excretion of albumin and retinol-binding protein (RBP) in different types of proteinuria. The urinary excretion of albumin is plotted on a double logarithmic scale against that of RBP of the same person. The distributions separate four different zones of normal (N), glomerular (G), tubular (T), and mixed glomerular-tubular (M) types of proteinuria. "N" includes distribution area of the results of normal persons, "G" that of the results of patients with common renal diseases, "T" that of patients with "Itai-itai" disease (O), and other etiologies (□). Uremic patients in each group are shown by the symbols (⊗, O) and those with multiple myeloma by the symbol (✻). (Redrawn from Kanai, M., Sasaoka, S., and Naiki, M., *Proc. Symp. Chem. Physiol. Pathol.,* 12, 325—330, 1972b.)

As is seen in Figure 13, there are some cases with prominent glomerular proteinuria components. This occurs also in control groups, however, and may in individual cases be unrelated to cadmium exposure (see below in this section).

In cases of high cadmium exposure from the general environment a similar relationship between urinary albumin and retinol-binding protein was found.[102] The authors outlined different zones of urinary protein findings (Figure 14) and defined these as "tubular", "glomerular", or "mixed" proteinurias.

Recent cases of cadmium-induced proteinuria have been diagnosed as "glomerular" in a series of studies of Belgian cadmium-exposed workers.[22,23,35,131] These are discussed in some detail here, because the diagnosis of the proteinuria is fundamental for the evaluation of the mechanism of cadmium toxicity. Lauwerys and co-workers[131] studied a group of 22 workers exposed to cadmium for 21 to 40 years. Urine specimens were dialyzed and the urine proteins were concentrated by ultrafiltration under negative pressure with Visking dialysis tubing 8/32 or by lyophylization after salt removal by passage through a dextran gel column (35 × 2 cm) with deionized water as eluting solvent. Unfortunately, this method for protein concentration is likely to cause considerable losses of low molecular weight proteins, such as β_2-microglobulin.[195] The urines were concentrated about 100 times only, which is not sufficient to give clear patterns in electrophoresis.[197,200] In addition, β_2-microglobulin is unstable at urinary pH levels below 5.6 (Section IV.B.5), a factor which was not considered in these studies. Lauwerys and co-workers concluded that the electrophoretic patterns were abnormal in 15 of the 22 workers and that 7 of these patterns were "glomerular" type and that the other eight were mixed "glomerular-tubular" type. Because of the methods used it is difficult to interpret the results.

In another study,[22] immunological methods for quantification of albumin and other proteins and radioimmunoassay for β_2-microglobulin were used in addition to the other analyses. This time the authors reported 10 abnormal electrophoretic patterns in 18 of the exposed workers. Two of these patterns were of "selective glomerular" type and eight were "mixed glomerular-tubular" type. There were no "pure low molecular weight protein patterns",[22] but the authors did not define the appearance of this pattern. It is interesting to note, though, that in the whole cadmium-exposed group (n = 18) the urinary total protein concentration as compared to a reference group (n = 21) increased by 6.9 times; urinary amino acid concentration increased by 2.6 times; the urinary albumin concentration increased by 1.2 times; and the urinary β_2-microglobulin concentration increased by 34 times.[22]

In a more extensive study of 42 cadmium-exposed workers,[23] the same type of electrophoresis as that performed earlier was used and 5 cases (12%) were classified as "glomerular" type patterns, 2 (5%) were "tubular" type patterns, and 8 (19%) were "mixed" type patterns. In a control group of 77 men, 4 (5%) "glomerular" type patterns and no other abnormal patterns were found. Some of the "glomerular" pattern cases found among the cadmium-exposed workers may thus be unrelated to cadmium. The prevalence of an increased functional clearance (above the control group mean + 2 SD) of individual proteins (C protein/C creatinine) was about 20 to 30% for all of the proteins analyzed (β_2-microglobulin, orosomucoid, albumin, transferrin, IgG).[23] In the control group, however, no subject had an increased β_2-microglobulin clearance, 1.3% had an increased IgG and transferrin clearance, 3.9% had an increased orosomucoid clearance, and 5.2% had an increased albumin clearance. There was a good correlation between urinary cadmium concentration and clearance of individual proteins. The relative increase of β_2-microglobulin clearance was higher than for the other proteins, which confirms the earlier studies.[81,226] The increase of the average proteinuria of the cadmium-exposed group as compared to the control group[23] was 31 times for β_2-microglobulin, whereas orosomucoid, albumin, transferrin, and IgG increased 11, 5.4, 4.7, and 3.6 times, respectively.

Bernard and co-workers[23] concluded that these findings can be explained by an effect of cadmium on the renal tubular reabsorption of proteins, but they point out that a direct effect of cadmium on the glomerulus should also be considered as the high molecular weight proteins are the most important quantitative components of the proteinuria caused by cadmium.

It is our opinion that taking the data as a whole, the calculated relative increases of excretion of high molecular weight and low molecular weight proteins do fit with the calculated expected increases based on a tubular reabsorption defect (Figure 1). Thus, the findings above can be explained mainly by renal tubular damage. Clinical studies support this view. Primary damage to the renal tubule gives rise to renal functional changes and is very similar to the Fanconi syndrome.[30] Kazantzis and co-workers[112] diagnosed one of their cadmium-poisoning cases as Fanconi syndrome and in a later study of the same workers[110] this case had developed renal osteomalacia. His biochemical abnormalities included tubular proteinuria, renal glycosuria, abnormal aminoaciduria, with particularly high hydroxyproline excretion (at the time he developed osteomalacia), polyuria with low osmolality, inappropriate urinary pH in response to an acid load, low plasma potassium and phosphate, high normal chloride, low normal bicarbonate, a low creatinine clearance, and high phosphate and uric acid clearances.[110] The reports on renal function changes show that, depending on the degree of poisoning, findings of glomerular damage as well as tubular damage can be seen in some cadmium workers. The glomerular damage may be secondary to the tubular damage.

It was shown by Friberg[67] that some of the more heavily cadmium-exposed workers had renal glomerular damage (as indicated by the low inulin clearance) as well as tubular damage. Proteinuria due to an increased glomerular filtration of proteins can be caused by a number of not so uncommon factors, such as, e.g., hypertension.[140,190,191] Thus, some proteins in the urine of a cadmium worker may be due to causes other than cadmium.

A late indicator of glomerular damage, which is not confused by the tubular reabsorption, is serum creatinine. Adams[1] showed that the vast majority of serum tests carried out on a group of 23 cadmium-poisoned workers with proteinuria showed increased serum creatinine levels. He defined proteinuria as a total protein excretion above 1 g/ℓ (or micrograms per gram creatinine) which corresponds to a β_2-microglobulin excretion above 100,000 μg/g creatinine (Figure 12) indicating a rather severe renal dysfunction. At that stage, glomerular involvement may thus be present in addition to tubular damage. An increased serum creatinine above the normal range only occurs when the GFR has decreased considerably.

There is no doubt that in severely poisoned cadmium workers and inhabitants of polluted areas, both tubular and glomerular damage may be seen. At an early stage of poisoning the renal function tests and urinary findings are in agreement with a renal tubular reabsorption defect. The urine will show an increased excretion of both high and low molecular weight proteins, but the relative increase of the latter is much greater than the relative increase of the former. At later stages of poisoning, more or less severe renal glomerular function defects may be seen.

4. Comparison of Methods for Measuring Proteinuria

The methodology and technology for analysis of proteins in urine have improved considerably since the first studies of cadmium poisoning. Princi[208] concluded that prolonged cadmium exposure did not cause proteinuria. However, he used the boiling test for his studies (Princi, cited by Friberg, 1950[67]) and this test is not sensitive to the urinary proteins of cadmium workers.[67,193] Other qualitative precipitation tests for proteinuria are more sensitive. Friberg[67] showed with repeated Heller tests (more than 10 times each) constant proteinuria in 5 out of 16 workers with long exposure, and in an additional 5 workers this test showed intermittent proteinuria in more than half of the tests. When Friberg[67] tested workers with both the Heller test and the trichloracetic acid (TCA) test, the latter was more sensitive.

A detailed comparison of the different qualitative methods[193] showed that the TCA test was more accurate and sensitive than Albustix, Esbach's test, or the boiling test.

In another comparison of urines from 32 suspected Itai-itai patients[226] it was found that all 32 had "tubular proteinuria" as defined by three different types of urine electrophoresis, but only 5 were clearly positive in Albustix (25 were +/−), 16 were clearly positive in the TCA test (12 were +/−), and 21 were clearly positive in the sulfosalicylic acid test (SA) (6 were +/−). Thus, the SA test may be even more sensitive than the TCA test.

Piscator[193] showed that neither TCA nor SA totally precipitates all the protein in urine. The best precipitation was achieved by the "Tsuchiya" phosphotungstic acid reagent. This was confirmed by Shiroishi and co-workers[226] who also showed that the sulfosalicylic acid precipitation is insufficient when the total urinary protein concentration is below 250 mg/ℓ.

The development of immunoassay methods has made it possible to measure specific proteins quantitatively. Nephelometric immunoassay[210] can be used for, e.g., albumin, orosomucoid, transferrin and IgG, and radioimmunoassay (RIA) for β_2-microglobulin[60] and retinol-binding protein (RBP).[102] Single radial immunodiffusion (SRI) has been used to measure quantitatively β_2-microglobulin,[114] albumin, and RBP.[101] The SRI with a detection limit of about 1 mg/ℓ is much less sensitive than the RIA with a detection limit of about 1 μg/ℓ. More recently a lateximmunoassay (LIA) method has been developed by Bernard and co-workers for analysis of β_2-microglobulin,[26] RBP,[27] and albumin.[20] This method is similar in sensitivity to the RIA method and has recently been developed for automated analysis.[19,21]

It is clear from the findings presented in Section IV.B.3 that measurement of total proteinuria alone is not a specific or sensitive way of diagnosing cadmium-induced proteinuria. It is the relative increase from normal excretion of low molecular weight proteins (such as β_2-microglobulin or RBP) and of high molecular weight proteins (such as albumin) that shows whether tubular proteinuria occurs and to what degree it has developed in an individual.

The relative increase of different proteins can also be measured by electrophoresis of a concentrate of the urinary proteins[194] or by gel filtration of the protein concentrate.[196] As a diagnostic procedure, electrophoresis has been most widely employed. Piscator[195] refined the method for paper electrophoresis of the proteins. Clear tubular and glomerular patterns could be seen with this method, but the separation of the proteins is not good enough for a semiquantitative estimate of the concentration of each protein. It was also shown by Piscator[195] that the often utilized Visking 8/32 dialysis tubes had pores which were too large to keep low molecular weight proteins in the concentrate. This has been confirmed by Kanai and co-workers.[103] A collodion membrane or a Visking 24/32 dialysis tube is better because they are not permeable for the proteins of interest.[195]

Polyacrylamide disc gel electrophoresis (Figure 11) and agarose-electrophoresis give a better separation of the proteins than paper electrophoresis (Figure 11), and these have been widely used in diagnosis of cadmium poisoning in Japan.[226] Even better separation is achieved by isoelectric focusing (Figure 11).[81]

When the cadmium damage to the kidney is severe (as in the 32 suspected Itai-itai disease cases mentioned above), the electrophoresis methods agree well and quantitative analysis of a low molecular weight protein also shows a pathologically high value.[226] In cases of less severe damage there is some overlap and cases declared "normal" with one method would be declared abnormal with another.[226] This may explain some of the findings in the Belgian cadmium workers (Section IV.B.3.).

It appears that the quantitative measurement of low molecular weight proteins is the most sensitive for cadmium-induced proteinuria and this method has the advantage of not being influenced by the technique for protein concentration. Kjellström and co-workers[121] studied 138 women with high cadmium exposure via food in a Japanese area; 63 had urinary β_2-microglobulin excretion above the 97.5 percentile of a reference group, 13 of them had clear and 48 had suspected tubular patterns in disc electrophoresis; 13 of them had clear proteinuria (SA method +) and 20 suspected proteinuria (SA method +/−).

The ideal method to diagnose cadmium-induced proteinuria would be to measure the clearance of a low molecular weight (LMW) protein, the clearance of albumin and GFR. This involves measuring the LMW protein, albumin and creatinine, inulin or Cr-EDTA in urine as well as in serum. With this information a rather clear picture of the function of the renal glomeruli and tubuli would be obtained.

The low molecular weight proteinuria in cadmium poisoning is relatively stable over different 6 hr periods during 1 day[136] so the timing of sampling can be made for convenience and to avoid low pH or long storage of the urine.[59] The common radioimmunoassay method to measure β_2-microglobulin is so sensitive that large volumes are not needed. The ideal may be to ask the participants to void urine in the morning and then collect a sample 4 hr later.

5. Some Features of β_2-Microglobulin and Retinol-Binding Protein

The main reason why β_2-microglobulin and retinol-binding protein (RBP) have been used as indicators of cadmium proteinuria is that sensitive and accurate RIA methods of analysis have been developed and also that they normally occur in serum at relatively constant concentrations and even a relatively limited increase in urine would be due to a change in renal function.

Human β_2-microglobulin has a molecular weight of 11,600.[15] The amino acid sequence is identical to a chain in IgG.[16] β_2-microglobulin has cytophilic properties and is found on the surface of most nonmalignant cells. On the surface of each mammalian lymphocyte there are about 100,000 β_2-microglobulin molecules. Malignant human lymphoma cell lines, however, have lower amounts of β_2-microglobulin on their surface. β_2-microglobulin is identical to the small invariant polypeptide chain of the serologically defined strong histocompatibility antigens, and it appears to be involved in the mechanism of lymphocyte activation and other immunological mechanisms.[206] It has been suggested that β_2-microglobulin and immunoglobulins evolved from the same ancestral gene, which would explain an evolutionary link between histocompatibility antigens and immunoglobulins. In spite of a wealth of research on this topic in recent years, the function of β_2-microglobulin is still not clear. The data point to β_2-microglobulin and β_2-microglobulin-associated molecules playing a role in immunological cell-cell recognition.[16]

The serum level of β_2-microglobulin in an individual stays relatively constant at about 2 mg/ℓ,[16] but in cases of lymphomas it may increase up to 10 to 15 mg/ℓ.[207] The serum level will increase when glomerular filtration rate decreases in the same way that serum creatinine increases.[16] Serum β_2-microglobulin and β_2-microglobulin clearance have therefore been used to monitor glomerular function in cases of renal failure. The kidney is a major organ in the catabolism of β_2-microglobulin.[44] Much of the β_2-microglobulin that passes through the kidney in plasma is filtered into primary urine as the sieving coefficient is high, 70% (Table 1). Almost all β_2-microglobulin in primary urine is reabsorbed in the renal tubules (Table 1) and catabolized into peptides and amino acids. In "normal" urine the β_2-microglobulin concentration is on average about 75 μg/24 hr (Table 1) or 100 μg/g creatinine.[120]

Saito et al.[217] studied 25 cadmium-exposed people in a Japanese polluted area. When urinary β_2-microglobulin was less than 10 μg/min (or 14,400 μg/day), plasma β_2-microglobulin varied between 1.5 and 4.5 mg/ℓ (Figure 15) and the increases in urinary β_2-microglobulin were not matched at all by similar increases in serum β_2-microglobulin. The figure shows the typical increase of urinary β_2-microglobulin explained mainly by tubular damage. There was also a very good correlation between urine β_2-microglobulin and β_2-microglobulin clearance (as a percentage of creatinine clearance). The reason for the somewhat higher serum β_2-microglobulin values in some cases could be the involvement of glomerular damage in these cases with severe cadmium-induced renal dysfunction.

Studies by Buchet and co-workers of workers exposed to cadmium, lead, or mercury;[35] workers exposed to lead and cadmium simultaneously;[36] and control workers showed that even when the blood cadmium or urine cadmium levels were increased on average 10 to 15 times over controls (Table 8) there was no increase of serum β_2-microglobulin or creatinine and no decrease of creatinine clearance. No functional glomerular changes were seen in these measures and still both urinary β_2-microglobulin and albumin were increased (Table 8). The relative increase of β_2-microglobulin was the greatest, as expected when tubular reabsorption damage occurs (Section II). The findings in Table 8 cannot be explained by an exceeded tubular maximum reabsorption for β_2-microglobulin or albumin. There was no difference in the renal functional changes between cadmium workers and cadmium-lead workers (Table 7). Animal data in support of these findings in people exposed in the general environment[217] and cadmium workers[35,36] have been provided by Piscator and co-workers.[203] Rabbits were exposed

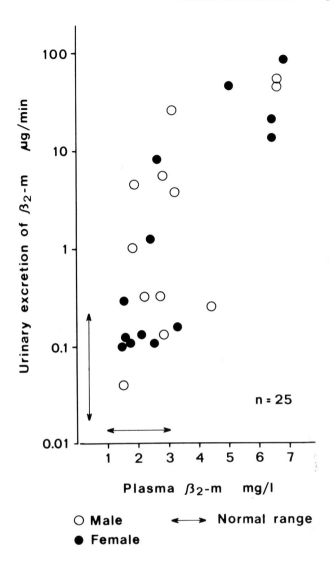

FIGURE 15.   Serum and urine β_2-microglobulin in 25 cadmium-exposed people from an area in Akita Prefecture, Japan. (Redrawn from Saito, H., Nakano, A., Tohyama, C., Mitane, Y., Sugihira, N., and Wakisaka, I., *Jpn. J. Hyg.,* 37, 245, 1982.)

to s.c. injections of cadmium chloride (0, 0.25, or 0.5 mg Cd per kilogram body weight) three times per week for up to 16 weeks. Blood specimens were collected at weeks 0, 3, 7, and 19 and urine specimens at weeks 0, 3, and 7. Even though the blood cadmium increased dramatically (Table 9), there was no increase in serum creatinine or β_2-microglobulin at any sampling time or any dose. After 7 weeks the urinary β_2-microglobulin had increased about 5 times in the rabbits given injections of 0.25 mg Cd per kilogram body weight and about 30 times in the higher dose rabbits (Table 9).

A correlation between blood cadmium (range 12 to 27 μg/ℓ) and serum β_2-microglobulin (range 0.1 to 3.9 mg/ℓ) in 18 cadmium workers was reported by Tsuchiya and co-workers.[259] No data on serum creatinine were given. The findings may be in agreement with the correlation between serum creatinine and serum β_2-microglobulin (range 0.9 to 3.9 mg/ℓ) reported by Kjellström and Piscator[120] in 33 cadmium workers. As

Table 8

PLASMA AND URINE LEVELS OF β_2-MICROGLOBULIN
(β_2-m) AND ALBUMIN (Alb) IN WORKERS WITH
EXPOSURE TO CADMIUM OR A COMBINATION OF
CADMIUM AND LEAD[35,36]

	Mean values for three groups of workers		
	Control	Cadmium	Cadmium-lead
No.	88	148	62
Blood-cadmium (μg/ℓ)	2.3	18.8	15.3
Urine-cadmium (μg/g creatinine)	0.88	15.8	15.5
Plasma-β_2-m (mg/ℓ)	1.70	1.78	1.91
Plasma-Alb (g/ℓ)	41.5	41.9	42.0
Plasma-creatinine (mg/ℓ)	9.4	9.1	8.8
Urine-β_2-m (μg/g creatinine)	71[a]	739	752
Urine-Alb (μg/g creatinine)	4950[a]	11650	10040
Creatinine clearance (mℓ/min) (GFR)	138	141	140
Urine-β_2-m (relative increase)	1	10	11
Urine-Alb (relative increase)	1	2.4	2.0

[a] One extreme outlier excluded.

Table 9

CADMIUM IN BLOOD (μg/ℓ), CREATININE (mg/ℓ),
β_2-MICROGLOBULIN-CREATININE RATIO IN SERUM IN CONTROLS
AND EXPOSED RABBITS (MEAN VALUES, RANGE WITHIN
PARENTHESES)[203]

				Cd-exposed			
Weeks		Controls		0.25 mg/kg		0.5 mg/kg	
	n	3		5		4	
0	CdB	<1		<1		<1	
	cr	9.4	(7.2—11.8)	9.0	(7.5—10.8)	8.1	(7.4—9.2)
	β_2-m	3.7	(2.52—5.02)	3.1	(2.64—3.40)	3.2	(2.42—3.4)
	β_2-m/cr	0.39	(0.35—0.43)	0.35	(0.29—0.37)	0.40	(0.26—0.45)
3	CdB	<1		381	(312—583)	717	(389—1102)
	cr	9.1	(8.4—10.1)	8.9	(7.2—9.7)	7.6	(7.2—8.1)
	β_2-m	3.5	(2.96—4.48)	3.2	(3.02—3.32)	3.2	(3.16—3.56)
	β_2-m/cr	0.38	(0.29—0.50)	0.36	(0.31—0.44)	0.42	(0.36—0.49)
7	CdB	<1		450	(259—726)	805	(614—1124)
	cr	8.1	(5.3—9.4)	8.3	(6.3—9.8)	7.6	(7.2—7.9)
	β_2-m	2.6	(2.30—2.90)	3.0	(2.64—3.32)	3.3	(2.64—3.82)
	β_2-m/cr	0.34	(0.28—0.43)	0.36	(0.33—0.42)	0.43	(0.35—0.48)
19	CdB	<1		520	(374—784)	804	(506—1327)
	cr	9.5	(8.9—10.3)	9.5	(7.2—11.2)	9.7	(9.0—11.6)
	β_2-m	3.2	(2.76—3.4)	3.3	(2.90—3.56)	3.4	(2.58—4.00)
	β_2-m/cr	0.33	(0.27—0.37)	0.35	(0.31—0.40)	0.35	(0.28—0.38)

Urinary β_2-microglobulin (mg/g creatinine)

	n	2		5		3	
7	β_2-m-u	140	(110—170)	730	(110—1570)	4380	(2620—7220)

mentioned earlier (Section IV.B.1.), workers with severe cadmium poisoning may have glomerular damage as well as tubular damage.

There are also some conflicting data indicating the existence of a tubular maximum reabsorptive capacity for β_2-microglobulin, but this can be seen only at very high serum β_2-microglobulin levels. Wibell and Evrin[269] suggested such a threshold existed at around 4.5 mg/ℓ in serum. Karlsson and co-workers[106] proposed a threshold at 8 mg/ ℓ. The filtered load would increase with the serum level and when the serum level increases above the threshold level no more filtered β_2-microglobulin can be reabsorbed by the tubuli and the urinary excretion increases rapidly. This seems unlikely as an explanation for the findings among cadmium-exposed people, because the urinary levels do not increase suddenly after a serum level of 4.5 mg/ℓ is reached (Figure 15). High urinary levels were found at much lower serum levels. Similarly, the cadmium workers studied by Buchet and co-workers[35] had very high increases of urinary β_2-microglobulin, even though their average plasma β_2-microglobulin was the same as in the control group (Table 8) and the range of plasma values was the same (0.9 to 3.64 mg/ℓ in 86 controls, 0.7 to 4.27 mg/ℓ in 148 cadmium workers).[35] The serum β_2-microglobulin levels may be as high as 10 to 30 mg/ℓ among certain blood cancer patients.[207,227] However, such patients were found to have a lower clearance of β_2-microglobulin than controls,[105] which does not support the tubular maximum theory.

Retinol-binding protein (RBP) is a "carrier protein" for vitamin A (retinol) in serum.[101] It has a molecular weight of 21,000, is filtered easily through the kidney glomeruli, and is reabsorbed in the tubuli to the same degree as β_2-microglobulin. The transport of retinol and metabolism of RBP is schematically shown in Figure 16. Vitamin A (retinol) and its precursors are absorbed from the intestines into epithelial cells, where they are converted into retinolesters and transported to the liver via chyle. In the liver RBP is synthesized. The retinol-free RBP is called apo-RBP and the retinol-RBP complex is called holo-RBP. Retinol and RBP bind in the liver and holo-RBP is transferred to the blood serum. Holo-RBP has a strong affinity for prealbumin (PA) (molecular weight = 61,000) and a complex between the two is formed in the blood. To a great extent this prevents the glomerular filtration of the retinol. In peripheral tissues the complex is dissociated to form free holo-RBP from which retinol is released and deposited into target cells. The apo-RBP is formed, and because of its lesser affinity for prealbumin than holo-RBP, free uncomplexed apo-RBP in blood will be available for glomerular filtration. The filtered apo-RBP is then reabsorbed and catabolized in the tubuli in a similar way as β_2-microglobulin (Figure 16). RBP is located in the lysosomes of the proximal tubules of patients with surgically removed kidneys.[260] This is in agreement with a reabsorption of RBP in this part of the kidney.

One disadvantage of β_2-microglobulin as compared to RBP is that it is unstable at urinary pH below 5.6.[59] Such low pH levels occur normally in a proportion of urine samples. The β_2-microglobulin will be degraded in the upper urinary tract, the bladder, and in the sample collection bottle, leading to measured concentrations which are erroneously low. Bernard and co-workers[26] compared the effect of urinary pH on measured concentrations of β_2-microglobulin and RBP in spot samples of urine from 150 people. A latex immunoassay method was used for the analysis.[25]

Above pH 6 there was no apparent variation in the concentrations of either of the two proteins, but below pH 6 the β_2-microglobulin decreased progressively with decreasing pH. In the urines of patients with renal diseases and urinary pH greater than 5.6, the concentrations of the two proteins were similar.[26] The urinary pH can be raised in vivo by consumption of sodium bicarbonate and it can be raised in the urine collection bottle by addition of an alkaline buffer. Still, the need to check urinary pH when β_2-microglobulin is used as an indicator of cadmium-induced renal tubular damage is a disadvantage.

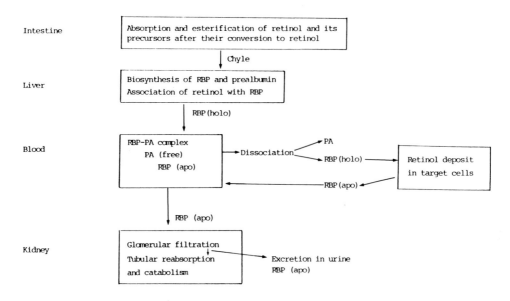

FIGURE 16.   Transport of retinol and metabolism of retinol-binding protein in human body. (Modified from Kanai, N., Nomoto, S., Sasaoka, S., and Muto, Y., *Proc. Symp. Chem. Physiol. Pathol.*, 12, 319—324, 1972a.)

β_2-microglobulin in urine can also be degraded by bacterial urinary infection.[185] On the other hand, β_2-microglobulin in urine has been reported to be increased in febrile patients[233] and in acute pancreatitis patients.[104] Circulating immune complexes (CIC) were found in the former patients and in the latter the serum analyses levels were very high. It may be that the renal handling of these substances could influence the tubular reabsorption of β_2-microglobulin.[104,233] Retinol-binding protein levels were not reported in any of these studies, so it is not known whether changes in urinary levels similar to those for β_2-microglobulin occur for RBP.

It has been suggested that urinary β_2-microglobulin increases with age to such a degree that its use in screening for cadmium poisoning is compromised.[117,169,259] It has been shown that in elderly people from low cadmium exposure areas (above age 70) a higher proportion of increased urinary β_2-microglobulin is seen than in younger age groups,[120,129] but the individual variations are great and also very low levels are more common.[120] Therefore, the average level (107 μg/ℓ) in the oldest age group studied,[129] 70 to 74 years, was only marginally higher than in the other age groups between age 20 and 70 (69 to 84 μg/ℓ). No age effect on urinary β_2-microglobulin was seen up to age 70, which agrees with the data by Kjellström and Piscator.[120] Nomiyama and co-workers[169] reported no changes in the age range 60 to 80 in a nonpolluted area in Tochigi prefecture (Nomiyama, personal communication), but after that the average urinary β_2-microglobulin increased with age. More complete data from the same area were reported by Nomiyama and co-workers.[175] In one of the less-polluted areas (Annaka) no age trend in average urinary β_2-microglobulin was seen for men,[169] but there was a tendency for an increase after age 60 in women. The average levels (about 100 to 200 μg/ℓ) were similar to what has been found in control groups.[120] The age-effect reported in control groups[174,259] is likely to disappear if the most elderly people are excluded. The age-effect in exposed groups[174,259] is to be expected as the cadmium dose increases with increasing age (Section V.B).

C. Mechanism for the Development of Renal Damage

In this section an attempt is made to describe the probable mechanism for cadmium-induced renal damage. To a great degree reference will be made to previous sections in this chapter and Volume I, Chapters 4 and 6, rather than to original data. Some aspects of this mechanism need further study.

It has been shown that during long-term chronic exposure, cadmium will accumulate in the kidneys. Only small amounts of cadmium are excreted in the urine and the half-time of cadmium in the kidneys is several decades (Chapter 6, Section V.D). The urinary excretion has been shown to be related to body burden, on a group basis (Chapter 6, Section VI.C). The accumulation of cadmium in the kidney takes place mainly in proximal tubular cells (Chapter 6, Section IV.B.1). This is due to the fact that cadmium in plasma is transported bound to metallothionein, which is readily filtered through the glomeruli and almost completely reabsorbed in the proximal tubules, as are other proteins (Section II). The cadmium-methallothionein complex will be transported by pinocytosis across the membrane of the renal proximal tubular cells (Section III.A.2) and will be found in the lysosomes. There the complex is broken down into "free" cadmium ions and peptides and amino acids (Chapter 6, Section IV.C). The "free" cadmium ions are likely to rapidly bind to other ligands. This catabolism of metallothionein is likely to be similar to the renal metabolism of a great number of proteins in the body.[30]

Presumably, nonmetallothionein-bound cadmium inside the renal tubular cells stimulates local metallothionein production in these cells and this metallothionein normally binds the cadmium (Chapter 4, Section V). If this binding does not take place, it is envisaged that the nonmetallothionein-bound cadmium will bind to metalloenzymes, particularly those with zinc, causing them to become inactive. Other destructive chemical reactions with the "free cadmium" may also occur. The newly produced metallothionein has a very strong affinity not only to cadmium, but also to zinc and copper. These are essential elements with homeostatically controlled levels in the cells. Zinc will bind to this metallothionein and new zinc will enter the cell to replace the bound zinc. Thus, the renal zinc level increases, but as has been shown by Elinder[53] in horses and by Iwao and co-workers[94,95] in humans, the zinc increase may follow a second-order function with a maximum. This is likely to be due to a limited capacity of the renal proximal tubular cell to produce metallothionein. If so much cadmium enters the cell that the available metallothionein cannot bind any more cadmium, a rapid increase in the concentration of "free cadmium" inside the cell is likely to follow and cadmium interference with renal tubular cell function occurs.

The first sign of cadmium damage will be a decrease of proximal tubular reabsorption capacity, as has been discussed in detail above (Sections IV.A.3, IV.B and IV.B.3). Low molecular weight proteins will be excreted at a relatively much greater degree than the major protein constituents in urine (Sections IV.A.1, IV.B.1, and IV.B.3). A large increase in the excretion of metallothionein is therefore also expected. An increase has been seen in animals (Section IV.A.5) and in humans (Sections IV.B.1.b and IV.B.2.b), but it is not as great as for β_2-microglobulin and is very similar to the increase of urinary cadmium studied in people (Chapter 6, Section V.A.3). Metallothionein in urine can come from two sources: nonreabsorbed part of primary urine from plasma and glomeruli or excretion from the tubular cells of the metallothionein produced there. The proportion of metallothionein in urine from these sources is not known.

The mechanism responsible for the loss of tubular reabsorption is not known. The tubular cells may be intact but the reabsorptive capacity is decreased because enzymes involved in this process are inactivated by cadmium. After more severe damage, the whole cells are destroyed by cadmium and the cell contents are excreted into urine (Sections III.A.2, IV.A.4, IV.B.2.c, and IV.B.1.d).

The animal data reviewed earlier (Section IV.A.5) show that urinary cadmium excretion increases when renal damage starts to occur and at some point in time a very rapid and large increase in excretion occurs. The latter finding supports the theory of excretion of the contents of the whole cell, bringing the accumulated cadmium with it. Further support is provided by the histological findings (Sections IV.A.4, IV.B.1.d. and IV.B.2.c) and the decreasing renal cadmium levels in spite of continued exposure. Experimental renal damage due to an agent other than cadmium has been shown to increase urinary cadmium dramatically (Section IV.A.5). In humans it is also clear that renal cadmium concentrations decrease when severe damage has occurred (Sections IV.B.1.d and IV.B.2.c). An alternative explanation for the increased urinary cadmium levels at the time damage occurred could be a rapid increase in the outflow of cadmium-metallothionein from the liver due to liver damage. There may be a maximum metallothionein production capacity in the liver, which limits the storage of cadmium in that tissue. Cadmium is in itself toxic to the liver (Chapter 11, Section III) and in animal experiments when renal damage has occurred the liver cadmium levels often decrease (Section IV.A). Liver damage caused by another agent increases urinary cadmium (Section IV.A.5). The situation after a massive release of cadmium-metallothionein from the liver may be similar to the acute experimental situations described in Section III.A.2.

As the tubular damage progresses the glomeruli become affected (Sections IV.A. and IV.B). The mechanism for this involvement of the glomeruli is not known. Even with very severely damaged tubuli though, the function of the glomeruli is maintained to some extent. Cadmium-induced renal damage rarely leads to uremia in humans.

A question which has not yet been resolved is the relationship between degree of cadmium damage, plasma cadmium levels, and renal cadmium levels. In experiments with parenteral injection exposure the plasma, kidney, and liver levels of cadmium build up rapidly, whereas peroral exposure can only manage to bring the levels up much slower. This is due to the limits of oral exposure levels related to acute effects on the intestines (Chapter 11, Section II). Injection exposures may cause very high peak cadmium levels in plasma and it has been suggested by Bernard and co-workers[28] that at the same renal cadmium level the relative damage to the kidney is more severe due to the peak exposure effect. This hypothesis needs further confirmation.

D. Prognosis of the Cadmium-Induced Renal Dysfunction

The reversibility of cadmium-induced renal dysfunction has not been studied extensively in animals. One animal experiment which indicated that cadmium-induced renal damage may be reversible is a study by Axelsson and Piscator[10] (Section IV.A.1). Rabbits were exposed to s.c. injections of $CdCl_2$ (0.25 mg Cd per kilogram body weight, 5 days/week) for 24 weeks. They had significant increases of urinary protein excretion from week 12 and onwards. The rabbits were followed up for 30 weeks after end of exposure. The proteinuria first increased and then decreased to the value at end of exposure. Due to individual variations, there was no difference in proteinuria between exposed and control rabbits 30 weeks after exposure ended. After 23 weeks of exposure there was severe damage to the renal tubules[12] and in some rabbits there was also mild glomerular damage. Thirty weeks after exposure was discontinued the tubular epithelium had regenerated in most animals, whereas mild glomerular changes remained in some rabbits.

Nomiyama and Nomiyama[176] gave rabbits pelleted food containing cadmium chloride (300 mg Cd per kilogram food) for 44 or 19 weeks. A maximum cadmium concentration in renal cortex at 300 mg Cd per kilogram wet weight was reached already after 9 weeks' exposure. After that the concentration decreased in spite of continued exposure. The rabbits with the long exposure period (44 weeks) showed "only little recovery" from cadmium-induced proteinuria and aminoaciduria. Nomiyama and Nomiyama[176] concluded that the rabbits of the second group recovered from the renal effects of cadmium, as the "proteinuria and aminoaciduria in most animals disappeared soon after the end of cadmium exposure". The data are not presented in such a way that quantitative comparisons can be made and the authors use no statistical methods to evaluate any changes with time. The proteinuria in the short exposure groups was intermittent before and after cadmium exposure ended. The report states that "urinary protein was elevated again after the 23rd week of recovery" and that "it took 26 weeks for the *depressed* urinary amino acids to recover to the control level". These data do not show conclusively that renal effects of cadmium are reversible.

Most studies of cadmium-exposed human beings report no improvement in the proteinuria when exposure ceases and one study (of five workers) reports the opposite. Friberg and Nyström[73] repeated qualitative proteinuria tests on some of the workers with cadmium-induced proteinuria first diagnosed by Friberg.[67] The proteinuria was either the same or in some cases worse than in the previous study. Piscator[193] studied a group of 40 workers of whom 39 had been included in the report by Friberg.[67] By using a quantitative method for the determination of the total urinary protein, Piscator[193] showed that in these workers, who had not been exposed to cadmium for the last 10 years, there was still a relationship between the total cadmium dose and urinary protein excretion.

Half a year later, Piscator[193] examined urine samples from 14 workers in the same group. Essentially, the same protein excretion values were obtained, indicating that this proteinuria was constant. In additional follow-ups it was found that in 18 workers reexamined in 1963[195] and in 26 workers reexamined in 1969,[198] from the above-mentioned group of 40 workers, the protein excretion in most cases remained at a high level 3 and 10 years after the first quantitative determinations in 1959. Most of these workers had been removed from cadmium exposure or had experienced a dramatic decrease in the exposure level.[122] As Friberg had not used a quantitative method for the determination of protein, a similar comparison could not be made with his original results. However, a comparison was made between the results of nitric acid and trichloracetic acid tests from the two investigations. No marked changes had occurred in the group.[193]

Piscator[200] reported on further follow-up investigations in this factory. Four workers, whose GFR had been measured in 1947 (range 78 to 132 mℓ/min), had been retested in 1969 (range 61 to 85 mℓ/min) and in 1978 (range 34 to 68 mℓ/min). There is a clear indication of a decreasing glomerular function with time and in each case the GFR is lower than "normal" values. Total proteinuria was first measured in 1959 and in these four and another three workers concentrations between 280 and 1300 mg/24 hr were found[200] ("normal" value 60 mg/24 hr; Table 1). Between 1969 and 1978 the concentrations tended to increase, particularly in the workers with the lowest starting values. (The range in 1978 was 370 to 1400 mg/24 hr.[200])

Piscator[200] also compared renal function and urinary concentration variables for 16 workers from the same factory tested in 1969 and 1982. All of them had much fewer renal effects of cadmium than the workers mentioned above. Piscator divided them into three groups: those with β_2-microglobulin clearance less than 0.1% of GFR, those with clearance above that, and those who had been on heavy medication for hypertension or heart congestion. The "normal" β_2-microglobulin clearance can be calculated from Table 1 as 0.027% of GFR. The cut-off level for the two groups would therefore be at about a fourfold increase of β_2-microblobulin excretion. The first group with low β_2-microglobulin clearance in 1982 had "normal" values of serum creatinine, GFR, serum β_2-microblobulin (2.1 mg/ℓ), urinary total protein (70 mg/g creatinine), and urinary β_2-microglobulin (99 mg/g creatinine) in 1982, and there was no deterioration of renal function since 1969. Their urinary cadmium was 6 μg/g creatinine. In the group with high β_2-microglobulin clearance, the urinary cadmium was almost twice as high (10.7 μg/g creatinine). In 1969 their urinary total protein (128 mg/g creatinine), and GFR (106 mℓ/min) were already indicating a lower renal function than the first group and in 1982 these variables had deteriorated further: 251 mg total protein per gram creatinine and GFR = 99 mℓ/min.[200] The group on heavy medication had a similar trend to the latter group, even though their average urinary cadmium was only 6 μg Cd per liter.

A study of another Swedish factory[202] found no improvement in some workers with proteinuria in cadmium-copper alloy factory. One of the workers, who had no signs of cadmium effects when he left the cadmium-exposed job 6 years earlier, developed a tubular proteinuria. His maximum blood cadmium just before he left cadmium exposure in 1969 was 72 μg Cd per liter, and at that time his urinary β_2-microglobulin was normal. In 1972 his blood cadmium had decreased to a third, his urinary cadmium was 21 μg Cd per gram creatinine, and his urinary β_2-microglobulin was five times the "normal" average.[202] After this his blood cadmium continued to decrease to 14 μg Cd per liter in 1981,[200] his urinary cadmium decreased to 12 μg Cd per gram creatinine, but his urinary β_2-microglobulin kept increasing to 5 mg/g creatinine (50 times the "normal" average).[200]

Kjellström and co-workers[123,124] reported repeated results of β_2-microglobulin analysis in urine of 23 workers with exposure durations mainly between 6 and 15 years in the cadmium battery factory which had been studied some years earlier.[122] The same RIA method for β_2-microglobulin analysis was used both times. Seven of these workers had in the first study β_2-microglobulin levels higher than the 97.5 percentile of the reference group (Figure 17). There was no apparent systematic change in the urinary β_2-microglobulin between 1974 and 1976 and this applied to workers with and without continued cadmium exposure. There is a scatter in the results but the correlation coefficient between log β_2-microglobulin in 1974 and log β_2-microglobulin in 1976 is high (r = +0.88; n = 23).

Even in cadmium poisoning cases in the general environment there is a tendency for irreversible proteinuria. Nogawa and co-workers[157] repeated urine tests on 39 Itai-itai disease patients and 26 suspected patients in 1967 and 1975. In spite of a decreasing trend for average urinary cadmium concentrations (31 μg Cd per gram creatinine among Itai-itai cases in 1967 and 27 μg Cd per gram creatinine in 1975; 30 μg and 28 μg Cd per gram creatinine, respectively, for the suspected cases), the average urinary protein excretions almost doubled in the same period. For Itai-itai cases the proteinuria increased from 660 mg/g creatinine in 1967 to 1272 mg/g creatinine in 1975. The corresponding values for the suspected cases were 683 and 1064 mg/g creatinine.[157]

A report which has been widely quoted as evidence for the reversibility of the cadmium damage to the kidneys is the one by Tsuchiya[256] which dealt with 16 workers exposed to cadmium fume in a factory where the exposure levels had decreased by a factor of 10 in 1971. Five of these workers had been working in the factory since 1957 to 1964. From 1965 onwards their urine was tested each month by the SA or TCA method. Detailed data were not reported, but the trends are indicated in a summarizing table (Table 10). As pointed out by Friberg,[67] there is a great variation in the results using these methods. There is no indication in the report of the comparibility of the laboratory results before and after 1970. Furthermore, in two cases it is doubtful whether the proteinuria had disappeared and in one case, the proteinuria tests were negative at a number of times all through the period. In four of the cases the concentration of urinary β_2-microglobulin concentrations ranged between 2.6 and 9.7 mg/ℓ (or mg/day) which is 30 to 320 times higher than the "normal" average (Table 1 and Section IV.B.5). The average concentration was 5.3 mg/ℓ or 70 times the "normal" value, and the corresponding figure for albumin was 74 mg/ℓ or 7.4 times "normal".

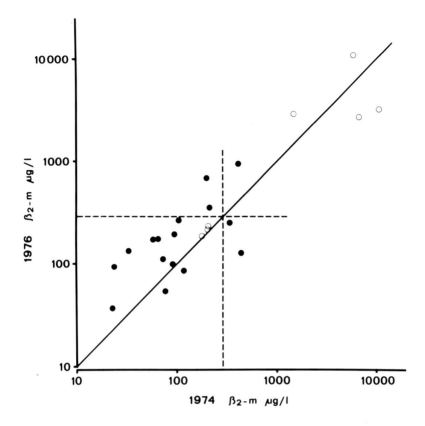

FIGURE 17. Comparison of urinary β_2-microglobulin measured at two times in a group of cadmium battery workers with 6 to 15 years' exposure. (O) Workers whose exposure had discontinued. (●) Workers with continuing exposure. (From Kjellström, T., Elinder, C.-G., and Friberg, L., *Environ. Res.*, 33, 284—295, 1984. With permission.)

These data fit the relative increases due to loss of tubular reabsorption expressed in Figure 1. The study by Tsuchiya[256] does not show that the cadmium damage is reversible. Adams[2] noted that removal from exposure was not followed by a complete loss of proteinuria in any, even mild, cases. Stewart and Hughes[238] also reported persistent proteinuria in four workers whose exposure had been discontinued for up to 12 years.

Roels and co-workers[213] measured urinary protein in 19 cadmium workers who had been removed from exposure. All had more than 15 years' exposure and the last examination took place up to 8 years after removal. Before being removed they all showed signs of cadmium-induced renal dysfunction. The cadmium level in blood and urine decreased considerably after removal, but there was no change in total urinary protein or albumin. In 11 workers β_2-microglobulin was measured before and after removal. In all cases the urinary β_2-microglobulin had increased 1 to 5 years after removal (Figure 18).

The available data indicate that the cadmium-induced renal damage is permanent even if exposure ceases, at least when it has reached an easily measureable level using proteinuria tests. In some cases a decrease in the daily urinary protein excretion may occur, but there is no evidence of total remission in affected human beings. In other cases a progression of the renal dysfunction is seen, even though cadmium exposure has ceased.

Table 10
TREND OF PROTEINURIA[256]

Worker	1965—1967	1968—1970	1971—1973	1974—1975
O.K.	(−)	(+) — (++)	(±)[a] — (−)	(−)
A.O.	(−) — (++)	(+++) — (++)	(++) — (±)	(+) — (±)
T.S.	(+) — (++)	(+) — (±)	(±) — (−)	(−)
S.O.	(−) — (+++)	(++) — (+++)	(++)[a] — (+)	(+)
N.I.	(+)	(+) — (+++)	(−) — (+)	(±) — (−)
Estimated cadmium in air (µg/m³)	0.123	0.2	0.02	(0.02)?

[a]   Translocation of job → to non-Cd exposure.

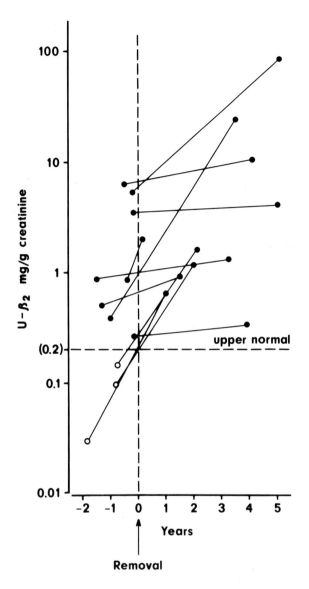

FIGURE 18. β_2-microglobulinuria before and after removal from cadmium exposure (○: none of the biological parameters of renal function abnormal; ●: at least one of the renal biological parameters abnormal). Paired t-test (n = 11): $p < 0.005$ (β_2-microglobulin: 1 mg = 0.0847 μmol). (Redrawn from Roels, H., Djubgang, J. Buchet, J.-P., Bernard, A., and Lauwerys, R., *Scand. J. Work Environ. Health*, 8, 191—200, 1982.)

V. EPIDEMIOLOGICAL STUDIES OF PROTEINURIA

A. Occupational Exposure

Several of the studies referred to in Section IV.B.1 aimed more at describing the clinical aspects of individual cadmium-poisoned workers than an epidemiological analysis of, e.g., the individual variations within groups of workers or the differences between exposed and control groups. Still, the data from some of the studies can be collated to produce prevalence rates of proteinuria in subgroups of the workers studied stratified according to exposure level and exposure duration (Table 11). In those stud-

Table 11

PREVALENCE OF PROTEINURIA IN CADMIUM WORKERS: SUMMARY OF STUDIES WITH DATA ON ENVIRONMENTAL LEVELS AS WELL AS ON HEALTH EFFECTS

Cadmium compounds	Estimated concentrations in the air (μg/m³)	Exposure time range and average (years)[a]	No. of examinees	Prevalence of proteinuria (%)	Method of detecting proteinuria[b]	Ref.
Cadmium oxide fume	40—50	Control	60	2	SA and TCA	116
		1—9	37	24		
		>9	63	46		
	64—241[c]	Control	11	0	TA	31, 32
		<1	4	0		
	123 (time-weighted average)	1—4	4	50	>100 mg/ℓ	255
		5—	4	100		
Cadmium oxide dust	3000—15000	1—4	15	0	Nitric acid (Heller test, positive in more than half the tests)	67
		9—15	12	33		
		16—22	17	41		
		23—34	14	64		
	500	5—44	116	28	SA	3
	300—5000	5—44	13	15	SA	
	50—1000+	5—44	43	51	SA	
	100—1000	5—44	40	20	SA	
	50—200	5—44	20	0	SA	
	31 (1.4)[c]	Control	31	0	Abnormal electrophoretic pattern as defined by the authors	131
		1—12 (4)	31	0		
	134 (88)	Control	27	4		
		0.6—19.7(9)	27	15		
	66 (21)	Control	22	0	β₂-microglobulin (RIA)	122
		21—40(28)	22	68		
	0	Control	87[a]	3.4	290 μg/ℓ (s.g. = 1.023)	
	50	0—3	50	6.0		
		3—6	26	7.7		
		6—12	18	17		
Cadmium stearate dust	30—690	Control	24	17	TCA	242
		3 (average)	19	58		

Cadmium sulfide dust	114^d		<1—5	12	17	EP	83
			5—21	7	100	
			<1—5	12	8	TCA
			5—21	7	43	
			Control	203	1	β_2-microglobulin (RIA)	237
	80		0—5	105	0	
	100		6—11	41	0	>765 µg/ℓ	238
	100—600		11—19	13	7.7	(s.g. = 1.016)
	100—600		20+	14	57	
Mixed exposures (12 factories, mainly zinc smelters)	Not reported but mean Cd-B levels after 1 year exposure were about 15 µg/ℓ		Controls	642	0(0.8)	β_2-microglobulin (RIA)	85
			<18 months	121	0(10)	
			19 months-5 years	168	1.8(8.3)	
			6—10	170	1.8(16)	>1000 µg/ℓ (or >200 µg/ℓ)
			11—15	82	7.3(22)	
			16—20	33	24(45)	
			20+	68	25(56)	

Note: + indicates different factory.

a   Numbers in parentheses are averages.

b   SA = sulfosalicylic acid method (qualitative); TCA = trichloroacetic acid method (qualitative); TA = tungstate method (quantitative); EP = electrophoresis; RIA = radioimmunoassay.

c   Measured in breathing zone — numbers in brackets refer to respirable (<5 µm) fraction.

d   Calculated average exposure for the worker with the most pronounced effect.

e   Including only those workers who were cadmium exposed during their whole employment.

ies where the exposed group has been divided into subgroups with different exposure durations, clear dose-response relationships are seen. The prevalence rates cannot be directly compared between the studies, because different methods were used to diagnose proteinuria. Depending on the type and duration of exposure and the criteria for diagnosis of proteinuria, prevalence rates as high as 100% were measured.

The β_2-microglobulin excretion tests were usually evaluated on the basis of the number of workers with excretion above the 97.5 percentile in the control group or above some other operational limit for "abnormal" values. By definition the prevalence of "abnormal" values would be about 2.5% in the control group. It is seen that for cadmium oxide fumes or dust, there is an apparently increased prevalence of proteinuria after even less than 10 years' exposure to concentrations of cadmium in air at 40 to 50 $\mu g/m^3$. The common exposure levels for the other cadmium compounds were higher (Table 11) and increases in the prevalence of proteinuria occur, but the data are not sufficient to show whether, e.g., cadmium sulfide dust is less toxic than cadmium oxide dust.

The effect of using different operational definitions of "abnormal" values is seen in the study by Holden[85] (Table 11). The prevalence of β_2-microglobulin above 200 $\mu g/\ell$ appears to increase already at exposure durations of 6 to 10 years, whereas the prevalence of levels above 1000 $\mu g/\ell$ increases at exposure durations of 11 to 15 years. These data also indicate the dose-effect relationship whereby the proteinuria becomes more severe the higher the dose. A more detailed picture of the dose-effect relationship is given by Kjellström and co-workers[122] (Figure 19). With increasing exposure duration (employment time) the average urinary β_2-microglobulin excretion increased and the maximum values increased, but even for the longest durations there are some workers with normal β_2-microglobulin. The individual variation in effect at the same exposure duration is great, but in each dose group a log-normal distribution of β_2-microglobulin excretion was found. This is likely to reflect variations in both the actual individual doses and in the sensitivity of the kidneys to a given cadmium dose. It should be pointed out that the vast majority of workers studied and referred to in Table 10 and Figure 19 were under age 70. Thus, their expected urinary β_2-microglobulin levels are more or less the same regardless of exposure duration (Section IV.B.4).

A number of epidemiological studies on cadmium workers (Sections IV.B.1, IV.B.5, IV.C, and IV.D) have earlier been referred to. Many studies lack environmental exposure data but often instead include urinary cadmium or blood cadmium data. As the urinary cadmium may increase as a sign of the renal tubular damage (Section IV.A.5), it is not possible to interpret relationships between urinary cadmium and occurence of proteinuria as true dose-response relationships. Blood cadmium data suffer from the lack of accuracy of blood cadmium analysis in the past (Chapter 2). Even in some recent studies no quality control data are presented. Epidemiological studies using in vivo neutron activation to establish amounts of cadmium in the liver and the kidney will be discussed in Section VI.B.

B. General Environment Exposure

As mentioned in Chapter 3, Section III, high general environment exposure has been observed mainly in Japan, but some other limited areas with unusually high cadmium intake have been found in Belgium and New Zealand.

A large number of epidemiological studies have been carried out in more than ten polluted areas in Japan. The first studies in the 1960s and early 1970s were concerned with Itai-itai disease (Appendix and Chapter 10, Section V) and rather insensitive methods to measure renal effects were used. At that time the main concern was the serious bone diseases. Friberg and co-workers in 1974[74] reviewed all the epidemiological studies in Japan and found dose effect and dose-response relationships for proteinuria in

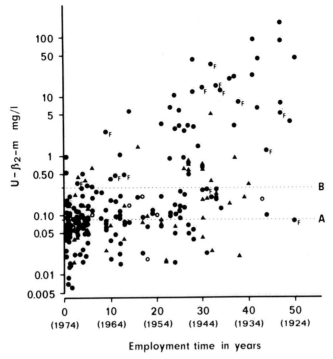

FIGURE 19.   Urinary β_2-microglobulin of workers in Swedish battery factory as a function of employment time and exposure continuity. A = geometric average β_2-microglobulin concentration in the reference group (84 $\mu g/\ell$). B = 97.5 percentile level in the reference group (290 $\mu g/\ell$). (Redrawn from Kjellstrom, T., Evrin, P.-E., and Rahnster, B., *Environ. Res.*, 13, 303—317, 1977b.)

several polluted areas. In the review by Tsuchiya in 1978,[257] further polluted areas and more recent studies were added. The current work of the governmental cadmium research committees in Japan was reviewed by Tsuchiya in 1982.[258] We will not attempt to make a complete review of the studies in this chapter, but will instead evaluate the more recently published studies that were designed to measure dose effect or dose-response relationships. The earlier studies have been summarized in Table 12 which includes the major original epidemiological studies from Japan. The table includes the areas where clear effects have been seen. In the other areas listed in Chapter 3, Table 6, the exposure levels are generally lower. The epidemiological methods used may not have been sufficiently sensitive to evaluate whether some effects occur also in these areas.

A reanalysis of earlier published data from the Fuchu area in Toyama prefecture[152] has shown very clear relationships between age or cadmium exposure level as measured by the average concentration of cadmium in rice for each village and prevalence of proteinuria or proteinuria with glucosuria (Table 13). In each age group there is a dose-response relationship between cadmium in rice and proteinuria prevalence. The prevalence of combined proteinuria and glucosuria in the control group does not change in the age range 30 to 59 years, but there is a tendency for it to increase at older age. This

Table 12
REPORTED OCCURRENCE OF PROTEINURIA AND GLUCOSURIA
IN EPIDEMIOLOGICAL STUDIES OF JAPANESE CADMIUM-
POLLUTED AREAS

Area	Range of cadmium concentration in rice[a]	Duration of high probable emissions from pollution source[b]	Effect reported to be increased above control group[c]	Ref.
Fuchu	0.5—2.0	86	Proteinuria (K-C)	77
			Glucosuria (Benedict)	
			LMW proteinuria (Disc)	225
			B2M (RIA)	121
Ikuno	0.6—1.0	105	Proteinuria (SA)	265
			LMW proteinuria (Disc)	117
			B2M (SRI)	265
Tsushima	0.75	30	Tendency for increased protein-uria (test tape)	145
			B2M (RIA)	218
Kakehashi	0.2—0.8	45	RBP (SRI)	90
			B2M (RIA)	218
Kosaka	0.5—0.6	91	Proteinuria (T-B)	127
			Glucosuria (no increase)	
			B2M (RIA)	128

Note: Areas listed in Volume I, Chapter 3, Table 6, but where statistically significant increases or proteinuria or glucosuria were not reported, have been excluded from this table.

[a]   See Chapter 3, Table 6; Friberg et al., 1974 and Tsuchiya, 1978.
[b]   According to Japan Mining Association, 1970 and 1973, National Institute of Geology, 1956, Dowa Mining Company, 1973, Toho Zinc Company, 1970, and Committee on Cadmium Pollution Mechanism in Yoshinokawa Area, 1974.
[c]   K-C = Kingsbury Clark method; LMW = low molecular weight; Disc = disc electrophoresis; RIA = radioimmunoassay; SA = sulfosalicylic acid method; SRI = single radial immunodif-fusion method; RBP = retinol-binding protein; B2M = beta-2-microglobulin; T-B = Tsuchiya-Biuret method.

has been noted also by Nomiyama[163] and Kitamura and Koizumi,[117] but it cannot be precluded that even in Japanese ''control'' groups the total cadmium dose in older age groups is sufficient to cause renal damage in some cases. On average, the cadmium exposure in Japanese ''nonpolluted'' areas is at least three to four times higher than the average exposure in European countries (Chapter 3, Section III). Because of the strong accumulation of cadmium in the kidneys, the exposure level and the exposure duration (equal to age of the exposed person living permanently in exposed area) are both equally important for the total dose.

The proteinuria prevalence also increases with age in the control group above age 60 (Table 13), but slower than the increase in the exposed groups. It is clear that the confounding influence of age on the prevalence rates must be taken into consideration in all epidemiological studies, but the high proteinuria rates in the exposed areas cannot be explained by age alone. One can say that in terms of proteinuria a person exposed to 0.6 mg Cd per kilogram in rice (Table 13) at age 55 has kidney function similar to a 75-year-old in the control area.

Nomiyama and co-workers[169] and Kitamura and Koizumi[117] first raised the issue of age as a more important explanation of increased urinary β_2-microglobulin than cadmium pollution. The relationship between cadmium exposure, age, and urinary β_2-

Table 13
PREVALENCE OF PROTEINURIA AND PROTEINURIA WITH
GLUCOSURIA IN RELATION TO AGE AND VILLAGE AVERAGE RICE-
CADMIUM CONCENTRATION

		Age								
		30—39		40—49		50—59		60—69		70—	
Cd in rice (μg/g)		N[a]	%[b]	N	%	N	%	N	%	N	%
				Proteinuria							
Control											
0.05—0.32	(0.15)[c]	655	11.8	465	13.1	467	11.6	362	20.4	201	36.3
Polluted village											
0.30—0.49	(0.41)	318	1.9	231	1.3	228	7.5[f]	210	21.0[d]	100	31.0[d]
0.50—0.69	(0.60)	346	2.6	245	3.3	234	14.5[d]	174	31.6[d]	102	46.1[d]
0.70—	(0.93)	147	2.0	159	6.9	111	20.7[d]	88	47.7[d]	48	50.0[d]
				Proteinuria with Glucosuria							
Control											
0.05—0.32	(0.15)	655	2.6	465	3.7	467	3.0	362	4.1	201	8.5
Polluted village											
0.30—0.49	(0.41)	318	1.9	231	1.3	228	7.5[f]	210	21.0[d]	100	31.0[d]
0.50—0.69	(0.60)	346	2.6	245	3.3	234	14.5[d]	174	31.6[d]	102	46.1[d]
0.70—	(0.93)	147	2.0	159	6.9	111	20.7[d]	88	47.7[d]	48	50.0[d]

[a] N, number of persons examined.
[b] Percentage of people with indicated condition.
[c] Mean in parentheses.
[d] $p < 0.001$.
[e] $p < 0.05$.
[f] $p < 0.01$.

From Nogawa, K. and Ishizaki, A., *Environ. Res.*, 18, 410—420, 1979. With permission.

microglobulin has been analyzed by multiple regression[174, 259] and age was found to be a strong determinant of urinary β_2-microglobulin. This should come as no surprise, as in polluted areas the residence time (or age) is directly related to the total cadmium dose in the kidney. This is due to the very long half-time in the kidney (Chapter 6, Section IV.A.2). In nonpolluted areas there is hardly any change with age of average urinary β_2-microglobulin excretion until about age 70 (Section IV.B.5). After that age the individual levels vary greatly. Occasional high levels occur in individuals with very low cadmium exposures due to factors other than cadmium.

Shigematsu and co-workers[223] pooled the data from seven Japanese polluted areas and compared the prevalence of various indicators of renal dysfunction. A total of 1826 exposed and 1611 control people over age 50 was included. Urinary total protein (Kingsbury-Clark method) and urinary glucose were analyzed at the primary screening stage for all participants. An increased prevalence with age was seen for both proteinuria (less than 100 mg/ℓ) and glucosuria, and there was a difference between the exposed and control groups, particularly for the combined proteinuria and glucosuria.[224] Such differences were reported by Nogawa and Ishizaki[152] for a population with, probably, a higher average cadmium exposure (Table 13).

The methods used by Shigematsu and co-workers[224] were insensitive and this study cannot be used to assess the occurrence of early effects. The subjects who were found to have either proteinuria or glucosuria according to the insensitive methods were tested for urinary β_2-microglobulin, RBP, cadmium, and amino acids. From an epidemiological point of view these tests have limited value, if they are not applied to the whole target population.

The combination of individual exposure level (cadmium level in family rice) and exposure duration was used to calculate a dose index (product of cadmium level in rice and exposure duration) in a study of women aged 50 to 59 in the Fuchu area.[121] A dose-response and a dose-effect relationship were found and there was also a correlation between urinary cadmium level and degree of tubular proteinuria. As mentioned earlier (Section IV.A.5), urinary cadmium in itself will increase when tubular damage occurs. This correlation can therefore not be interpreted directly as a dose-effect relationship.

The prevalence of proteinuria (as well as other clinical measures of the renal effects) was shown to correlate closely with average urinary cadmium concentration in groups of exposed Japanese farmers.[155] Average urinary cadmium in groups also correlated with prevalence of increased urinary RBP, amino acids,[154] and β_2-microglobulin.[156]

The method of estimating cadmium dose from the product of cadmium in rice and exposure duration has been used also in the Kosaka area.[128,216] Correlations showing the dose-effect relationship between cadmium dose and urinary β_2-microglobulin were found. Further studies in the Kosaka area[217] have confirmed the increased urinary β_2-microglobulin levels in exposed areas as compared to control areas, particularly in the older age groups. The average β_2-microglobulin excretion in different age groups is very similar in the exposed Tsushima area and the Kosaka area.[217] Fecal cadmium has also been used[128,259] to measure current exposure level. For Kosaka, estimates of 139 mg Cd per day[128] and 158 μg Cd per day[259] were given. In Tsushima the estimate was 255 μg Cd/day[259] and in another area (Kakehashi River) 149 μg Cd per day. In the three control areas the fecal cadmium was 60, and 102 μg Cd per day, respectively.[259]

Detailed studies have been carried out in the Kakehashi river basin in Ishikawa Prefecture. Elevated cadmium levels in rice were found along the river downstream from several mines all the way to Komatsu city and the sea (Figure 20). All people over age 50 living in villages along the river were studied for proteinuria using RBP as an indicator of tubular proteinuria. With a semiquantitative immunodiffusion method it was considered that an "increase" had occurred when the concentration was above 4 mg/ℓ.[90] There was a clear dose-response relationship between cadmium in rice and prevalence of increased RBP (Figure 21).

In one area outside Japan renal effects in people exposed in the general environment were reported by Roels and co-workers.[211] They defined three areas with different cadmium concentrations in air and different cadmium dustfall. Women over 60 years of age were studied. Almost all of the women were nonsmokers. The median cadmium concentration in urine was 0.79 μg/24 hr in one area (n = 70), 1.3 μg/24 hr in another area (n = 45), and 2.0 μg/24 hr in a third area (n = 60). The dustfall and air data followed the same pattern with the third area having more than twice as high levels as the first area. The daily cadmium intake via food was 15 μg/day in both the third and first area, but the average renal cortex cadmium concentration at autopsy was twice as high in the third area.[132] The prevalence of total proteinuria (above 360 mg/24 hr) increased from 11% in the low cadmium area to 37% in the high cadmium area. For β_2-microglobulin in urine above 2.4 mg/24 hr, the prevalence increased from 2.6 to 10%[211] with increasing exposure level. In the most highly exposed Belgian area a doubling of the age-standardized mortality rate in nephritis-nephrosis has also been reported.[130]

Nogawa and co-workers[156] studied women over age 20 in a Japanese polluted area. At a median urine cadmium level of 3.6 μg/g creatinine, 3.9% of the women had increased β_2-microglobulin (above 5 mg/ℓ urine). The prevalence increased to 85% at

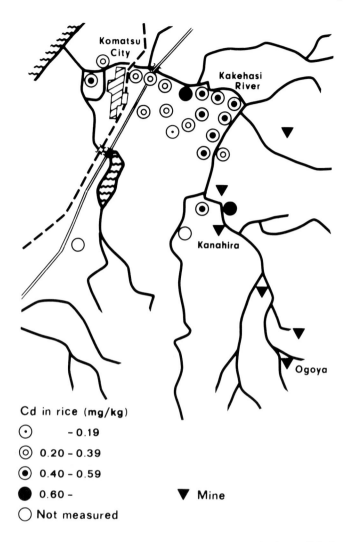

Cd in rice (mg/kg)

⊙   - 0.19

◎   0.20 - 0.39

◉   0.40 - 0.59

●   0.60 -   ▼ Mine

○ Not measured

FIGURE 20.   Average cadmium concentration in fresh unpolished rice in villages studied in the Kakehashi river basin. (Modified from Ishikawa Prefecture, *Report on the Health Examination of Residents of the Kakehashi River Basin,* Health and Welfare Department, Kanazawa, Japan, 1976.)

a median urinary cadmium of 47.8 $\mu g/g$ creatinine. The two sets of data are not exactly comparable because Nogawa and co-workers used a higher level of β_2-microglobulin per liter urine as an operational definition of "increased" values. In addition, they studied younger women than Roels and co-workers.[213]

Nomiyama and co-workers[174] reported urinary cadmium and β_2-microglobulin levels of inhabitants in three polluted areas. There were some elderly cases in two areas with β_2-microglobulin above 1000 $\mu g/\ell$ even though urinary cadmium was below 5 $\mu g/\ell$. Kjellstrom and co-workers[121] also reported some very high β_2-microglobulin values at such low urinary cadmium values in 50 to 59-year-old women from a polluted area. It is thus possible that a urinary cadmium level as low as 2 $\mu g/24$ hr can be associated with a slightly increased prevalence of cadmium-induced renal damage.

Kobayashi[125,126] and Honda and co-workers[86] studied more than 500 people over 5 years of age in nine cadmium-polluted villages of Fuchu area. About 400 controls from

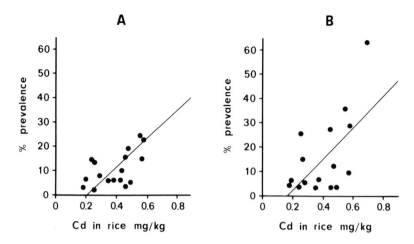

FIGURE 21.   Prevalence of increased urinary RBP above 3 mg/ℓ for inhabitants in villages of the Kakehashi river basin as a function of village average cadmium concentration in rice. (A) Men and women over 50 years of age. (B) Women over 50 years of age. (Redrawn from Ishikawa Prefecture, *Report on the Health Examination of Residents of the Kakehashi River Basin,* Health and Welfare Department, Kanazawa, Japan, 1976.)

Ishikawa Prefecture were also included. The urinary concentrations of total protein, glucose, total amino acids, proline, calcium, phosphorus, retinol-binding protein, and β_2-microglobulin were on average higher in the polluted area than in the control area for each 10-year age group above 40 years. After that age the averages of these variables in the exposed groups all increased with age.[125] Urinary cadmium increased with age all through life. There were statistically significant correlations between the variables in the exposed group.[86] Using 4 mg/ℓ as an operational definition of "increased" urinary retinol-binding protein or β_2-microblobulin,[125] it was seen that the prevalence figures were higher (up to double) for β_2-microglobulin than for retinol-binding protein. The analytical method used had a high detection limit of about 1 mg/ℓ.[86]

Kobayashi[125] analyzed the data for relationships between the individuals' residence time in the polluted area and occurrence of urinary findings. This time duration is likely to be a better indicator of accumulated dose than age. Not everybody at age 50 in a polluted area had lived in the area all their lives.[121] Dose-response relationships were seen for all urinary variables.[126] When the analysis was only based on people who had lived all their lives in the polluted area (Figure 22), a dose-response curve very similar to the theoretical s-curve was seen for β_2-microglobulin (above 4 mg/ℓ).

Of particular interest is the calculation by Kobayashi[126] of separate average urinary cadmium concentrations in the groups with high β_2-microglobulin and low β_2-microglobulin (Figure 23). The differences seen between people with high and low β_2-microglobulin may be explained partly by the increased excretion of cadmium after renal damage has occurred (Section IV.A.5) and partly by a higher average dose among those with damaged kidneys.

VI. ORGAN DOSE-EFFECT AND DOSE-RESPONSE RELATIONSHIPS

The data in Section IV.A.2 and Chapters 4 and 6 indicate that the concentration of nonmetallothionein-bound cadmium in the renal cell may be crucial for the development of cadmium damage. Few studies have measured this fraction of cadmium accurately in the cells. After a single dose of cadmium-methallothionein,[236] rats had initially

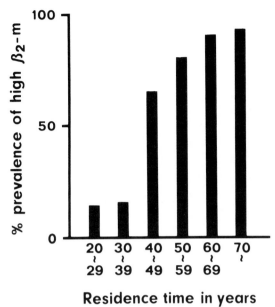

FIGURE 22. The prevalence of increased urinary β_2-microglobulin (above 4 mg/ℓ) as a function of residence time in the polluted area. Includes only men who have lived all their lives in the polluted area. (Redrawn from Kobayashi, E., *Sangyo Igaku*, 3, 316—317, 1982.)

a very high nonmetallothionein fraction, but after 8 hr this fraction had decreased to 16% of total cadmium in cytoplasma (Table 2). One third of the whole kidney amount of cadmium is in cytoplasma, so the proportion of cytoplasma nonmetallothionein-cadmium out of whole kidney-cadmium is about 5% at 8 hr after a single dose.

There are no data on how this fraction in the kidney changes with age or other factors. Even at "steady state" for cadmium accumulation in the kidney a redistribution between different fractions may occur. However, it may be that the fraction of nonmetallothionein cadmium is relatively stable.

Renal tubular damage has been seen at nonmetallothionein-cadmium concentrations of about 10 mg/kg.[181,236] Using the 5% figure mentioned above, this level would be equivalent to a total cadmium concentration of 200 mg/kg. This is a level at which renal damage starts developing in several studies (Sections VI.A and VI.B).

In a theoretical scientific sense one should not apply the term critical concentration to the total cadmium concentration in renal cortex. It is more correct to use the term for nonmetallothionein cadmium concentrations. As will be seen below, the total cadmium levels and the renal tubular effects are related, and we believe that at present total cadmium concentrations in renal cortex are the best available organ dose measures.

A. Animals

The effect of cadmium that occurs at the lowest renal cortex cadmium level appears to be an increase in the renal cortex zinc concentration (Section IV.C) related to the stimulation of local metallothionein production (Chapter 4, Section V). This may in itself be a "physiological adaptation" to the cadmium exposure, but its mechanism serves as a valuable explanation of the toxicity of cadmium to the kidney. The increase of zinc concentration starts already at the lowest cadmium levels.

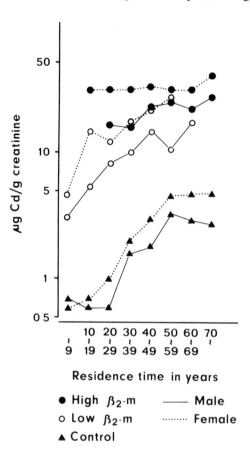

FIGURE 23.   Average urinary cadmium excretion as a function of residence time in the polluted area and renal tubular effects. (Redrawn from Kobayashi, E., *Sangyo Igaku*, 3, 316—317, 1982.)

Elinder and co-workers[54] found dose-response relationships between cadmium concentration in horse renal cortex and tubular dilatation and casts, as well as renal interstitial changes (Figure 24). Some horses also had renal glomerular changes (mesangial sclerosis), but the prevalence of this was quite high even in low-cadmium horses, so the dose-response relationship was not clear. All these horses had long-term "natural" cadmium exposure. Unfortunately, proteinuria could not be measured, so it is not known how tubular and interstitial changes related to any pathological urinary findings. In some horses morphological changes occurred already at renal cortex cadmium concentration in the range 30 to 70 mg Cd per kilogram and a 50% prevalence of slight interstitial changes was reached at about 70 to 80 mg Cd per kilogram (Figure 24). At that level more severe changes occurred in 25% of the horses (Figure 24).

The study by Elinder and co-workers[54] was unusual in that it clearly presented the individual variation of sensitivity to cadmium among the animals. Most animal studies simply present the average concentrations for a group of animals, and the groups usually contain inbred animals with very similar genetic structure and are, therefore, not similar to the conditions for humans. However, the horses studied by Elinder and co-workers[54,55] would be rather similar to humans in this respect.

Changes in the rabbit urinary protein electrophoresis pattern with a progressive increase of the excretion of proteins at molecular weights of about 12,000, 25,000 and

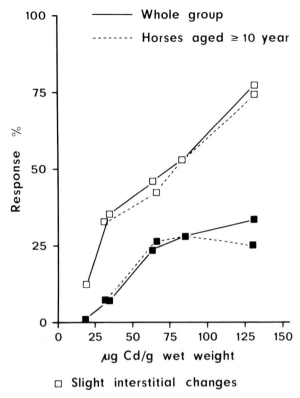

FIGURE 24.   Frequency of slight and severe interstitial changes in relation to cadmium concentration and in relation to age of horse. (From Elinder, C.-G., Nordberg, M., Palm, B., and Piscator, M., *Environ. Res.*, 26, 22—32, 1981b.)

67,000 (Figure 6) have been reported by Nomiyama and co-workers.[173] The increase started after about 5 weeks of exposure, when the renal cortex cadmium concentration was close to 200 mg/kg. More massive total protein excretion started between weeks 10 and 15 (Figure 5), when the renal cortex cadmium concentration reached a peak at about 430 mg/kg.[173] Similar data were presented by Suzuki[243] for rats. After five weeks of exposure, the total proteinuria increased (Figure 2) and at the same time urinary cadmium rapidly increased. At this stage, the renal cortex cadmium concentration was on average 125 mg Cd per kilogram,[243] assuming 25% higher cadmium levels in cortex than in whole kidney. After 10 weeks of exposure the urinary cadmium reached a peak and at that stage there was no further increase in the renal cortex cadmium concentration (190 mg/kg), in spite of continued exposure. Feeding rats cadmium via drinking water produces similar results.[24] After 8 to 9 months of exposure an increase of total proteinuria starts (Figure 3) and at about the same time there is a great increase of urinary cadmium excretion (Figure 4). At that stage the renal cortex cadmium concentration was about 220 mg Cd per kilogram (Figure 4).

In several other studies quoted in Section IV.A.1, it was reported that proteinuria occurred when the renal cortex cadmium concentration was in the range 100 to 300 mg/kg. Low molecular weight proteinuria is likely to occur earlier than total proteinuria, but both of these are not always measured. Piscator and Larsson[201] found, for instance, that an increase in ribonuclease excretion could be seen in the renal cortex of calcium-deficient, cadmium-exposed rats at 90 mg Cd per kilogram. Tohyama et al.[252] found an up to tenfold increase of urinary metallothionein excretion of rats at about 200 to 250 mg Cd per kilogram.

It appears that glucosuria, aminoaciduria, and other changes in renal function develop at about the same or possibly at a somewhat higher dose as the total proteinuria (Sections IV.A.2 and IV.A.3). However, these findings (as well as those of proteinuria) are very much influenced by the operational definition of upper "normal" limit used by each individual researcher.

Table 14 summarizes the data from animal studies. Generally the cadmium levels quoted are average levels at which the majority or all of the animals showed the effects. Prevalence rates are seldom given, and one has to assume that they were generally high (50% or more) at the points where clear effects were reported. In one study[24] the prevalence rates were reported (Figure 3).

After 8 to 9 months of exposure there is an approximate 20% increase of proteinuria prevalence (Figure 3) when the average kidney cortex cadmium level is about 220 mg/kg (Figure 4). This has been compared with human data in Chapter 13 (Figure 2).

In conclusion, most animal data indicate that at a renal cortex cadmium concentration of 200 mg Cd per kilogram renal tubular damage with a decreased tubular reabsorption of proteins will be evident if proper diagnostic procedures are used. In some studies an increase of low molecular weight proteinuria or morphological changes have been seen in sensitive animals at renal cortex cadmium levels of about 100 mg/kg.

The interindividual variation of dose-effect relationships is seldom reported in animal experimental studies. Most of the data include only averages for groups of animals and, therefore, the lowest renal cortex level at which effects occur may be considerably lower than the values mentioned above.

B. Humans
1. Estimated from Autopsy and Biopsy Findings

As mentioned earlier, there is evidence from animal experiments (Section IV.A.1) as well as humans (Sections IV.B.1.d and IV.B.2.c) that the renal cortex cadmium concentration will decrease as the cadmium-induced renal damage progresses, after the initial damage has occurred. In individuals the time development of the cadmium concentration may be as shown in Figure 25. These changes in the renal cadmium levels make it difficult to interpret autopsy data in straight dose-response terms. Only longitudinal studies of biopsies (or in vivo neutron activation analysis) carried out several times for each exposed person can directly show the dose-effect and dose-response relationships, but no such studies have been carried out.

A plot (Figure 26) of the autopsy and biopsy data in Tables 4, 6, and 7 shows the paradoxically low cadmium levels in the kidney in the groups with the most severe renal damage. Few of the exposed subjects have levels above 300 mg/kg. In the group with proteinuria only, about half have levels above 200 and half below 200 mg/kg. In the most severely affected group no subjects had levels above 200 mg/kg.

Figure 27 shows that the renal cortex cadmium concentrations in almost all cases where renal damage was found are lower than would be "expected" from the liver cadmium concentrations in the same individual based on the model calculations in Volume I, Chapter 7. The majority of the affected people have liver values in the range 40 to 120 mg/kg, "corresponding to" expected renal cortex values in the range 100 to 300 mg Cd per kilogram (Chapter 7). For general environmental exposure, eight out of nine cases with both proteinuria and morphological changes had liver values in the range 60 to 150 mg/kg, "corresponding to" renal cortex values in the range 150 to 350 mg/kg, and the cases with proteinuria only or morphological changes only have somewhat lower liver values. The cases of industrial exposure have a much greater scatter

Table 14

SUMMARY OF ANIMAL STUDIES WITH DATA ON BOTH RENAL CADMIUM LEVELS AND EFFECTS

Species	Administration	Exposure level	Duration (months) before effect	Average cadmium level in renal cortex (mg/kg wet weight)	Reported findings of renal damage	Ref.
Mouse	SC	0.25 mg/kg bw	6	140—210ᵃ	No effects	180
		0.5 mg/kg bw	6	210ᵃ	Tubular protein patterns in urine	
Rat	IP	0.75 mg/kg bw	3	250ᵃ	No effects	33
			4	370ᵃ	Histological change in 50% of animals	
	SC	0.5 mg/kg bw	1.5	125ᵃ	Total proteinuria	243
	Water	10 mg/ℓ	8.5	14ᵃ	No histological change	107
		50 mg/ℓ	8.5	44ᵃ	Slight histological change	
		100 mg/ℓ	8.5	115ᵃ	Histological change	
		200 mg/ℓ	8.5	180ᵃ	Histological change	
		7.5 mg/ℓ	12	90	Increase in urinary ribonuclease	201
		200 mg/ℓ	11	200	Total proteinuria	24
		50 mg/ℓ	3	125ᵃ	Decreased insulin and PAH clearance	108
					Histological changes	
Rabbit	SC	0.25 mg/kg bw	2.5	235	Slight histological changes; proximal tubules	10
			4	460	More severe histological changes	12
		0.5 mg/kg bw	2.5	300	Total proteinuria	173
		0.5 mg/kg bw	0.7	200	Proteinuria, glucosuria, aminoaciduria, decrease in C_{IN} and Tm_{PAH}	165
		1.5 mg/kg bw	1	50—200	Decreased tubular reabsorption	160
						169
	Water	0.5 mg/ℓ	2	200ᵃ	Slight histological change	107
		160 mg/ℓ	6	210ᵃ	Extensive fibrosis, pronounced changes	239
	Water	50 mg/ℓ	10	58	Slight tubular atrophy	107
		200 mg/ℓ	10	200	Severe interstitial and tubular fibrosis	107

Table 14 (continued)

SUMMARY OF ANIMAL STUDIES WITH DATA ON BOTH RENAL CADMIUM LEVELS AND EFFECTS

Species	Administration	Exposure level	Duration (months) before effect	Average cadmium level in renal cortex (mg/kg wet weight)	Reported findings of renal damage	Ref.
	Diet	300 mg/kg	4	200	Aminoaciduria, enzymuria	167
			10	300	Proteinuria, glucosuria	
Swine	Diet	50—1350 mg/kg		<15[a]	Equimolar increase in zinc in kidney	45
				92[a]	Decrease in renal leucine amino peptidase	
Monkey	Diet	3—300 mg/kg	Up to 24 months	160—240	Slight swelling of tubular epithelium	171
				380	Low molecular weight proteinuria	
				450	Total proteinuria	
		30 mg/kg	30	800	Beta$_2$-microglobulin	115
Horse	Normal diet	No Cd added	Up to 24 months	75	Renal tubular interstitial changes and fibrosis in 25% of animals	54
Bird	SC	0.16 mg/kg bw	1.5	20[b]	Histological change	148

Note: SC = subcutaneous; IP = intraperitoneal; bw = body weight.

[a] Denotes concentrations calculated as 1.25 times renal average concentration (Volume I, Chapter 6, Section IV.B.2).

[b] Denotes concentrations (wet weight) calculated as 0.2 times dry weight concentrations.

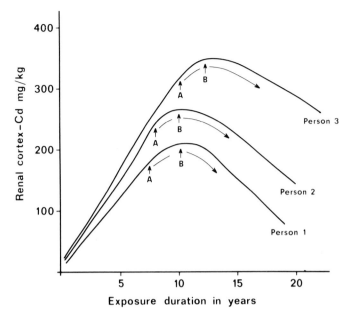

FIGURE 25. Schematic presentation of the time development of the renal cortex cadmium concentration in three individuals who develop early cadmium-induced renal damage at (A) and this progresses to more severe damage at (B).

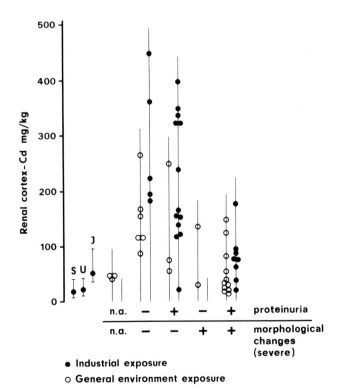

FIGURE 26. Relationship between cadmium concentrations in renal cortex and degree of renal damage in people with ''excessive'' exposure to cadmium. Data from Tables 4, 6, 7, and Appendix. Comparison data for ''normal'' groups. (From Kjellstrom, T., *Environ. Health Perspect.*, 28, 169—197, 1979. With permission.)

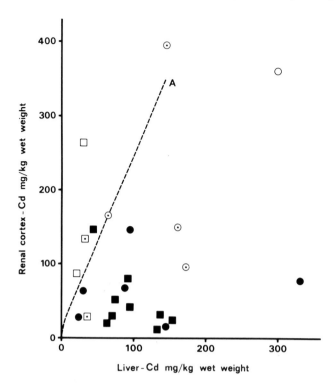

FIGURE 27.   Renal cortex cadmium and liver cadmium measured at autopsy. People with occupational or general environment exposure to cadmium and with different signs of renal damage as indicated in Figure 26. Data from Tables 4 and 6. (A) Relationship between liver cadmium and renal cortex cadmium of actively exposed people as calculated by the model explained in Chapter 7.

Occupational exposure	General environmental exposure	
●	■	both proteinuria and morphological changes
⊙	⊡	only one of the above reported (usually protein- uria)
○	□	none of the above

and are more difficult to interpret, but 10 of 12 liver values are above 60 mg/kg. The kidney biopsy cases reported in Table 4 had mostly proteinuria and no or slight morphological changes. For these cases no liver cadmium values were available. Kidney cortex values varied between 150 and 450 mg Cd per kilogram except in one case. Taking these data as a whole it can be concluded that at renal cortex cadmium concentrations in the range 100 to 300 mg/kg, effects on the kidney leading to proteinuria begin to occur. It appears that when morphological changes occur these are almost always associated with proteinuria and sometimes other urinary findings. The morphological changes are also associated with high liver cadmium concentrations but low renal cortex cadmium concentrations, indicating that the kidney has lost cadmium due to the damage (Section IV.A.5).

2. *Estimated from In Vivo Analyses of Liver and Kidney*

Recently, some studies have utilized in vivo neutron activation analysis (Chapter 2, Section V) to study the dose-response relationships for cadmium-induced renal effects. Ellis and co-workers[56] measured the cadmium concentrations in the liver of 82 industrially exposed subjects and 10 nonexposed controls in the U.S. No quality assessment of the analytical method was reported, but the limits of detection were reported as 2.2 mg for kidney amount and 1.5 mg/kg for liver concentration. Based on a kidney weight of 145 g and 1.25 times higher cadmium concentration in the cortex than in whole kidney (Chapter 6, Section IV.B.2), 2.2 mg in whole kidney corresponds to 19 mg Cd per kilogram in kidney cortex. Ellis and co-workers[56] used a ratio of 1.5 between cortex and whole kidney for their calculations. This may give too high values.[123] The reported ranges and geometric means of values for the control group are 3 to 102 mg Cd per kilogram in renal cortex (mean = 32) and 0.6 to 7.9 mg Cd per kilogram in liver (mean = 2.6).[56] Autopsy data for a group of 92 male Americans in the age range 30 to 59 years[118] found renal cortex values in the range 3 to 90 mg Cd per kilogram (geometric mean = 20) and liver values in the range 0.1 to 4 mg Cd per kilogram (geometric mean = 1.1), which are comparable with the data by Ellis and co-workers. It is therefore likely that, at least at higher levels than the detection limit (19 mg Cd per kilogram), the in vivo neutron activation method used and chemical analysis of autopsy specimens would give comparable data. Of the 82 workers, 14 had urinary β_2-microglobulin levels above 25,000 μg/g creatinine and 25 had levels above 300 μg/g creatinine, indicating the prevalence of different degrees of renal tubular damage.

The scatter of the individual cadmium concentrations is also great for these in vivo data (Figure 28). This is in agreement with the autopsy data (Figure 27). Ellis and co-workers[56] concluded that the data best fitted a two-component curve with an inflection point at approximately 40 mg Cd per kilogram in liver and 30 mg cadmium in one kidney. Ellis and co-workers[56] calculated the inflection point to have an average at 31 mg Cd in the kidney (95% confidence interval within ±9 mg) (Ellis, personal communication), but did not explain how this calculation was carried out. They assumed that the inflection point corresponded to the renal cadmium concentration at which renal damage started to occur. Using the ratio 1.25 for cortex/whole kidney cadmium concentrations, 31 mg Cd would correspond to 270 mg Cd per kilogram in renal cortex. It was pointed out by Kjellström and co-workers[123] and Friberg and Kjellström[72] that this is the average concentration at which renal damage occurs. If the cadmium concentration at the inflection point followed a "normal distribution", about 10% of the workers would have developed renal damage at kidney cortex cadmium concentrations below 200 mg/kg.[123]

Tohyama and co-workers[252] measured liver and renal cadmium concentrations by in vivo neutron activation in 16 smelter workers. Liver values up to 500 mg/kg were found but there were no cases of very high β_2-microglobulin. There was a correlation between liver or kidney cadmium concentration and urinary metallothionein concentration, but there was no threshold for sudden increase of metallothionein. At the highest renal cortex cadmium concentrations of about 300 mg/kg, the urinary metallothionein excretion was about 300 μg/g creatinine,[252] which is 20 to 100 times higher than the lowest values reported among cadmium workers without renal effects.[182,215]

Roels and co-workers[212] studied 264 workers in Belgian zinc-cadmium production plants with different degrees of cadmium exposure. No quality assessment of the in vivo neutron activation analysis was reported. The detection limits were 4.5 mg cadmium in one kidney and 7 mg Cd per kilogram in liver. No control group was studied, but for one group of workers with low cadmium exposure most values were below the detection limit. The accuracy of the data is questioned by the authors themselves who added a note in proof saying that all the kidney cadmium concentrations may be 35%

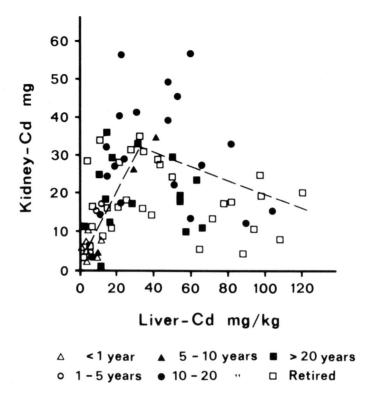

FIGURE 28.   Relation between kidney cadmium and liver cadmium for exposed American workers. Values based on neutron activation analysis. The dashed line represents the two components of a linear model fit to the data (Modified from Ellis, K. J., Morgan, W. D., Zanzi, Il., Yasamura, S., Vartsky, D., and Cohn, S. H., *J. Toxicol. Environ. Health,* 7, 691—703, 1981.)

too low. This error was explained by a lack of adjustment for the depth of the kidney below the skin surface.[6] Revised data were published by Roels and co-workers.[214] An error due to the use of the cortex/whole kidney ratio 1.5 still remains in these data.[123] For those workers whose kidney and liver values were above the detection limits, there was a similar relationship between kidney and liver (Figure 29) to those seen in Figures 27 and 28. In the report by Roels and co-workers,[212] the data are expressed in a way which makes it possible to evaluate dose-effect and dose-response relationships. None of the workers had an increased urinary β_2-microglobulin (>200 μg/g creatinine) if the renal cortex cadmium concentration was less than 104 mg Cd per kilogram (recalculated according to Kjellström and co-workers),[123] or if the liver cadmium concentration was less than 24 mg Cd per kilogram (Figure 29).

In the range 25 to 39 mg Cd per kilogram liver there were 4 cases with increased urinary β_2-microglobulin among 28 workers (14%) and 3 cases of increased total proteinuria (11%). In the range 40 to 50 mg Cd per kilogram liver (n = 24), there were 5 cases with increased β_2-microglobulin (21%) and 2 total proteinuria cases (8%) and in the range over 60 mg Cd per kilogram (n = 13), there were 10 cases with increased β_2-microglobulin (77%) and 6 total proteinuria cases (46%).

The regression line for workers who had no abnormal protein findings in the urine (Figure 29) does not go through the origin as the low level results have all been excluded. It appears, though, that the average renal cortex cadmium concentration at the liver cadmium level (30 mg Cd per kilogram liver), where effects (β_2-microglobulin

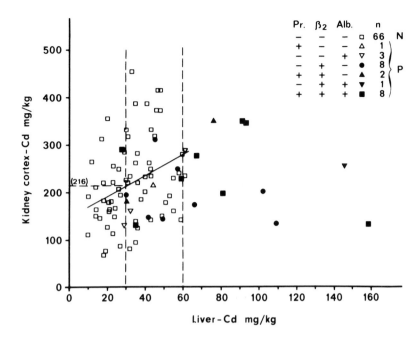

FIGURE 29.   Cadmium level in renal cortex as a function of cadmium level in liver for cadmium workers (active, retired, removed from cadmium exposure) with liver and renal cortical cadmium above the detection limits (CdL ≥ 10 ppm, CdKc ≥ 50 ppm). Regression line for subjects without renal dysfunction (□——□), group N). The individual values of the subjects with renal dysfunction(s) (group P, n = 23) are plotted with different symbols to distinguish between those with increased proteinuria (Prot-U) or β_2-microglobulinuria (β_2m-U) or albuminuria (Alb-U) or a combination of these. (From Roels, H. A., Lauwerys, R. R., Buchet, J. P., Bernard, A., Chettle, D. R., Harvey, T. C., and Al-Haddod, I. K., *Environ. Res.*, 26, 217—240, 1981b; with subsequent corrections from Ref. 214.)

excretion) start to occur (response rate about 10%), is just above 200 mg Cd per kilogram. By using the new cadmium concentration ratio renal cortex to whole kidney of 1.25 (Chapter 6, Section IV.B.2), this value can be recalculated to 167 mg Cd per kilogram.

Roels and co-workers[212] concluded that the "critical concentration" (Chapter 13, Section II) for renal cortex cadmium (the level where effects start to occur in the individual) is found in the range 160 to 285 mg Cd per kilogram (recalculated with the factor 1.25 to 138 to 238 mg/kg). Beyond 285 mg/kg (238 mg/kg recalculated) the probability is very high that all persons will show signs of renal dysfunction. Based on the revised data, Roels and co-workers[214] concluded that 10% of workers would have renal dysfunction at 216 mg Cd per kilogram in renal cortex (can be recalculated with the factor 1.25 to 180 mg/kg).

The in vivo data and the autopsy and biopsy data included a mixture of currently exposed and formerly exposed people. This makes the interpretation of the data more complicated, as the liver levels start to decrease as soon as exposure ceases (Chapter 7, Section V.B) and the renal cortex levels stay at the same level for several years in spite of cessation of exposure, unless the kidney is already severely damaged by cadmium (Sections IV.B.1.d and IV.B.2.c).

Ellis and co-workers[57] reanalyzed their in vivo neutron activation analysis data and separated the workers in the analysis into active and retired. Based on data from 30 controls and 31 active workers a dose-response relationship was calculated (Figure 30).

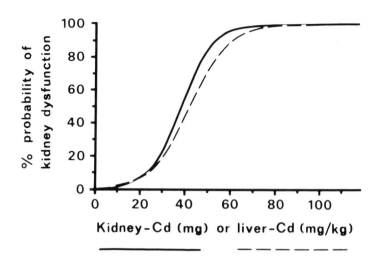

FIGURE 30.   The percentage probability of having kidney dysfunction as a function of kidney or liver cadmium burden. (Modified from Ellis, K. J., Yuen, K., Yasamura, S., and Cohn, S. H., *Environ. Res.,* 33, 216—226, 1984.)

It is seen that a 10% response rate occurs at about 22 mg cadmium in the whole kidney or about 190 mg Cd per kilogram in renal cortex. These and other dose-response data are discussed further in Chapter 13.

The human and animal data are in good agreement. At an average renal cortex cadmium concentration of 200 mg Cd per kilogram, a considerable proportion (10% plus) of the group will have developed renal tubular damage with low molecular weight proteinuria. Sensitive individuals will have a measurable renal functional change in the form of low molecular weight proteinuria at lower levels than 200 mg/kg.

VII. SUMMARY AND CONCLUSIONS

Cadmium exposure can induce both acute and chronic effects in the kidneys of animals and humans. The major damage is done to the renal tubules, but at higher doses the glomeruli and the renal blood vessels may also be affected. The mechanism of the renal effects is related to the fact that cadmium in plasma is bound to metallothionein, which is freely filtered through the renal glomeruli and reabsorbed in the proximal tubuli. After entering the lysosomes of proximal tubular cells, the metallothionein is catabolized and "free cadmium" (or nonmetallothionein-bound cadmium) is released, which probably interferes with zinc-dependent enzymes and causes damage to the metabolism of the renal tubular cell.

The dose-effect relationship for the acute renal effects is such that humans only risk acute effects after high accidental or intentional ingestion of a cadmium salt.

The chronic effects may occur after long-term (years) occupational exposure to cadmium compounds in air or long-term (years) exposure to cadmium in contaminated food. Hundreds of workers have been poisoned in Europe, U.S., and Australia, and thousands of farmers in Japan have been poisoned after consuming cadmium-contaminated rice and other foods. The signs of the chronic renal effects are proteinuria, glucosuria, and aminoaciduria. In severe cases other signs of impaired renal tubular function tests are seen. Histopathological changes in the renal tubules are commonly seen in animal experiments and in poisoned humans.

The cadmium-induced proteinuria is the most characteristic sign of renal tubular damage. It involves an increase of all plasma proteins in urine and particularly high relative increase of low molecular weight proteins. It appears that this proteinuria is mainly due to a renal tubular reabsorption deficiency, which similarly affects the reabsorption of all proteins, glucose, amino acids, phosphate, etc. from the primary urine. The relative increase of the urinary excretion of each compound depends very much on the "normal" reabsorption rate. A high "normal" reabsorption rate leads to a high relative increase when tubular reabsorption is reduced.

Studies of animals, cadmium-exposed workers, and people exposed in the general environment show that the renal tubular damage is not likely to be reversible. In some workers the renal dysfunction has in fact progressed even after exposure ceased.

The concentration of cadmium in renal cortex at which the first signs of dysfunction occur, in about half of exposed animals, is about 100 to 150 mg/kg. At 200 mg/kg proteinuria is common in most species. Autopsies and in vivo neutron activation analysis studies of humans show that the first effects occur in about 10% of individuals at renal cortex levels in the range 170 to 200 mg/kg.

REFERENCES

1. Adams, R. G., Clinical and biochemical observations in men with cadmium nephropathy. A twenty-year study, *Arh. Hig. Rada Toksikol.,* 30, 219—231, 1979.
2. Adams, R. G., Osteopathy associated with tubular nephropathy in employees in an alkaline battery factory, in *Cadmium Induced Osteopathy,* Shigematsu, I. and Nomiyama, K., Eds., Japan Public Health Association, Tokyo, 1980, 66—73.
3. Adams, R. G., Harrison, J. F., and Scott, P., The development of cadmium-induced proteinuria, impaired renal function, and osteomalacia in alkaline battery workers, *Q. J. Med.,* 38, 425—443, 1969.
4. Ahlmark, A., Axelsson, B., Friberg, L., and Piscator, M., Further investigations into kidney function and proteinuria in chronic cadmium poisoning, *Int. Congr. Occup. Health,* 13, 201—203, 1961.
5. Akahori, F., Nomiyama, K., Masaoka, T., Nomiyama, H., Nomura, Y., Kobayashi, K., and Suzuki, T., Effects on monkeys of long-term dietary exposure to cadmium chloride, in *Kankyo Hoken Report No. 49,* Japan Public Health Association, Tokyo, 1983, 1—27.
6. Al-Haddad, I. K., Chettle, D. R., Fletcher, J. G., and Fremlin, J. H., A transportable system for measurement of kidney cadmium *in vivo, Int. J. Appl. Radiat. Isot.,* 32, 109—112, 1981.
7. Aughey, E., Fell, G. S., Scott, R., and Black, M., Histopathology of early effects of oral cadmium in the rat kidney, *Environ. Health Perspect.,* 53, 153—161, 1984.
8. Aukland, K. J., Stop flow analysis of renal protein excretion in the dog, *Scand. J. Clin. Lab. Invest.,* 12, 300—310, 1960.
9. Axelsson, B., Urinary calculus in long-term exposure to cadmium, in *Proc. 14th Int. Congr. Occup. Health, Madrid, 1963,* Sangro, P., Akoun, G., and Beate, H. L., Eds., Int. Congr. Ser. 62, Excerpta Medica Foundation, Amsterdam, 1963.
10. Axelsson, B. and Piscator, M., Renal damage after prolonged exposure to cadmium. An experimental study, *Arch. Environ. Health,* 12, 360—373, 1966.
11. Axelsson, B. and Piscator, M., in *Cadmium in the Environment,* Friberg, L., Piscator, M., and Nordberg, G. F., Eds., Chemical Rubber Co., Cleveland, 1971, 85.
12. Axelsson, B., Dahlgren, S. E., and Piscator, M., Renal lesions in the rabbit after long-term exposure to cadmium, *Arch. Environ. Health,* 17, 24—28, 1968.
13. Baader, E. W., Chronic cadmium poisoning, *Dtsch. Med. Wochenschr.,* 76, 484—487, 1951 (in German).
14. Berggaard, I., Plasma proteins in normal urine, in *Protein in Normal and Pathological Urine,* Manuel, Y., Revillard, J. P., and Betnel, H. C., Eds., S. Karger, Basel, 1970, 7—19.
15. Berggaard, I. and Bearn, A. C., Isolation and properties of a low molecular weight globulin occurring in human biological fluids, *J. Biol. Chem.,* 243, 4095—4101, 1968.

16. Berggaard, I., Björck, L., Cigen, R., and Lögdberg, L., Beta$_2$-microglobulin, *Scand. J. Clin. Lab. Invest.,* 40(Suppl. 154), 13—25, 1980.

17. Bernard, A., Evaluation of renal dysfunction induced by cadmium in man by determination of enzymes and specific proteins in urine, *Arch. Toxicol.,* 4(Suppl.), 223—232, 1980.

18. Bernard, A. M. and Lauwerys, R. R., The effects of sodium chromate and carbon tetrachloride on the urinary excretion and tissue distribution of cadmium in cadmium-pretreated rats, *Toxicol. Appl. Pharmacol.,* 57, 30—38, 1981.

19. Bernard, A. M. and Lauwerys, R. R., Comparison of tubidimetry with particle counting for the determination of human b-2-microglobulin by latex immunoassay (LIA), *Clin. Chim. Acta,* 119, 335—339, 1982.

20. Bernard, A. and Lauwerys, R. R., Latex immunoassay of urinary albumin, *J. Clin. Chem. Clin. Biochem.,* 21, 25—30, 1983a.

21. Bernard, A. M. and Lauwerys, R. R., Continuous-flow system for automation of latex immunoassay by particle counting, *Clin. Chem.,* 29, 1007—1011, 1983b.

22. Bernard, A., Roels, H., Hubermont, G., Buchet, J. P., Masson, P. L., and Lauwerys, R. R., Characterization of the proteinuria in cadmium-exposed workers, *Int. Arch. Occup. Environ. Health,* 38, 19—30, 1976.

23. Bernard, A., Buchet, J. P., Roels, H., Masson, P., and Lauwerys, R., Renal excretion of proteins and enzymes in workers exposed to cadmium, *Eur. J. Clin. Invest.,* 9, 11—22, 1979.

24. Bernard, A., Lauwerys, R., and Gengoux, P., Characterisation of the proteinuria induced by prolonged oral administration of cadmium in female rats, *Toxicology,* 20, 345—357, 1981a.

25. Bernard, A., Vyskocil, A., and Lauwerys, R., Determination of beta$_2$-microglobulin in human urine and serum by latex immunoassay, *Clin. Chem.,* 27, 832—837, 1981b.

26. Bernard, A., Moreau, D., and Lauwerys, R., Comparison of retinol-binding protein and beta-2-microglobulin determination in urine for the early detection of tubular proteinuria, *Clin. Chim. Acta,* 126, 1—7, 1982a.

27. Bernard, A., Moreau, D., and Lauwerys, R., Latex immunoassay (LIA) of human retinol-binding protein, *Clin. Chem.,* 28, 1167—1171, 1982b.

28. Bernard, A., Viau, C., and Lauwerys, R., Renal handling of human β-2-microglobulin in normal and cadmium poisoned rats, *Arch. Toxicol.,* 53, 49—57, 1983.

29. Beton, D. C., Andrews, G. S., Davies, H. J., Howells, L., and Smith, G. F., Acute cadmium fume poisoning, five cases with one death from renal necrosis, *Br. J. Ind. Med.,* 23, 292—301, 1966.

30. Black, D. and Jones, N. F., *Renal Disease,* 4th ed., Blackwell Scientific, Oxford, 1979.

31. Bonnell, J. A., Emphysema and proteinuria in men casting copper-cadmium alloys, *Br. J. Ind. Med.,* 12, 181—197, 1955.

32. Bonnell, J. A., Kazantzis, G., and King, E., A follow-up study of men exposed to cadmium oxide fume, *Br. J. Ind. Med.,* 16, 135—145, 1959.

33. Bonnell, J. A., Ross, J. H., and King, E., Renal lesions in experimental cadmium poisoning, *Br. J. Ind. Med.,* 17, 69—80, 1960.

34. Buchet, J. P., Roels, H., Lauwerys, R., Bruaux, P., Clays-Thoreau, F., Lafontaine, A., and Verduyn, G., Repeated surveillance of exposure to cadmium, manganese and arsenic in school-age children living in rural, urban and non-ferrous smelter areas in Belgium, *Environ. Res.,* 22, 95—108, 1980a.

35. Buchet, J. P., Roels, H., Bernard, A., and Lauwerys, R., Assessment of renal function of workers exposed to inorganic lead, cadmium or mercury vapor, *J. Occup. Med.,* 22, 741—749, 1980b.

36. Buchet, J. P., Roels, H., Bernard, A., and Lauwerys, R., Assessment of renal function of workers simultaneously exposed to inorganic lead and cadmium, *J. Occup. Med.,* 23, 348—352, 1981.

37. Bulmer, F. M. R., Rothwell, H. E., and Frankish, E. R., Industrial cadmium poisoning, *J. Can. Public Health,* 29, 19—26, 1938.

38. Butler, E. A. and Flynn, F. V., The proteinuria of renal tubular disorders, *Lancet,* 2, 978—980, 1958.

39. Butler, E. A. and Flynn, F. V., The occurrence of post-gamma protein in urine: a new protein abnormality, *J. Clin. Pathol.,* 14, 172—178, 1961.

40. Butler, E. A., Flynn, F. V., Harris, H., and Robson, E. B., A study of urine proteins by two dimensional electrophoresis with special reference to the proteinuria of renal tubular disorders, *Clin. Chim. Acta,* 7, 34—41, 1962.

41. Chang, C. C., Lauwerys, R., Bernard, A., Roels, H., Buchet, J. P., and Garvey, J. S., Metallothionein in cadmium-exposed workers, *Environ. Res.,* 23, 422—428, 1980.

42. Clarkson, T. W. and Kench, J. E., Urinary excretion of amino acids by men absorbing heavy metals, *Biochem. J.,* 62, 361—372, 1956.

43. Committee on Cadmium Pollution Mechanism in Yoshinokawa Area, *Pollution Source of Cadmium in Yoshinokawa Area, Yamagata, Yamagata Prefecture,* 1974, 2—3 and 17—19 (in Japanese).

44. Conway, T. P. and Poulik, M. D., Catabolism of rat beta-2-microglobulin in the rat, *J. Lab. Clin. Med.,* 89, 1208—1214, 1977.

45. Cousins, R. J., Barber, A. K., and Trout, J. R., Cadmium toxicity in growing swine, *J. Nutr.,* 103, 964—972, 1973.
46. Dalhamn, T. and Friberg, L., Dimercaprol (2,3-dimercaptopropanol) in chronic cadmium poisoning, *Acta Pharmacol. Toxicol.,* 11, 68—71, 1955.
47. Dalhamn, T. and Friberg, L., Morphological investigations on kidney damage in chronic cadmium poisoning, *Acta Pathol. Microbiol. Scand.,* 40, 475—479, 1957.
48. Dance, N. and Price, R. G., The excretion of N-acetyl-beta-glucosaminidase and beta-glactosidase by patients with renal disease, *Clin. Chim. Acta,* 27, 87—92, 1970.
49. de Silva, P. E. and Donnan, M. B., Chronic cadmium poisoning in a pigment manufacturing plant, *Br. J. Ind. Med.,* 38, 76—86, 1981.
50. Diem, K. and Lentner, C., Eds., *Documenta Geigy, Scientific Tables,* 7th ed., Ciba-Geigy, Basel, 1975.
51. Dowa Mining Company Ltd., *Outline of Kosaka Mining Station* (pamphlet), 1973 (in Japanese).
52. Druet, P., Bernard, A., Hirsch, F., Weening, J. J., Gengoux, P., Mahieu, P., and Birkeland, S., Immunologically mediated glomerulonephritis induced by heavy metals, *Arch. Toxicol.,* 50, 187—194, 1982.
53. Elinder, C.-G., Early Effects of Cadmium Accumulation in Kidney Cortex, with Special Reference to Cadmium and Zinc Interactions, Doctoral thesis, Karolinska Institute, Stockholm, Sweden, 1979.
54. Elinder, C.-G., Jönsson, L., Piscator, M., and Rahnster, B., Histopathological changes in relation to cadmium concentration in horse kidneys, *Environ. Res.,* 26, 1—21, 1981a.
55. Elinder, C.-G., Nordberg, M., Palm, B., and Piscator, M., Cadmium, zinc and copper in horse liver and in horse liver metallothionein: comparison with kidney cortex, *Environ. Res.,* 26, 22—32, 1981b.
56. Ellis, K. J., Morgan, W. D., Zanzi, Il., Yasamura, S., Vartsky, D., and Cohn, S. H., Critical concentrations of cadmium in human renal cortex: dose effect studies in cadmium smelter workers, *J. Toxicol. Environ. Health,* 7, 691—703, 1981.
57. Ellis, K. J., Yuen, K., Yasamura, S., and Cohn, S. H., Dose-response analysis of cadmium in man: body burden vs. kidney dysfunction, *Environ. Res.,* 33, 216—226, 1984.
58. Engstrom, B., Norin, H., Jawaid, M., and Ingman, F., Influence of different Cd-EDTA complexes on distribution and toxicity of cadmium in mice after oral or parenteral administration, *Acta Pharmacol. Toxicol.,* 46, 219—234, 1980.
59. Evrin, P.-E. and Wibell, L., The serum levels and urinary excretion of beta$_2$-microglobulin in apparently healthy subjects, *Scand. J. Clin. Lab. Invest.,* 29, 69—74, 1972.
60. Evrin, P.-E., Peterson, P. A., Wide, L., and Berggaard, I., Radioimmunoassay of beta$_2$-microglobulin in human biological fluids, *Scand. J. Clin. Lab. Invest.,* 28, 440—443, 1971.
61. Eybl, V., Sykora, J., and Mertl, F., Effects of CaEDTA and CaDTPA in cadmium intoxication, *Acta Biol. Med. Ger.,* 17, 178—185, 1966 (in German).
62. Favino, A. and Nazari, G., Renal lesions induced by a single subcutaneous injection of cadmium chloride in rat, *Lav. Um.,* 19, 367—372, 1967 (in Italian).
63. Foster, C. L. and Cameron, E., Observations on the histological effects of sub-lethal doses of cadmium chloride in the rabbit. II. The effect on the kidney cortex, *J. Anat.,* 97, 281—288, 1963.
64. Foulkes, E. C., Nature of inhibition of renal aspartate reabsorption in experimental Fanconi syndrome, *Toxicol. Appl. Pharmacol.,* 71, 445—450, 1983.
65. Foulkes, E. C. and Gieske, T., Specificity and metal sensitivity of renal amino acid transport, *Biochim. Biophys. Acta,* 318, 439—445, 1973.
66. Fowler, B. A., Jones, H. S., Brown, H. W., and Haseman, J. K., The morphological effects of chronic cadmium administrations on the renal vasculature of rats given low and normal calcium diets, *Toxicol. Appl. Pharmacol.,* 34, 233—252, 1975.
67. Friberg, L., Health hazards in the manufacture of alkaline accumulators with special reference to chronic cadmium poisoning. Doctoral thesis, *Acta Med. Scand.,* 138(Suppl. 240), 1—124, 1950.
68. Friberg, L., Further investigations on chronic cadmium poisoning; a study on rabbits with radioactive cadmium, *AMA Arch. Ind. Hyg. Occup. Med.,* 5, 30—36, 1952.
69. Friberg, L., Iron and liver administration in chronic cadmium poisoning and studies on the distribution and excretion of cadmium. Experimental investigations in rabbits, *Acta Pharmacol.,* 11, 168—178, 1955.
70. Friberg, L., Edathamil calcium-disodium in cadmium poisoning, *AMA Arch. Ind. Health,* 13, 18—23, 1956.
71. Friberg, L., Deposition and distribution of cadmium in man in chronic cadmium poisoning, *AMA Arch. Ind. Health,* 16, 27—29, 1957.
72. Friberg, L. and Kjellström, T., Toxic metals, pitfalls in risk estimation, in *Int. Conf. Heavy Metals in the Environ.,* Commission of the European Communities, Luxembourg, 1981, 1—11.
73. Friberg, L. and Nyström, Å., Aspects on the prognosis in chronic cadmium poisoning, *Läkartidningen,* 49, 2629—2639, 1952 (in Swedish).

74. Friberg, L., Piscator, M., Nordberg, G. F., and Kjellstrom, T., *Cadmium in the Environment,* 2nd ed., CRC Press, Boca Raton, Fla., 1974.

75. Fukushima, I., Nakajima, T., Higuchi, Y., Tsuchiya, T., Nakajma, K., Tanaka, T., Fukamachi, M., Arai, K., and Schiono, T., Results of screening examination on tubular dysfunction in aged persons confined to bed, in *Kankyo Hoken Report No. 36,* Japanese Public Health Association, Tokyo, 1975, 138—142 (in Japanese).

76. Fukushima, M., Kobayashi, S., and Sakamoto, M., Urinary free amino acids in Itai-itai patients and among inhabitants in a cadmium polluted area, in *Kankyo Hoken Report No. 24,* Japanese Public Health Association, Tokyo, 1973, 53—57 (in Japanese).

77. Fukushima, M., Ishizaki, A., Nogawa, K., Sakamoto, M., and Kobayashi, E., Epidemiological studies on renal failure of inhabitants in "Itai-Itai" disease endemic district. I. Some urinary findings of inhabitants living in and around the endemic district of the Jinzu river basin, *Jpn. J. Public Health,* 21, 65—73, 1974 (in Japanese with English summary).

78. Gieske, T. H. and Foulkes, E. C., Acute effects of cadmium on proximal tubular function in rabbits *Toxicol. Appl. Pharmacol.,* 27, 292—299, 1974.

79. Goyer, R. A., Tsuchiya, K., Leonard, D. L., and Kahyo, H., Aminoaciduria in Japanese workers in the lead and cadmium industries, *Am. J. Clin. Pathol.,* 57, 635—642, 1972.

80. Hagino, N., About investigations on Itai-itai disease, *J. Toyama Med. Assoc.,* Dec. 21, 7, 1957 (in Japanese).

81. Hansén, L., Kjellström, T., and Vesterberg, O., Evaluation of different urinary proteins excreted after occupational Cd exposure, *Int. Arch. Occup. Environ. Health,* 40, 273—282, 1977.

82. Harada, A., Findings on urinary protein electrophoresis, in *Kankyo Hoken Report No. 11,* Japanese Public Health Association, Tokyo, April 1972, 87 (in Japanese).

83. Harada, A. and Shibutanni, E., Medical examination of workers in a cadmium pigment factory, in *Kankyo Hoken Report No. 24,* Japanese Public Health Association, Tokyo, 1973, 16—22 (in Japanese).

84. Harada, A., Hirota, M. and Kono, K., Surveillance of cadmium workers, in *Cadmium 79, Proc. 2nd Int. Cadmium Conf., Cannes,* Metal Bulletin Ltd., London, 1980, 183—193.

85. Holden, H., Health status of European cadmium workers, in *Occupational Exposure to Cadmium,* Cadmium Association, London, 1980, 330—337.

86. Honda, R., Yamada, Y., Tsuritani, I., Kobayashi, E., Ishizaki, M. and Nogawa, K., Significance of low molecular proteinuria in cadmium poisoning, *J. Kanazawa Med. Univ.,* 7, 142—151, 1982 (in Japanese with English summary).

87. Honda, R., Yamada, Y., Kido, T., Tsuritani, T., Ishizaki, M., and Nogawa, K., Urinary N-acetyl-beta-D-glucosaminidase as a biological indicator of renal damage due to environmental cadmium exposure, *J. Kanazawa Med. Univ.,* 8, 248—256, 1983 (in Japanese with English summary).

88. Hörstebrock, A., The pathological anatomy of chronic cadmium poisoning, *Bundesarbeitsblatt,* 10(7), 1951 (in German).

89. ICRP, *Report of the Task Group on Reference Man,* Report No. 23, International Commission on Radiological Protection, Pergamon Press, Oxford, 1975.

90. *Report on the Health Examination of Residents of the Kakehashi River Basin,* Health and Welfare Department, Ishikawa Prefecture, Kanazawa, 1976 (in Japanese).

91. Ishizaki, A., About the Cd and Zn concentrations in organs of Itai-itai disease patients and in inhabitants of Hokuriku area, in *Kankyo Hoken Report No. 11,* Japanese Public Health Association, Tokyo, April 1972, 154 (in Japanese).

92. Ishizaki, A., Fukushima, M. and Sakamoto, M., On the accumulation of cadmium in the bodies of Itai-itai disease patients, *Jpn. J. Hyg.,* 25, 86, 1970 (in Japanese).

93. Iwao, S., Tsuchiya, K. and Sakurai, H., Serum and urinary beta-2-microglobulin among cadmium-exposed workers, *J. Occup. Med.,* 22, 399—402, 1980.

94. Iwao, S., Tsuchiya, K., and Sugita, M., Variation of cadmium accumulation among Japanese, *Arch. Environ. Health,* 38, 156—162, 1983a.

95. Iwao, S., Kodama, Y., and Tsuchiya, K., Associations of cadmium with zinc and copper in cancer cases and controls, *Biol. Trace Element Res.,* 5, 383—388, 1983b.

96. Japan Mining Association, *General Review of Ore Deposits in Japan,* Vol. 1, Japan Mining Association, Tokyo, 1970 (in Japanese).

97. Japan Mining Association, *General Review of Ore Deposits in Japan,* Vol. 2, Japan Mining Association, Tokyo, 1973 (in Japanese).

98. Joshi, B. C., Dwivedi, C., Powell, A., and Holscher, M., Immune complex nephritis in rats induced by long-term oral exposure to cadmium, *J. Comp. Pathol.,* 91, 11—14, 1981.

99. Kajikawa, K., Pathogenesis of Itai-itai disease based on post-mortem studies, in *Cadmium Studies in Japan — A Review,* Tsuchiya, K., Ed., Elsevier, New York, 1978, 286—295.

100. Kajikawa, K., Kitagawa, M., Nakanishi, I., Ueshima, H., Katsuda, S., and Kuroda, K., A pathological study of Itai-itai disease, *J. Juzen Med. Soc.,* 83, 309—347, 1974 (in Japanese with English summary).

101. Kanai, M., Nomoto, S., Sasaoka, S., and Muto, Y., Retinol-binding protein levels in blood and urine from patients with "Itai-itai" disease: pathological mechanism for its increased excretion, *Proc. Symp. Chem. Physiol. Pathol.*, 12, 319—324, 1972a.

102. Kanai, M., Sasaoka, S., and Naiki, M., Radioimmunoassay of retinol-binding protein and its use for an early detection of renal tubular disorders and differentiation of proteinuric patterns, *Proc. Symp. Chem. Physiol. Pathol.*, 12, 325—330, 1972b.

103. Kanai, M., Nomoto, S., Sasaoka, S., and Naiki, M., Clinical significance of urinary excretion of retinol-binding protein in patients with "Itai-itai" disease, *Proc. Symp. Chem. Physiol. Pathol.*, 11, 194—199, 1972c.

104. Karlsson, F. A. and Jacobson, G., Renal handling of beta-2-microglobulin amylase and albumin in acute pancreatitis, *Acta Chir. Scand.*, 145, 59—63, 1979.

105. Karlsson, F. A., Sege, K., Beauduin, M., Pluygers, E., Wibell, L., Groth, T., and Peterson, P. A., Turnover studies of beta$_2$-microglobulin in normal persons and in patients with increased serum levels of the protein, in *Beta$_2$-m in Proliferative Disorders and Heavy Metal Intoxication*, Peterson, P. A. and Lauwerys, R., Eds., 1978, 9—21.

106. Karlsson, F. A., Wibell, L., and Evrin, P. E., Beta-2-microglobulin in clinical medicine, *Scand. J. Clin. Lab. Invest.*, 40(Suppl. 154), 27—37, 1980.

107. Kawai, K., Fukuda, K., and Kimura, M., Morphological alterations in experimental cadmium exposure with special reference to the onset of renal lesions, in *Effects and Dose-Response Relationships of Toxic Metals*, Nordberg, G. F., Ed., Elsevier, Amsterdam, 1976, 343—370.

108. Kawamura, J., Yoshida, O., Nishino, K., and Itokawa, Y., Disturbances in kidney functions and calcium and phosphate metabolism in cadmium-poisoned rats, *Nephron*, 20, 101—110, 1978.

109. Kazantzis, G., Industrial hazards to the kidney and urinary tract, in *Sixth Symposium on Advanced Medicine*, Slater, J. D. H., Ed., Pitman & Sons, London, 1970, 263—274.

110. Kazantzis, G., Some long-term effects of cadmium on the human kidney, in *Cadmium 77, Proc. 1st Int. Cadmium Conf. San Francisco, 1977*, Metal Bulletin Ltd., London, 1978, 194—198.

111. Kazantzis, G., Cadmium nephropathy, *Contrib. Nephrol.*, 16, 161—166, 1979.

112. Kazantzis, G., Flynn, F. V., Spowage, J. S., and Trott, D. G., Renal tubular malfunction and pulmonary emphysema in cadmium pigment workers, *Q. J. Med.*, 32, 165—192, 1963.

113. Kennedy, A., The effect of L-cysteine on the toxicity of cadmium, *Br. J. Exp. Pathol.*, 49, 360—364, 1968.

114. Kimura, M., A method to analyse beta$_2$-microglobulin, in *Kankyo Hoken Report No. 36*, Japanese Public Health Association, Tokyo, 1976, 185—187.

115. Kimura, M., Watanabe, M., and Otaki, N., Determination of heavy metals and beta-2-microglobulin, in *The Tertiary Monkey Experiment Team, Effects of Nutritional Factors on Cadmium-Administered Monkeys*, Japanese Public Health Association, Tokyo, 1983, 66—82.

116. King, E., An environmental study of casting copper-cadmium alloys, *Br. J. Ind. Med.*, 12, 198—205, 1955.

117. Kitamura, M. and Koizumi, N., Results of urine tests for bedridden old persons in Hyogo Prefecture (Report 2), in *Kankyo Hoken Report No. 36*, Japanese Public Health Association, Tokyo, 1975, 135—137 (in Japanese).

118. Kjellström, T., Exposure and accumulation of cadmium in populations from Japan, the United States and Sweden, *Environ. Health Perspect.*, 28, 169—197, 1979.

119. Kjellström, T. and Elinder, C.-G., Interpretation of the urinary excretion of different proteins in cadmium poisoning, in press.

120. Kjellström, T. and Piscator, M., *Quantitative Analysis of β$_2$-Microglobulin in Urine as an Indicator of Renal Tubular Damage Induced by Cadmium*, Phadedoc No. 1, Diagnostic Communications, Pharmacia Diagnostics AB, Uppsala, Sweden, 1977, 3—21.

121. Kjellström, T., Shiroishi, K., and Evrin, P.-E., Urinary β$_2$-microglobulin excretion among people exposed to cadmium in the general environment. An epidemiological study in cooperation between Japan and Sweden, *Environ. Res.*, 13, 318—344, 1977a.

122. Kjellström, T., Evrin, P.-E., and Rahnster, B., Dose-response analysis of cadmium-induced tubular proteinuria. A study of urinary beta$_2$-microglobulin excretion among workers in a battery factory, *Environ. Res.*, 13, 303—317, 1977b.

123. Kjellström, T., Elinder, C.-G., and Friberg, L., Conceptual problems in establishing the critical concentration of cadmium in human kidney cortex, *Environ. Res.*, 33, 284—295, 1984.

124. Kjellström, T., et al., submitted.

125. Kobayashi, E., An epidemiological study on the health effects of environmental cadmium. II. Some urinary findings by residence time in Cd-polluted area, *Jpn. J. Public Health*, 29, 201—207, 1982a (in Japanese with English summary).

126. Kobayashi, E., Delta aminolevulinic acid dehydratase activities in inhabitants in a cadmium-polluted area, *Sangyo Igaku*, 3, 316—317, 1982, (in Japanese).

127. Kojima, S., Haga, Y., Kurihara, T., and Yamawaki, T., Report from Akita Prefecture, in *Kankyo Hoken Report No. 36,* Japanese Public Health Association, Tokyo, 1975, 114—123 (in Japanese).

128. Kojima, S., Haga, Y., Kurihara, T., Yamawaki, T., and Kjellstrom, T., A comparison between fecal cadmium and urinary β_2-microglobulin, total protein, and cadmium among Japanese farmers. An epidemiological study in cooperation between Japan and Sweden, *Environ. Res.,* 14, 436—451, 1977.

129. Kowal, N. E. and Kraemer, D. F., Urinary cadmium and beta-2-microglobulin levels of persons aged 20—74 years from a subsample of the national HANES II survey, 1976—1980, in *Cadmium 81, Proc. 3rd Int. Cadmium Conf. Miami,* February 3 to 5, 1981, Cadmium Association, London, 1982, 119—122.

130. Lauwerys, R. R. and De Wals, P., Environmental pollution by cadmium and mortality from renal diseases, *Lancet,* 1, 383, 1981.

131. Lauwerys, R. R., Buchet, J.-P., Roels, H. A., Brouwers, J., and Stanescu, D., Epidemiological survey of workers exposed to cadmium, *Arch. Environ. Health,* 28, 145—148, 1974.

132. Lauwerys, R. R., Bernard, A., Roels, H. A., Buchet, J.-P., and Viau, C., Characterization of cadmium proteinuria in man and rat, *Environ. Health Perspect.,* 54, 147—154, 1984.

133. Levinsky, N. G. and Levy, M., Clearance techniques, in *Handbook of Physiology. VIII. Renal Physiology,* Orloff, J., Berliner, R. W., and Geiger, S. R., Eds., American Physiology Society, Washington, D.C., 1973, 103—117.

134. Maack, T., Johnson, V., Kan, S. T., Figueiredo, J., and Sigulem, D., Renal filtration transport and metabolism of low-molecular weight proteins: a review, *Kidney Int.,* 16, 251—270, 1979.

135. McKenzie, J. M., Kjellström, T., and Sharma, R., *Cadmium Intake, Metabolism and Effects in People with a High Intake of Oysters in New Zealand,* U.S. Environmental Protection Agency, Health Effects Research Laboratory, Cincinatti, 1982.

136. Mitane, Y., Saito, H., Nakano, A., Tohyama, E., Sugihira, N., and Wakisaka, I., Diurnal variation in urinary excretion of various elements, *Jpn. J. Hyg.,* 37, 264, 1982 (in Japanese).

137. Miyasaki, K., Bone lesions in inhabitants of the cadmium polluted areas in Hyogo Prefecture: five autopsy cases, in *Cadmium-Induced Osteopathy,* Shigematsu, I. and Nomiyama, K., Eds., Japanese Public Health Association, Tokyo, 1980, 106—123.

138. Mogensen, C. E. and Sølling, K., Studies on renal tubular protein reabsorption: partial and near complete inhibition by certain amino acids, *Scand. J. Clin. Lab. Invest.,* 37, 477—486, 1977.

139. Mogensen, C. E., Vittinghus, E., and Sølling, K., Increased urinary excretion of albumin, light chains and beta-2-microglobulin after intravenous arganine administration in normal man, *Lancet,* 2, 581—583, 1975.

140. Mogensen, C. E., Gjöde, P., and Christensen, C. K., Albumin excretion in operating surgeons and in hypertension, *Lancet,* 1, 774—775, 1979.

141. Mudge, G. H., Berndt, W. O., and Valtin, H., Tubular transport of urea, glucose, phosphate, uric acid, sulfate, and thiosulfate, in *Handbook of Physiology. VIII. Renal Physiology,* Orloff, J., Berliner, R. W., and Geiger, S. R., Eds., American Physiology Society, Washington, D.C., 1973, 587—652.

142. Murakami, M., Tohyama, C., Sano, K., Kawamura, R., and Kubota, K., Autoradiographical studies on the localisation of metallothionein in proximal tubular cells of the rat kidney, *Arch. Toxicol.,* 53, 185—192, 1983a.

143. Murakami, M., Cain, K., and Webb, M., Cadmium-metallothionein-induced nephropathy: a morphological and autoradiographic study of cadmium distribution, the development of tubular damage and subsequent cell regeneration, *J. Appl. Toxicol.,* 3, 237—244, 1983b.

144. Murase, H., Inoue, T., Sagai, M., Iwata, T., and Kubota, K., Microscopic pathology of dogs chronically exposed for cadmium *per os,* in *Materials of the Research Meeting Concerning Cadmium Poisoning,* sponsored by the Japanese Public Health Association, Tokyo, March 16, 1974, 72—73 (in Japanese).

145. Nagasaki Prefecture, Report of medical examinations of the population in the river Sasu and Shiine basin, Department of Health, March 30, 1970 (in Japanese).

146. Nakagawa, S., A study of osteomalacia in Toyama Prefecture (so-called Itai-itai disease), *J. Radiol. Phys. Ther. Univ. Kanizawa,* 56, 1—51, 1960 (in Japanese with English summary).

147. National Institute of Geology, *Historical Review of Mining Products in Japan,* Tokyo Geological Association, Japan, 1956 (in Japanese).

148. Nicholson, J. K. and Osborn, D., Kidney lesions in pelagic seabird with high tissue levels of cadmium and mercury, *J. Zool. London,* 200, 99—118, 1983.

149. Nicholson, J. K., Kendall, M. D., and Osborn, D., Cadmium and mercury nephrotoxicity, *Nature, (London),* 304, 633—635, 1983.

150. Niemeier, B., The influence of chelating agents on the distribution and toxicity of cadmium, *Int. Arch. Gewerbepathol. Gewerbehyg.,* 24, 160—168, 1967 (in German).

151. Nogawa, K., Itai-itai disease and follow-up studies, in *Cadmium in the Environment. II. Health Effects,* Niriagu, J. O., Ed., John Wiley & Sons, New York, 1981, 1—38.

152. Nogawa, K. and Ishizaki, A., A comparison between cadmium in rice and renal effects among inhabitants of the Jinzu River Basin, *Environ. Res.,* 18, 410—420, 1979.

153. Nogawa, K., Ishizaki, A., Fukushima, M., Shibata, I., and Hagino, N., Studies on the women with acquired Fanconi syndrome observed in the Ichi river basin polluted by cadmium, *Environ. Res.,* 10, 280—307, 1975.

154. Nogawa, K., Kobayashi, E., Inaoka, H., and Ishizaki, A., The relationship between the renal effects of cadmium and cadmium concentration in urine among the inhabitants of cadmium-polluted areas, *Environ. Res.,* 14, 391—400, 1977.

155. Nogawa, K., Ishizaki, A., and Kobayashi, E., A comparison between health effects of cadmium and cadmium concentration in urine among inhabitants of the Itai-itai disease endemic district, *Environ. Res.,* 18, 297—409, 1979a.

156. Nogawa, K., Kobayashi, E. and Honda, R., A study of the relationship between cadmium concentrations in urine and renal effects of cadmium, *Environ. Health Perspect.,* 28, 161—168, 1979b.

157. Nogawa, K., Kobayashi, E., Honda, R., and Ishizaki, A., A follow-up study on urinary findings of patients with and suspected of Itai-itai disease, *Jpn. J. Public Health,* 26, 25—31, 1979c (in Japanese with English summary).

158. Nogawa, K., Yamada, Y., Honda, R., Tsuritani, I., Ishizaki, M., and Sakamoto, M., Urinary N-acetyl-beta-d-glucosaminidase and beta-2-microglobulin in "Itai-itai" disease, *Toxicol. Lett.,* 16, 317—322, 1983.

159. Nogawa, K., Tamada, Y., Honda, R., Tsuritani, I., Kobayashi, E., and Ishizaki, M., Copper and zinc levels in serum and urine Sf cadmium-exposed people with special reference to renal tubular damage, *Environ. Res.,* 33, 29—38, 1984.

160. Nomiyama, K., Development mechanism and diagnosis of cadmium poisoning, in *Kankyo Hoken Report No. 24,* Japanese Public Health Association, Tokyo, 1973, 11—15 (in Japanese).

161. Nomiyama, K., Experimental studies on cadmium intoxication, *Jpn. Med. Assoc. J. (Nikon Isshikai Zasshi),* 72, 955—966, 1974 (in Japanese).

162. Nomiyama, K., Toxicity of cadmium-mechanism and diagnosis, in *Progress in Water Technology,* Krenkel, P. A., Ed., Pergamon Press, Oxford, 1975, 15—23.

163. Nomiyama, K., Recent progress and perspectives in cadmium health effects studies, *Sci. Total Environ.,* 14, 199—232, 1980.

164. Nomiyama, K. and Foulkes, E. C., Reabsorption of filtered cadmium-metallothionein in the rabbit kidney, *Proc. Soc. Exp. Biol. Med.,* 156, 97—99, 1977.

165. Nomiyama, K. and Nomiyama, H., Tissue metallothioneins in rabbits chronically exposed to cadmium, with special reference to the critical concentration of cadmium in the renal cortex, in *Biological Roles of Metallothionein,* Foulkes, E. C., Ed., Elsevier, New York, 1982, 47—67.

166. Nomiyama, K., Sato, C., and Yamamoto, A., Early signs of cadmium intoxication in rabbits, *Toxicol. Appl. Pharmacol.,* 24, 625—636, 1973.

167. Nomiyama, K., Sugata, Y., Yamamoto, A., and Nomiyama, H., Effects of dietary cadmium on rabbits. I. Early signs of cadmium intoxication, *Toxicol. Appl. Pharmacol.,* 31, 4—12, 1975.

168. Nomiyama, K., Foulkes, E. C., and Taguchi, T., *The Metabolism of Metallothionein, with Particular Reference to the Excretion and Reabsorption of Metallothionein in Rabbit Kidneys,* Special Report, Environment Agency, Tokyo, 1976 (in Japanese).

169. Nomiyama, K., Nomiyama, H., Yotoriyama, M., and Taguchi, T., Some recent studies on the renal effects of cadmium, in *Cadmium 77, Proc. 1st Int. Cadmium Conf., San Francisco, 1977,* Metal Bulletin Ltd., London, 1978, 186—194.

170. Nomiyama, K., Nomiyama, H., Nomura, T., Taguchi, T., Matsui, J., Yotoriyama, M., Akahori, F., Iwao, S., Koizumi, N., Masaoka, T., Kitamura, S., Tsuchiya, K., Suzuki, T., and Kobayashi, K., Effects of dietary cadmium on rhesus monkeys, *Environ. Health Perspect.,* 28, 223—243, 1979.

171. Nomiyama, K., Nomiyama, H., Akahori, F., and Masaoka, T., Further studies on effects of dietary cadmium on rhesus monkeys. V. Renal effects, in *Recent Studies on Health Effects of Cadmium in Japan,* Environment Agency, Tokyo, 1981, 59—104.

172. Nomiyama, K., Nomiyama, H., and Yotoriyama, M., Low-molecular weight proteins in urine from rabbits given nephrotoxic compounds, *Ind. Health,* 20, 1—10, 1982a.

173. Nomiyama, K., Nomiyama, H., Yotoriyama, M., and Matsui, K., Sodium dodecyl sulfate acrylamide gel electrophoretic studies of low-molecular-weight proteinuria, an early sign of cadmium health effects in rabbits, *Ind. Health,* 20, 11—18, 1982b.

174. Nomiyama, K., Yotoriyama, M., and Nomiyama, H., Dose-effect relationship between cadmium and beta-2-microglobulin in the urine of inhabitants of cadmium-polluted areas (Japan), *Arch. Environ. Contam. Toxicol.,* 12, 147—150, 1983a.

175. Nomiyama, K., Yotoriyama, M., and Nomiyama, H., Urinary beta-2-microglobulin and renal functions in elderly people in an area with no known cadmium pollution (Japan), *Arch. Environ. Contam. Toxicol.,* 12, 143—146, 1983b.

176. Nomiyama, K. and Nomiyama, H., Reversibility of cadmium-induced health effects in rabbits, *Environ. Health Perspect.*, 54, 201—211, 1984.

177. Nomura, Y., Masaoka, T., and Akahori, F., Further studies on effects of dietary cadmium on rhesus monkeys. VIII. Pathological findings, in *Recent Studies on Health Effects of Cadmium in Japan*, Environment Agency, Tokyo, August 1981, 129—138.

178. Nordberg, G. F., Cadmium metabolism and toxicity. Experimental studies on mice with special reference to the use of biological materials as indices of retention and the possible role of metallothionein in transport and detoxification of cadmium, *Environ. Physiol. Biochem.*, 2, 7—36, 1972.

179. Nordberg, G. F., in *Cadmium in the Environment*, 2nd ed., Friberg, L. F., Piscator, M., Nordberg, G. F., and Kjellström, E., Eds., CRC Press, Boca Raton, Fla., 1974.

180. Nordberg, G. F. and Piscator, M., Influence of long-term cadmium exposure on urinary excretion of protein and cadmium in mice, *Environ. Physiol. Biochem.*, 2, 37—49, 1972.

181. Nordberg, G. F., Goyer, R. A., and Nordberg, M., Comparative toxicity of cadmium-metallothionein and cadmium chloride on mouse kidney, *Arch. Pathol.*, 99, 192—197, 1975.

182. Nordberg, G. F., Garvey, J. S., and Chang, C. C., Metallothionein in plasma and urine of cadmium workers, *Environ. Res.*, 28, 179—182, 1982.

183. Nordberg, M. and Nordberg, G. F., Distribution of metallothionein-bound cadmium and cadmium chloride in mice: preliminary studies, *Environ. Health Perspect.*, 12, 103—108, 1975.

184. Nordberg, M., Elinder, C.-G., and Rahnster, B., Cadmium, zinc and copper in horse kidney metallothionein, *Environ. Res.*, 20, 341—350, 1979.

185. Norden, A. G. W. and Flynn, F. V., Degradation of beta-2-microglobulin in infected urine by leukocyte elastase-like activity, *Clin. Chim. Acta*, 134, 167—176, 1983.

186. Olhagen, B., Special investigations of urine protein, in Friberg, L., Health hazards in the manufacture of alkaline accumulators with special reference to chronic cadmium poisoning, *Acta Med. Scand.*, 138(Suppl. 240), 35—40, 1950.

187. Parizek, J., The destructive effect of cadmium ion on testicular tissue and its prevention by zinc, *J. Endocrinol.*, 15, 56—63, 1957.

188. Parizek, J., Sterilization of the male by cadmium salts, *J. Reprod. Fertil.*, 1, 294—309, 1960.

189. Parizek, J. and Zahor, Z., Effect of cadmium salt on testicular tissue, *Nature, (London)*, 177, 1036—1038, 1956.

190. Parving, H.-H., Jensen, J. A., Mogensen, C. E., and Evrin, P.-E., Increased urinary albumin excretion rate in benign essential hypertension, *Lancet*, 1, 1190—1192, 1974.

191. Pedersen, E. B., Mogensen, C. E., and Larsen, J. S., Effects of exercise on urinary excretion of albumin and beta-2-microglobulin in young patients with mild essential hypertension without treatment and during long-term propranol treatment, *Scand. J. Clin. Lab. Invest.*, 41, 493—498, 1981.

192. Peterson, P. A., Evrin, P. E., and Berggaard, I., Differentiation of glomerular, tubular and normal proteinuria: determinations of urinary excretion, beta-2-microglobulin, albumin and total proteins, *J. Clin. Invest.*, 1189—1198, 1969.

193. Piscator, M., Proteinuria in chronic cadmium poisoning. II. The applicability of quantitative and qualitative methods of protein determination for the demonstration of cadmium proteinuria, *Arch. Environ. Health*, 5, 325—332, 1962a.

194. Piscator, M., Proteinuria in chronic cadmium poisoning. I. An electrophoretic and chemical study of urinary and serum proteins from workers with chronic cadmium poisoning, *Arch. Environ. Health*, 4, 607—621, 1962b.

195. Piscator, M., Proteinuria in chronic cadmium poisoning. III. Electrophoretic and immunoelectrophoretic studies on urinary proteins from cadmium workers, with special reference to the excretion of low molecular weight proteins, *Arch. Environ. Health*, 12, 335—344, 1966a.

196. Piscator, M., Proteinuria in chronic cadmium poisoning. IV. Gel filtration and ion-exchange chromatography of urinary proteins from cadmium workers, *Arch. Environ. Health*, 12, 345—356, 1966b.

197. Piscator, M., Proteinuria in chronic cadmium poisoning, Beckmans, Stockholm, 1966c.

198. Piscator, M., in *Cadmium in the Environment*, Friberg, L., Piscator, M., and Nordberg, G., Eds., Chemical Rubber Co., Cleveland, 1971.

199. Piscator, M., Cadmium toxicity, industrial and environmental experience, Paper presented at the 17th Int. Congr. Occup. Health, Buenos Aires, 1972.

200. Piscator, M., The progress of renal dysfunction in cadmium-exposed persons, in *Cadmium 83, Proc. 4th Int. Cadmium Conf., Munich 1983*, Metal Bulletin Ltd., London, in press.

201. Piscator, M. and Larsson, S. E., Retention and toxicity of cadmium in calcium-deficient rats, in *Proc. of the 17th Int. Congr. Occup. Health*, Buenos Aires, 1972.

202. Piscator, M. and Pettersson, B., Chronic cadmium poisoning — diagnosis and prevention, in *Clinical Chemistry and Chemical Toxicology of Metals*, Brown, S. S., Ed., Elsevier, Amsterdam, 1977, 143—155.

203. Piscator, M., Björk, L., and Nordberg, M., Beta-2-microglobulin levels in serum and urine of cadmium exposed rabbits, *Acta Pharmacol. Toxicol.,* 49, 1—7, 1981.
204. Pollak, V. E., First, M. R., and Piesce, A. J., Value of the sieving coefficient in the interpretation of renal protein clearances, *Nephron,* 13, 82—92, 1974.
205. Potts, C. L., Cadmium proteinuria — the health of battery workers exposed to cadmium oxide dust, *Ann. Occup. Hyg.,* 8, 55—61, 1965.
206. Poulik, M. D. and Reisfeld, R. A., Beta₂-microglobulins, *Contemp. Top. Mol. Immunol.,* 4, 157—201, 1975.
207. Poulik, M. D., Farrah, D., and Smithies, O., Association of urinary beta₂-microglobulin with myeloproliferative diseases, *Fed. Proc.,* 31, 741, 1972.
208. Princi, F., A study of industrial exposures to cadmium, *J. Ind. Hyg. Toxicol.,* 29, 315—324, 1947.
209. Prodan, L., Cadmium poisoning. II. Experimental cadmium poisoning, *J. Ind. Hyg. Toxicol.,* 14, 174—196, 1932.
210. Richie, R. F., Alper, C., Graves, J., Pearson, N., and Larson, C., Automated quantitation of proteins in serum and other biological fluids, *Am. J. Clin. Pathol.,* 59, 151—159, 1973.
211. Roels, H. A., Lauwerys, R. R., Buchet, J.-P., and Bernard, A., Environmental exposure to cadmium and renal function of aged women in three areas of Belgium, *Environ. Res.,* 24, 117—130, 1981a.
212. Roels, H. A., Lauwerys, R. R., Buchet, J. P., Bernard, A., Chettle, D. R., Harvey, T. C., and Al-Haddad, I. K., *In vivo* measurement of liver and kidney cadmium in workers exposed to this metal: its significance with respect to cadmium in blood and urine, *Environ. Res.,* 26, 217—240, 1981b.
213. Roels, H., Djubgang, J., Buchet, J.-P., Bernard, A., and Lauwerys, R., Evolution of cadmium-induced renal dysfunction in workers removed from exposure, *Scand. J. Work Environ. Health,* 8, 191—200, 1982.
214. Roels, H., Lauwerys, R., and Dardenne, A. N., The critical level of cadmium in human renal cortex: a reevaluation, *Toxicol. Lett.,* 15, 357—360, 1983a.
215. Roels, H., Lauwerys, R., Buchet, J. P., Bernard, A., Garvey, J. S., and Linton, H. J., Significance of urinary metallothionein in workers exposed to cadmium, *Int. Arch. Occup. Environ. Health,* 52, 159—166, 1983b.
216. Saito, H., Shioji, R., Furukawa, Y., Nagai, K., Arikawa, T., Saito, T., Sasaki, Y., Furuyama, T., and Yoshinaga, K., Cadmium-induced proximal tubular dysfunction in a cadmium-polluted area, *Nephron,* 6, 1—12, 1977.
217. Saito, H., Nakano, A., Tohyama, C., Mitane, Y., Sugihira, N., and Wakisaka, I., Studies on cadmium-exposure and urinary beta-2-microglobulin, *Jpn. J. Hyg.,* 37, 245, 1982 (in Japanese).
218. Saito, H., Nakano, A., Tohyama, C., Mitane, Y., Sugihira, N., and Wakisaka, I., Report on a health examination of inhabitants of areas with cadmium pollution of soil, in *Kankyo Hoken Report No. 40,* Japanese Public Health Association, Tokyo, 1983, 91—95 (in Japanese).
219. Scott, R., Mills, E. A., Fell, G. S., Husain, F. E. R., Yates, A. J., Paterson, P. J., McKirdy, A., Ottoway, J. M., Fitzgerald-Finch, O. P., Lamont, A., and Roxburgh, S., Clinical and biochemical abnormalities in coppersmiths exposed to cadmium, *Lancet,* August 21, 396—398, 1976.
220. Scott R., Aughey. E., and Sinclair, J., Histological and ultrastructural changes in rat kidney following cadmium injection, *Urol. Res.,* 5, 15—20, 1977.
221. Scott, R., Patterson, P. J., Burns, R., Ottoway, J. M., Hussain, F. E. R., Fell, G. S., Dumbuya, S., and Iqbal, M., Hypercalciuria related to cadmium exposure, *Urology,* 11, 462—465, 1978.
222. Selenke, W. and Foulkes, E. C., The binding of cadmium-metallothionein to isolated renal brush border membranes, *Proc. Soc. Exp. Biol. Med.,* 167, 40—44, 1981.
223. Shigematsu, I., Minowa, M., and Fukutomi, S., Health survey of population in cadmium polluted areas according to the standard method by the Environment Agency (intermediate report), In *Kankyo Hoken Report No. 45,* Japanese Public Health Association, Tokyo, 1979a, 75—85 (in Japanese).
224. Shigematsu, I., Minowa, M., Yoshida, T., and Miyamoto, K., Recent results of health examinations on the general population in cadmium-polluted and control areas in Japan, *Environ. Health Perspect.,* 28, 205—210, 1979b.
225. Shiroishi, K., Anayama, M., Tanii, M., Fukuyama, Y., and Kubota, K., Urinary findings of different age groups in the inhabitants of Itai-itai disease districts, *Med. Biol.,* 85, 263—267, 1972 (in Japanese).
226. Shiroishi, K., Kjellström, T., Kubota, K., Evrin P.-E., Anayama, M., Vesterberg, O., Shimada, T., Piscator, M., Iwata, I., and Nishino, H., Urine analysis for detection of cadmium-induced renal changes, with special reference to beta₂-microglobulin, *Environ. Res.,* 13, 407—424, 1977.
227. Shuster, J., Gold, P., and Poulik, M. D., Beta₂-microglobulin levels in cancerous and other disease states, *Clin. Chim. Acta,* 67, 307—313, 1976.
228. Smith, J. C. and Kench, J. E., Observations on urinary cadmium and protein excretion in men exposed to cadmium oxide dust and fume, *Br. J. Ind. Med.,* 14, 240—249, 1957.
229. Smith, J. C., Kench, J. E., and Lane, R. E., Determination of cadmium in urine and observations on urinary cadmium and protein excretion in men exposed to cadmium oxide dust, *Biochem. J.,* 61, 698—701, 1955.

230. Smith, J. P., Smith, J. C., and McCall, A. J., Chronic poisoning from cadmium fume, *J. Pathol. Bacteriol.*, 80, 287—296, 1960.

231. Smith, J. P., Wells, A. R., and Kench, J. E., Observations on the urinary protein of men exposed to cadmium dust and fume, *Br. J. Ind. Med.*, 18, 70—78, 1961.

232. Snell, K. C., *Pathology of Laboratory Rats and Mice*, Cotchin, E. and Roe, F. J. C., Eds., Blackwell Scientific, Oxford, 1967, 105—148.

233. Sølling, J., Sølling, K., and Mogensen, C. E., Patterns of proteinuria and circulating immune complexes in febrile patients, *Acta Med. Scand.*, 212, 167—169, 1982.

234. Squibb, K. S., Ridlington J. W., Carmichael, N. G., and Fowler, B. A., Early cellular effects of circulating cadmium-thionein on kidney proximal tubules, *Environ. Health Perspect.*, 28, 287—296, 1979.

235. Squibb, K. S., Pritchard, J. B., and Fowler, B. A., Renal metabolism and toxicity of metallothionein, in *Biological Roles of Metallothionein*, Foulkes, E. C., Ed., Elsevier, New York, 1982, 181—192.

236. Squibb, K. S., Pritchard, J. B., and Fowler, B. A., Cadmium-metallothionein nephropathy: relationships between ultrastructural/biochemical alterations and intracellular cadmium binding, *J. Pharmacol. Exp. Ther.*, 229(1), 311—321, 1984.

237. Stewart, M., Use of beta₂-microglobulin as a parameter of biological monitoring in cadmium workers, in *Occupational Exposure to Cadmium*, Cadmium Association, London, 1980, 30—33.

238. Stewart, M. and Hughes, E. G., Urinary beta-2-microglobulin in the biological monitoring of cadmium workers, *Br. J. Ind. Med.*, 38, 170—174, 1981.

239. Stowe, H. D., Wilson, M., and Goyer, R. A., Clinical and morphological effects of oral cadmium toxicity in rabbits, *Arch. Pathol.*, 94, 389—405, 1972.

240. Strober, W. and Waldmann, T. A., The role of the kidney in the metabolism of plasma proteins, *Nephron*, 13, 35—66, 1974.

241. Sumpio, B. E. and Maack, T., Kinetics, competition and selectivity of tubular absorption of proteins, *Am. J. Physiol.*, 243, F379—F392, 1982.

242. Suzuki, S., Suzuki, T., and Ashizawa, M., Proteinuria due to inhalation of cadmium stearate dust, *Ind. Health*, 3, 73—85, 1965.

243. Suzuki, Y., The amount of cadmium bound to metallothionein in liver and kidney after long-term cadmium exposure, in *Proc. 47th Annu. Meet. Jpn. Assoc. Ind. Health*, Nagoya, March 29, 1974, 124—125 (in Japanese).

244. Svartengren, M., Elinder, C.-G., Friberg, L., and Lind, B., Distribution and concentration of cadmium in human kidney, *Environ. Res.*, in press.

245. Takebayashi, S., First autopsy case, suspicious of cadmium intoxication, from the cadmium-polluted area in Tsushima, Nagasaki Prefecture, in *Cadmium-Induced Osteopathy*, Shigematsu, I. and Nomiyama, K., Eds., Japanese Public Health Association, Tokyo, 1980, 124—138.

246. Taniguchi, N., Tanaka, M., Kishihara, C., Ohno, H., Kondo, T., Matsuda, I., Fujino, T., and Harada, M., Determination of carbonic anhydrase C and beta-2-microglobulin by radioimmunoassay in urine of heavy-metal-exposed subjects and patients with renal tubular acidosis, *Environ. Res.*, 20, 154—161, 1979.

247. Task Group on Metal Toxicity, in *Effects and Dose-Response Relationships of Toxic Metals*, Nordberg, G. F., Ed., Elsevier, Amsterdam, 1976.

248. Tepperman, H. M., The effect of BAL and BAL-glucoside therapy on the excretion and tissue distribution of injected cadmium, *J. Pharmacol.*, 89, 343—349, 1947.

249. Tertiary Monkey Experiment Team, *Effects of Nutritional Factors on Cadmium-Administered Monkeys*, Japanese Public Health Association, Tokyo, 1983.

250. Toho Zinc Company Ltd., Outline of Annaka Refinery, 1970 (pamphlet in Japanese).

251. Tohyama, C. and Shaikh, Z. A., Metallothionein in plasma and urine of cadmium exposed rats determinated by a single-antibody radioimmunoassay, *Fundam. Appl. Toxicol.*, 1, 1—7, 1981.

252. Tohyama, C., Shaikh, Z. A., Ellis, K. J., and Cohn, S. H., Metallothionein excretion in urine upon cadmium exposure: its relationship with liver and kidney cadmium, *Toxicology*, 20, 181—191, 1981a.

253. Tohyama, C., Shaikh, Z. A., Nogawa, K., Kobayashi, E., and Honda, R., Elevated urinary excretion of metallothionein due to environmental cadmium exposure, *Toxicology*, 22, 289—297, 1981b.

254. Toyoshima, I., Seino, A., and Tsuchiya, K., Urinary amino acids in cadmium workers, inhabitants in a cadmium-polluted area, and in Itai-itai disease patients, in *Kankyo Hoken Report No. 24*, Japanese Public Health Association, 1973, 65—71 (in Japanese).

255. Tsuchiya, K., Proteinuria of workers exposed to cadmium fume. The relationship to concentration in the working environment, *Arch. Environ. Health*, 14, 875—880, 1967.

256. Tsuchiya, K., Proteinuria of cadmium workers, *J. Occup. Med.*, 18, 463—466, 1976.

257. Tsuchiya, K., Ed., *Cadmium Studies in Japan — A Review*, Elsevier, Amsterdam, 1978.

258. Tsuchiya, K., Cadmium research in Japan, in *Cadmium 81, Proc. 3rd Int. Cadmium Conf., Miami*, February 3 to 5, 1981, Cadmium Association, London, 1982.

259. Tsuchiya, K., Iwao, S., Sugita, M., and Sakurai, H., Increased urinary beta-2-microglobulin in cadmium exposure: dose-effect relationship and biological significance of beta-2-microglobulin, *Environ. Health Perspect.*, 28, 147—153, 1979.
260. Usuda, N., Kameko, M., Kanai, M., and Nagata, T., Immunocytochemical demonstration of retinol-binding protein in the lysosomes of the proximal tubules of the human kidney, *Histochemistry*, 78, 487—490, 1983.
261. Vander Mallie, R. J. and Garvey, J. S., Production and study of antibody produced against rat cadmium thionein, *Immunochemistry*, 15, 857—868, 1978.
262. Vesterberg, O., Nise, G., and Hansen, L., Urinary proteins in occupational exposure to chemicals and in diseases, *J. Occup. Med.*, 18, 473—476, 1976.
263. Viau, C., Bernard, A., Lauwerys, R. R., Tulkens, P., Laurent, G., and Maldague, P., Gentamicin nephrotoxicity in cadmium, lead and mercury pretreated rats, *Toxicology*, 27, 15—25, 1983.
264. Watanabe, H. and Murayama, H., A study on health effect indices concerning population in cadmium-polluted area, in *Recent Advances in the Assessment of the Health Effects of Environmental Pollution, Vol. 1*, Commission of the European Communities, Luxembourg, 1975, 91—104.
265. Watanabe, H., Hasegawa, Y., Murayama, H., Matsushita, S., Nagakura, S., Okuno, T., Ono, K., Araka, K., Ogawa, T., and Teraoka, Y., Health effect indices of inhabitants in a cadmium-polluted area, *Kankyo Hoken Report No. 24*, Japanese Public Health Association, Tokyo, 1973, 122—130 (in Japanese).
266. Webb, M. and Etienne, A. T., Studies on the toxicity and metabolism of cadmium-thionein, *Biochem. Pharmacol.*, 26, 25—30, 1977.
267. Wellwood, J. M., Ellis, B. G., Prince, R. G., Hammond, K., Thompson, A. E., and Jones, N. F., Urinary N-acetyl-beta-D-glucosaminidase activity in patients with renal disease, *Br. Med. J.*, 3, 1975.
268. WHO, Environmental health criteria for cadmium, Interim Rep. No. EHE/EHC/79.20, World Health Organization, Geneva, 1979.
269. Wibell, L. and Evrin, P.-E., The urinary excretion of beta$_2$-microglobulin (b$_2$-u) in renal disorders, *Protides Biol. Fluids*, 21, 519—523, 1973.
270. Wilson, R. H., DeEds, F., and Cox, A. J., Effects of continued cadmium feeding, *J. Pharmacol. Exp. Ther.*, 71, 222—235, 1941.
271. Wisniewska-Knypl, J. M., Jablonska, J., and Myslak, Z., Binding of cadmium on metallothionein in man: an analysis of a fatal poisoning by cadmium iodide, *Arch. Toxikol.*, 28, 46—55, 1971.
272. Yamada, Y. K., Shimizu, F., Suzuki, K. T., Yamamura, M., and Kubota, K., A study on cadmium-induced nephropathy in rats pretreated with puromycin aminonucleoside, *Environ. Res.*, 32, 179—187, 1983.

Chapter 10

EFFECTS ON BONE, ON VITAMIN D, AND CALCIUM METABOLISM

Tord Kjellström

TABLE OF CONTENTS

I. INTRODUCTION

The cadmium concentrations in bone are much lower than the concentrations in several other tissues (Volume I, Chapter 5, Section VII and Chapter 6, Section IV.A). The average concentration in Japanese adults was about 0.1 mg Cd per kilogram at an average whole kidney concentration of 45 mg Cd per kilogram.[92] Bone can be affected by cadmium both directly and indirectly, via vitamin D metabolism. Some of the most severe clinical effects in cadmium poisoning, osteoporosis and osteomalacia, are due to effects on the bone.

In animal experiments cadmium effects on bones, in the form of decreased mineral content, were reported many years ago by Ceresa.[16] At about the same time Nicaud and co-workers[74] reported osteomalacia in six workers from a cadmium battery factory. It was not until the unraveling of the association between cadmium exposure and Itai-itai disease (Section IV) in the 1960s and 1970s that interest in the cadmium effects on bone became more widely developed.

Accumulating evidence shows that cadmium exposure can affect several mechanisms in the metabolism of calcium, vitamin D, and collagen. It is therefore to be expected that experimental as well as epidemiological studies of cadmium poisoning will reveal a combination of different types of bone effects, including osteoporosis and osteomalacia. One or the other will be more prominent, depending on the circumstances of exposure.

II. ASPECTS OF BONE PHYSIOLOGY AND PATHOLOGY OF RELEVANCE TO CADMIUM TOXICITY

A. Bone Metabolism in General

Hard bone tissue consists mainly of fibers of collagen in which hydroxyapatite (a compound of calcium and phosphate) is deposited as a solid matrix. There is a constant turnover of bone with replacement of collagen, calcium, phosphate, and other minor components.[9] Osteoblasts are the cells actively depositing hydroxyapatite in the collagen fibers. They turn into osteocytes within the hard bone structure where they serve as mediators for the metabolism of the bone structure. Osteoclasts are the cells which actively remove hydroxyapatite from the collagen fibers. This leads to a more rapid release of calcium and phosphorus from bone and the osteoclasts become active when there is a need for a large influx of calcium into plasma.

One of the functions of bone is to serve as a storage tissue for calcium. The plasma level of calcium is very closely maintained at about 100 mg Ca per liter. A complicated feedback mechanism involving intestines, bone and kidney, vitamin D and its metabolites, parathyroid hormone, and calcitonin maintains the calcium homeostasis.[9]

Osteoporosis is characterized by a reduction in trabecular bone mass in relation to the total area of a histological bone section[8] and a normal ratio of mineral to organic matter. In osteomalacia the ratio of mineral to organic matter is low and trabecular bone mass may be normal, decreased, or increased. Often histological analysis of bone sections is needed to distinguish between the two diseases. In osteomalacia the bone trabecular surfaces of newly forming haversian systems have wide osteoid seams, an increased number of surface seams per unit area of bone, and a decreased rate of osteoid mineralization (so-called appositional rate) at the calcification front.[8] Kanis and co-workers[44] concluded that the decreased mineralization is the main feature of osteomalacia and reviewed in detail the methods available to measure such a decrease.

Usually osteoporosis patients do not have changes in levels of calcium, inorganic phosphate, or alkaline phosphatase in serum or urine.[8] On the other hand, hypocalciuria, hypocalcemia, and elevations in serum alkaline phosphatase are often found in

Table 1
THE DIFFERENT FEATURES OF OSTEOPOROSIS AND
OSTEOMALACIA, WHICH SHOULD BE CONSIDERED IN
ASSESSING THE EFFECTS OF CADMIUM ON BONE. IN
INDIVIDUAL CASES, ALL OR ONLY SOME OF THE FEATURES
MAY OCCUR[9]

Features	Osteoporosis	Osteomalacia
Clinical	Fractures easily at impact	Spontaneous fractures, "bone pains"
Radiological	Extremely low density bone, uniformly	"Looser's zones" due to pseudofractures
Biochemical	No change in serum or urine calcium	Low serum calcium Low urine calcium High serum alkaline phosphatase
Histological	Low trabecular bone mass	Low mineralization of trabecular bone Wide osteoid seams Low rate of osteoid mineralization

osteomalacia.[20] The different features of osteoporosis and osteomalacia are listed in Table 1. The effects of cadmium on bone will be discussed with these basic aspects of bone metabolism and bone pathology in mind.

B. Vitamin D Metabolism

The role of vitamin D in calcium metabolism has been extensively reviewed by several authors.[19,57,61] The active metabolite of vitamin D, 1,25-dihydroxycholecalciferol (1,25-DHCC) is formed in the kidney and stimulates calcium absorption from the intestine and the resorption of calcium from bone. It is also necessary for the proper mineralization of bone matrix. This metabolite is now called Kalcitriol® [88] and is commercially available. The involvement of the kidney in vitamin D metabolism is of particular importance for the understanding of cadmium poisoning, because the kidney is one of the main target organs for cadmium effects (Chapter 9).

A feedback control system including the calcium concentration in serum and parathyroid hormone (PTH) is involved in vitamin D metabolism. A schematic drawing of these pathways is shown in Figure 1. The steps in the feedback control system where cadmium has been shown or suggested to interact (Section III.A) are also shown in Figure 1. For some time it was thought that the kidney was the only tissue where Kalcitriol® was produced and that this would form the basis for renal osteomalacia. However, it has been shown in tissue culture of human bone cells[34] that these cells can produce Kalcitriol® in amounts similar to those of kidney cells.

Further evidence for the fact that the renal hydroxylation of vitamin D is not crucial for bone mineralization in patients treated with vitamin D is given by Memmos and co-workers.[67] Patients with nutritional osteomalacia and osteomalacia of chronic renal failure were treated with 25-OH vitamin D3 (25-HCC) (5 μg p.o. per day) or vitamin D3 (50 μg p.o. per day). The rise of plasma 25-HCC was studied and it reached a steady state after about 3 months' treatment. The steady-state plasma levels were similar in the two patient groups, with a tendency for higher values among those with renal osteomalacia. The degree of improvement of bone mineralization was also similar, but with a tendency for a more rapid initial improvement in the nutritional osteomalacia cases. However, in some of these cases even 9 months' treatment was insufficient to completely reverse the defect in mineralization, although the cases had recovered clinically.

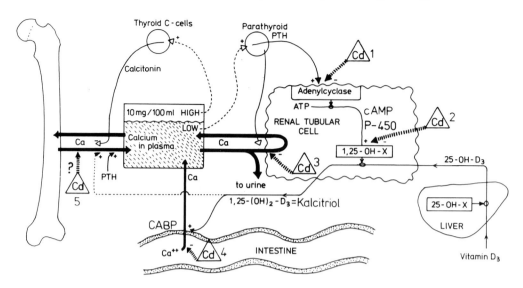

FIGURE 1.   Schematic drawing of those aspects of calcium and vitamin D metabolism that may be affected by cadmium. (1) Cadmium decreases PTH stimulation of adenylcyclase. (2) Cadmium inhibits hydroxylation of 25-OH-D₃. (3) Cadmium increases urinary calcium excretion. (4) Cadmium decreases gastrointestinal calcium absorption. (5) Cadmium affects bone mineralization directly(?).

All the patients had some degree of secondary hyperparathyroidism with low plasma calcium and high plasma PTH. This normalized in the cases with nutritional osteo-malacia (prompt rise in plasma calcium and fall in plasma PTH), whereas in the cases with renal osteomalacia there was no change.[67] The authors assumed that this was due to little or no rise in circulating Kalcitriol®.

The findings agree with an intraosseous production of Kalcitriol® sufficient for the needs of bone mineralization at high exposures to vitamin D3 or 25-HCC, and a lack of renal production of Kalcitriol®. The latter is needed for an increase of plasma Kalcitriol® and a subsequent stimulation of gastrointestinal calcium absorption.

A detailed discussion of the role of vitamin D metabolites in the osteomalacia of renal diseases was published by Kanis and co-workers.[44] Such osteomalacia is not un-common among patients with chronic renal failure and who are undergoing dialysis treatment. They point out that plasma levels of 25-HCC are often normal in patients with chronic renal failure osteomalacia, and that plasma levels of Kalcitriol® are usu-ally very low in end-stage chronic renal failure.[30] The administration of Kalcitriol® at doses close to the estimated daily endogenous production reverses the biochemical and radiographic features of renal osteomalacia,[17] whereas 25-HCC has to be given at much higher doses.[44] Chronic renal failure also involves disturbances in phosphate and calcium metabolism, which may exacerbate the effects on vitamin D metabolism.[44] Of particular importance for the interpretation of the data from Itai-itai disease patients (Section V) are the findings that delayed osteoid maturation in vitamin D-deficient rats with hypocalcemia can be corrected by dietary calcium supplements[33] and that in chronic renal failure severe calcium deficiency may render the patient unresponsive to Kalcitriol®, whereas healing of osteomalacia occurs when the diet is adequate in cal-cium.[18] Thus, the interaction between vitamin D and calcium in the restoration of the bone findings in renal osteomalacia may be complicated and in view of the possibility of multiple mechanisms for the effects of cadmium on bone (Figure 1), the available data have to be interpreted with caution.

C. Calcium Metabolism

Calcium is absorbed from the intestinal tract, stored in the bones, and excreted via urine. It is an essential element for a number of biochemical mechanisms and the plasma calcium level is controlled by an elaborate homeostatic mechanism (Figure 1).

As mentioned earlier, 1,25-DHCC (Kalcitriol®) stimulates calcium absorption in the intestine and the resorption of calcium from bone. This will increase the plasma calcium concentration. In the intestinal mucosa 1,25-DHCC activates a calcium-binding protein.[105] This protein probably facilitates passive diffusion as well as active transportation of calcium. Calcitonin from the thyroid C cells blocks the resorption of bone and this will lead to a decrease of plasma calcium (Figure 1).

Several interactions between homeostasis and cadmium are already known (Figure 1) and it is therefore to be expected that the clinical manifestations of bone effects due to cadmium poisoning are not identical to bone disease due to pure renal damage or other causes.

D. Collagen Metabolism

Most of the inorganic matrix in bone is stored in conjunction with collagen fibers. These are packed in an overlapping fashion which creates 40-nm gaps between the ends of adjoining fibers[83] in which the mineralization of the collagen fibers starts. The fiber structure of the bone is essential for proper mineralization.

Collagen has a molecular weight of about 95,000 and contains about 1,000 amino acids. There is a high percentage (11%) of hydroxyproline[83] and this is important for the helix fiber formation of collagen. Together with proline and glycine, hydroxyproline forms a "helical sequence". Figure 2 shows the different steps in collagen production.

The amino acids lysine and hydroxylysine and the enzyme lysyloxidase (which requires copper) are of particular importance for the intermolecular cross-linking of collagen molecules into fibrils.[83] A decrease in the production of mature collagen fibers in bone would eventually lead to loss of bone matrix. The data below indicate that cadmium may interfere with collagen production. This could be an additional mechanism in the development of bone disease after cadmium exposure.

It should be pointed out that there is a disease specifically caused by the inhibition of cross-linking of collagen and elastin fibers: lathyrism.[10] The disease earned its name from a certain species of poisonous "peas" from Lathyrus plants. The active ingredient is β-amino-propio-nitrile (BAPN) which inhibits lysyl-oxidase in vivo and in vitro.[10]

Osteolathyrism has been induced in animals by feeding of Lathyrus peas. Spinal curvatures, sternal deformities, severe abnormalities of the femurs and humeri, and enlargement of the costochondral junctions occurred. In regard to bone effects there is a strong resemblance between lathyrism and copper deficiency, but the BAPN toxicity is not affected by dietary copper addition.[10]

III. ANIMALS

A. Effects on Vitamin D Metabolism

Feldman and Cousins[22] first reported an effect of cadmium on the activity of 1,25-DHCC (Kalcitriol®) in kidney mitochondria of chicks. Different cadmium concentrations in the incubation solutions gave a clear dose-effect relationship (Table 2). They also showed that cadmium-binding protein from the chicks eliminated a large part of the inhibitory influence of cadmium on the chick kidney mitochondria formation of 1,25-DHCC. These effects of cadmium in vitro on vitamin D activation in the kidney have been confirmed by Suda and co-workers.[89]

Feldman and Cousins[22] also found an in vivo effect on renal vitamin D metabolism.

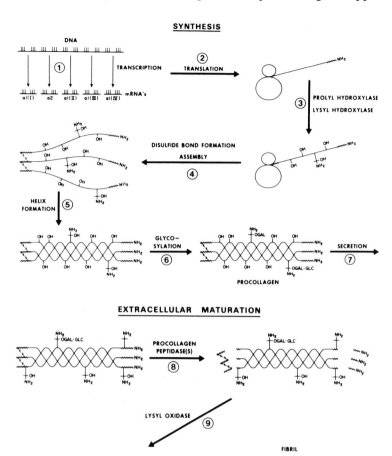

FIGURE 2.   An outline of the major steps in collagen synthesis and fibril formation. The steps shown represent a minimal number of control points. For example,
several processes are involved in forming collagen from procollagen; some may be
intracellular and some extracellular, and different enzymes are probably required to
remove the amino + carboxy terminals of procollagen (step 8). There are several
enzymatic and nonenzymatic processes involved in the formation of cross-links,
which include many different chemical species (step 10). Note that the collagen molecule is drawn with the amino terminal on the right and the carboxy terminal on the
left, opposite to the more usual presentation. (Figure kindly provided by Dr. Rich-
ard A. Salvador.) (From Raisz, L. G., *Metabolic Bone Disease*, Vol. 1, Avioli, L.
V. and Krane, S. M., Eds., Acdemic Press, New York, 1977, 1—49. With permission.)

Chicks on a low calcium diet were given 50 mg Cd per liter in drinking water for 2
weeks and at the same time 11 daily i.p. injections of 1.0 mg Cd (body weight not
mentioned). The mitochondria were isolated and it was found that the conversion of
25-HCC to 1,25-DHCC decreased to about 60% of the control value among the cadmium-exposed chicks. Kimura and co-workers[50] confirmed these findings in a similar
experiment with up to 8 weeks' exposure. It should be pointed out that even in the
highest dose group, the whole kidney cadmium concentration only reached about 36
mg Cd per kilogram wet weight.[50]

Table 2

EFFECT OF CADMIUM
CONCENTRATION ON 1,25-
DIHYDROXYCHOLECALCIFEROL
SYNTHESIS BY CHICK KIDNEY
MITOCHONDRIA

Cadmium (μM) [a]	1,25-DHCC (%)
0	65.0
25	38.0
50	17.2
55	14.5
60	13.3
65	11.1
70	10.5
75	9.7
100	6.0
125	2.3

[a] Cadmium solutions were incubated with the mitochondria for 3 min prior to addition of [³H]-25-HCC.

From Feldman, S. L. and Cousins, R. J., *Nutr. Rep. Int.*, 8, 251—259, 1973. With permission.

Higher doses were given in an experiment on rats by Lorentzon and Larsson.[62] For 3 months diets with normal or low calcium content were given in combination with different cadmium concentrations in drinking water. The dose reached in the low calcium, highest cadmium group was 170 mg Cd per kilogram wet weight in renal cortex.

Five fractions of vitamin D metabolites in chloroform extracts of the tissues were analyzed by chromatography on Sephadex LH-20. In Table 3 peak 2 represents vitamin D3 and peak 1, which is eluted with the void volume, is considered to contain vitamin D esters.[62] It can be seen (Table 3) that low calcium intake decreased peak 1 and peak 2 in serum, whereas cadmium exposure at normal calcium intake greatly increases peak 1 and peak 2 in serum in a dose-related manner. In kidneys the changes are less and appear opposite to those in serum (Table 3). There is no effect of low calcium on 25-HCC in serum or kidney, whereas cadmium exposure decreases 25-HCC in serum. The highest cadmium exposure at normal calcium intake causes the production of 1,25-DHCC to be very much depressed and in serum none of this metabolite is found, whereas in kidney about 15% of the control level was found. A decrease of kidney 1,25-DHCC was seen already in the lower cadmium dose group (Table 3). The serum levels of 24,25-DHCC change in an entirely opposite manner to 1,25-DHCC, which indicates that the calcium deficiency and cadmium exposure cause a shift from the production of 1,25-DHCC to the production of 24,25-DHCC.

Low calcium intake, on the other hand, causes an increase in 1,25-DHCC in serum and kidneys (Table 3) and also in the intestinal mucosa.[62] The combination of low calcium and cadmium exposure cancels out these effects, so that in serum and in kidneys the levels are the same as in the control group.

For the production of 1,25-DHCC in the renal tubular cells there is a sequential relationship between PTH, adenylcyclase, cyclic-AMP, P-450, and 1,25-DHCC (Figure 1). The mechanism underlying the decreased production of 1,25-DHCC can involve any of these steps. Cyclic AMP (cAMP) and P-450 (Figure 1) are involved in a number

Table 3
PERCENTAGE DISTRIBUTION OF VARIOUS FRACTIONS OF VITAMIN D METABOLITES IN CADMIUM-EXPOSED AND OTHER ANIMALS[62]

Group of animals	Peak I	Peak II	25-HCC	24,25-DHCC	1,25-DHCC
Serum					
Ca⁺(control)	3.4(2.6—4.2)	2.9(1.1—4.7)	84.3(82.9—85.7)	1.9(1.8—2.0)	7.6(6.9—8.3)
Ca⁻	0.7(0.5—0.9)	2.1(0.9—3.3)	84.9(82.4—87.4)	0	12.3(10.3—14.3)
Ca⁺, 0.22 mmol of Cd/ℓ	6.1(4.6—7.6)	6.5(4.3—8.7)	78.0(72.7—83.3)	5.4(4.5—6.3)	4.2(3.4—5.0)
Ca⁺, 0.67 mmol of Cd/ℓ	21.2(17.0—25.4)	16.8(13.9—19.7)	46.1(36.2—56.0)	11.3(9.7—12.8)	0
Ca⁻, 0.67 mmol of Cd/ℓ	1.5(1.0—2.0)	1.9(1.6—2.2)	88.4(87.1—89.6)	0.3(0.1—0.6)	8.0(7.2—8.7)
Kidneys					
Ca⁺(control)	1.6(1.5—1.6)	1.7(1.6—1.7)	84.5(83.2—85.8)	0.1(0.0—0.2)	12.2(10.8—13.6)
Ca⁻	3.3(2.2—4.4)	2.1(1.1—3.1)	74.4(69.7—79.1)	0	20.3(17.6—22.9)
Ca⁺, 0.22 mmol of Cd/ℓ	1.2(0.5—1.9)	1.9(0.2—3.5)	90.0(85.5—94.5)	0	6.7(4.8—8.6)
Ca⁺, 0.67 mmol of Cd/ℓ	1.2(1.1—1.2)	2.2(0.9—3.6)	94.0(91.9—96.0)	0	1.9(1.3—2.5)
Ca⁻, 0.67 mmol of Cd/ℓ	1.1(0.3—1.8)	1.9(1.6—2.2)	82.7(78.8—86.5)	0	13.9(11.6—16.1)

Note: For serum and kidneys, mean values and range are given, and for intestinal mucosa and liver, mean values are obtained from repeated analyses of pooled samples. Note the dose-related depression of the 1,25-DHCC fraction and the increase in 24,25-DHCC and less-polar metabolites in serum due to cadmium exposure at a normal calcium intake, which was completely prevented by a concomitant low calcium intake.

of biochemical oxidative processes. In renal cortex much of the cyclic AMP is generated by the PTH-adenylcyclase system and 30 to 50% of urinary cAMP originates from the PTH-stimulated reaction.[15]

Kjellström and co-workers[56] exposed rats for 17 weeks to s.c. injections of $CdCl_2$ (0.9 mg Cd per kilogram body weight per day), leading to kidney cortex concentrations of about 300 mg Cd per kilogram wet weight. In the cadmium-exposed rats the urinary adenylcyclase activity increased with PTH stimulation (10 μg per rat) and in the control rats it increased 433% ($p < 0.05$, student's t-test).

Merali and Singhal[68] found decreased cAMP levels in kidney cortex of rats after acute i.p. exposure (0.25 or 1 mg Cd per kilogram body weight). They suggested that the decrease could be due to enhanced breakdown by phosphodiesterase or an increased excretion of cAMP due to tubular damage.

A subacute experiment on rats (i.p. injections for 7 days to 2.0 mg Cd per kilogram per day) showed a decreasing renal cortex cAMP concentration over time.[43]

A cadmium-induced decrease in P-450 activity in the liver has also been shown in vitro and in vivo (Chapter 11, Section III.A.1). This is possibly associated with the cadmium-induced stimulation of metallothionein production in the liver (Volume I, Chapter 4). However, there are no reports on such cadmium effects in renal cortex.

The involvement of the parathyroid gland in cadmium effects on vitamin D and calcium metabolism has been shown by Jones and Fowler.[42] Rats exposed to cadmium had dose-relelated changes in the microscopic structure of parathyroid cells.

The evidence to date indicates that there is a dose-effect relationship for decreased production of 1,25-DHCC in renal cortex. Some decrease was already found at renal cortex cadmium concentrations of about 50 mg Cd per kilogram[22] and at 170 mg Cd per kilogram[62] the production was almost totally inhibited. At higher renal cortex levels adenylcyclase stimulation was also impaired.

B. Effects on Calcium Metabolism

The mineral content of bone and intestinal calcium absorption can be affected by cadmium. Long-term studies of rats exposed to cadmium in drinking water (25 to 50 mg Cd per kilogram water)[60,63] and fed a low calcium diet showed that cadmium caused decreased mineral content of the bone, whereas low calcium intake alone or cadmium exposure alone did not induce such changes. However, the low calcium intake increased cadmium absorption[104] and the amount of cadmium accumulated in tissues (Volume I, Chapter 6, Section II.B.3.a).

Sugawara and Sugawara[91] studied the influence of cadmium on calcium uptake from the intestine of rats. Three groups of 5 to 13 rats were given a diet *ad libitum* containing 0.22% calcium. In one group the food was mixed with a solution of cadmium in drinking water at a cadmium concentration of 10 mg/ℓ and in another group the cadmium concentration in drinking water was 50 mg/ℓ. The rats were studied for 287 days.

No significant differences in calcium or phosphorus absorption rates were seen between the control group and the group given 10 mg Cd per liter in drinking water. There was a significant decrease in the calcium absorption rate to about half and a 25% decrease in phosphorus absorption in the group given 50 mg Cd per liter. Eosinophilic cell infiltration and erosions in the duodenal mucosa were seen in some of the rats given 50 mg Cd per liter in drinking water.[91]

Kobayashi[58] gave rats a calcium-deficient diet with an addition of 30 to 200 mg Cd per kilogram. The cadmium-exposed rats accumulated less calcium in the body than rats on a calcium-deficient diet without added cadmium. Some of the cadmium-exposed rats even developed a negative calcium balance after 25 to 35 weeks, with almost all of the calcium excretion in feces.

Kawamura and co-workers[47] exposed four groups of female rats to combinations of

Table 4

CALCIUM AND CADMIUM LEVELS IN FEMURS AND KIDNEYS AS A FUNCTION OF CALCIUM AND CADMIUM EXPOSURE[47]

| | | Femur | | Kidney | | Serum |
	Exposure	Calcium (g/kg)	Cadmium (mg/kg)	Calcium (mg/kg)	Cadmium (mg/kg)	Calcium (mg/ℓ)
Group I (n = 6)	Ca$^+$Cd$^-$	138.2 ± 8.8^a	0.7 ± 0.4	89.0 ± 32.6	0	97 ± 7
Group II (n = 8)	Ca$^-$ Cd$^-$	130.7 ± 13.1^a	0.7 ± 0.4	94.7 ± 44.0	0	77 ± 10
Group III (n = 7)	Ca$^+$ Cd$^+$	124.8 ± 12.4^a	4.7 ± 1.6^a	90.8 ± 12.9	63.9 ± 24.9	80 ± 7
Group IV (n = 9)	Ca$^-$ Cd$^+$	106.9 ± 10.6^a	9.8 ± 4.2^a	112.3 ± 52.6	90.4 ± 26.5	66 ± 10

Note: In the paper, the values are given as per dry tissue weight, but this is a misprint according to Itokawa (personal written communication) and the values are instead wet weight based (mean ± SD). Cd$^+$ = 50 mg Cd/ℓ in water for 3 months.

^{a} $p = < 0.01$.

calcium adequate or deficient diet with or without 50 mg Cd per liter (Table 4). After 90 days exposure the renal clearances of inulin and PAH were studied with i.v. infusion (Chapter 9, Section IV.A.3) and the renal clearances of calcium and phosphorus were calculated from urine and plasma concentrations. After the clearance studies, the animals were sacrificed and histological examination as well as analysis of calcium and cadmium were carried out in kidneys and femurs. Cadmium concentration in kidney in group III (Table 4) was reported to be 64 mg/kg dry weight,[47] but there was a misprint in the publication. Itokawa (written personal communication) has confirmed that all the concentration values given in Table 4 were in fact based on wet weight.

In group IV with the lower calcium intake but the same cadmium intake, the cadmium concentration in kidney was 50% higher than in group III. This is in agreement with the studies by Larsson and Piscator[60] showing a higher cadmium absorption in animals with low calcium intake. Cadmium concentration in femur had increased about 150% in group IV compared to group III, indicating that there may be a difference between kidney and bone in the stimulation of cadmium accumulation by low calcium. Serum calcium concentration as well as the femur calcium concentrations had decreased (see Table 4). There was a dose-related relationship between cadmium in kidney and inulin clearance (Chapter 9, Section IV.A.3), and urine calcium was decreased in the cadmium-exposed groups with normal calcium intake as well as in the groups with low calcium exposure. In contrast to this Hirota[31] reported an increased urinary excretion of calcium in rabbits after long-term cadmium exposure in water. The rabbits had been exposed to cadmium in drinking water (200 mg/ℓ) for 1 year and in serum they had an increased calcium, decreased phosphate, and increased alkaline phosphatase.

A subacute dietary exposure study of chicks[26] showed that the calcium absorption in the intestines as well as the concentration of calcium-binding protein (CaBP) in the intestines decreased after cadmium exposure. The chicks were exposed for 3 weeks to different levels of cadmium in the diet (0, 3, 10, and 100 μg Cd per day). No data on the kidney concentrations of cadmium were given in the paper, but they are unlikely to have reached kidney cortex concentrations higher than about 10 mg/kg.

Fullmer and co-workers[26] also studied a group of chicks with low calcium intake with and without cadmium exposure at 80 mg Cd per kilogram in the diet for 24 weeks. The low calcium intake increased the intestinal calcium-binding protein levels three times, but in combination with cadmium exposure, the concentrations were as depressed as in the high calcium/high cadmium group (about 20% of the control values). There was no decrease in the 25 (OH)-D3-1-hydroxylase activity in the kidney in the low calcium/high cadmium group as compared to the low calcium/no cadmium group,[26] but this is not surprising as the cadmium levels in the kidney must have been far below toxic levels.

Tsuruki and co-workers,[101] using the everted gut sac technique in rats, showed a cadmium-induced inhibition of vitamin D-stimulated calcium transport in vitro. This was confirmed by the *in situ* ligated duodenal loop procedure during a 30-min absorption period.[26] A very rapid onset (within 4 hr) of the cadmium-induced decrease (down to about 40% of base value) of calcium absorption and intestinal calcium-binding protein was seen. As the concentration of calcium-binding protein in the intestine was decreased to the same degree as the decrease in calcium absorption,[26] it is not likely that competition between calcium and cadmium at the binding site would be the main reason for the effect of cadmium on calcium absorption. It seems more likely that cadmium affects the production of calcium-binding protein itself. In the study by Fullmer and co-workers,[26] this effect on calcium-binding protein production could not have been mediated via vitamin D metabolism in the kidney, because of the low kidney cadmium levels and the rapid onset of the effect.

Ando and co-workers[7] exposed rats to cadmium in water via gastric tube (10 mg Cd per kilogram body weight) for up to 90 days. During 3 days some of the rats were given subcutaneously 10 nmol 1-alpha-OH-vitamin D3. The intestinal calcium absorption, as measured by the everted intestinal sac procedure, was the same in cadmium-tested and control rats when 1-alpha-OH-D3 was not given. However, when it was given, the control rats showed a marked increase of calcium absorption and the cadmium-treated rats showed no change. The calcium-binding activity of the duodenal mucosa was measured in the rats given 1-alpha-OH-D3. The binding activity of the cadmium-treated rats had decreased to half after 30 days' cadmium exposure and to one sixth after 90 days' exposure.

Decalcification due to cadmium exposure has been shown in fish raised in soft water containing 0.01 mg Cd per liter.[58] The bones were low in density and had abnormal bends (Figure 3A, B). Such bends in animal bones have been seen in osteolathyrism (Section II.C). Similar bends in fish spinal columns due to vertebral damage were seen in fish (minnows) after 70 days in cadmium-contaminated water.[11] There was a clear dose-response relationship. At a concentration of 10 μg Cd per liter water, about 5% of the fish had spinal deformations and at 500 μg Cd per liter about 50% had deformations. Some of the exposed fish died, and the response rate was measured among survivors only.[11]

C. Effects on Collagen Metabolism

The first studies on this topic were carried out on cadmium-exposed humans and not on animals. Sano and Iguchi[84] discussed the possibility that a decline in the formation of matured collagen would cause, on the one hand, an increased urinary excretion of proline and, on the other hand, an impairment of polymerization of soluble collagen. This would cause an accelerated decomposition of this type of collagen and increased excretion of hydroxyproline in urine (Figure 4).

Sano and Iguchi[84] analyzed the excretion of proline and hydroxyproline in urine from seven Itai-itai patients in the Fuchu area of Japan, seven cadmium-exposed persons in the Ikuno area, five cadmium workers and ten control persons. Neither the exposure time nor the age of the persons was stated by the authors. A dramatic increase (up to about 100 times) in the urinary excretion of these amino acids was found among the Itai-itai patients, the exposed people from the Ikuno area, and the cadmium-exposed workers.

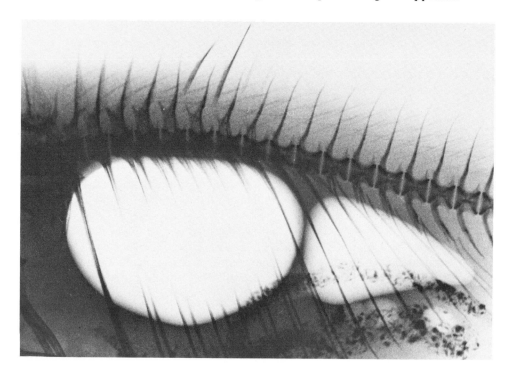

A

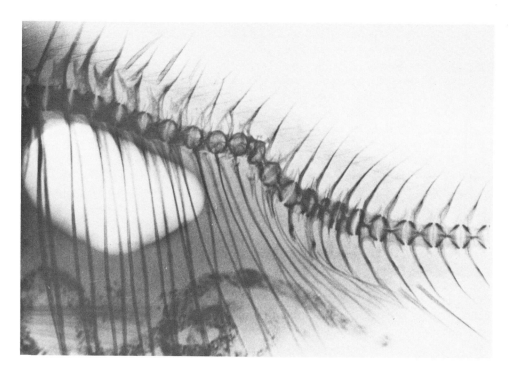

B

FIGURE 3.   (A) Normal carp bone. (B) Bone of cadmium-exposed carps. (From Kobayashi, J., *Trace Substances in Environmental Health — VIII*, Hemphill, D. D., Ed., University of Missouri, Columbia, 1974, 263—266. With permission.)

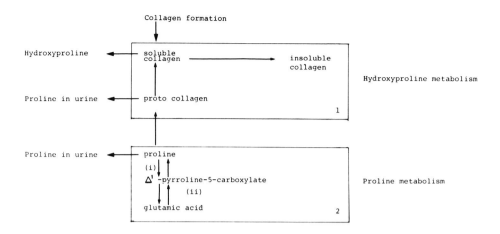

FIGURE 4.   Metabolic pathways of imino acids and collagen. (Modified from Sano, H. and Iguchi, H., Kankyo Hoken Rep. No. 31, Japanese Public Health Assoc., Tokyo, 1974, 95—96, in Japanese.)

Other studies on urinary excretion of amino acids in cadmium-exposed people with kidney and bone effects have also shown high excretion of proline and hydroxyproline (Chapter 9, Section IV.B.2.b).

The enzyme lysyl-oxidase, which is necessary for collagen formation (Section II.D), is a copper-dependent enzyme that can be affected by cadmium in vitro. Iguchi and Sano incubated purified lysyl-oxidase from chick embryo cartilage with cadmium and measured the copper content as well as the cadmium content. The higher the cadmium content in the incubation medium, the lower the copper content in the enzyme and the higher the cadmium content in the enzyme. Iguchi and Sano[35] reported that cadmium (5×10^{-4} M) inhibited lysyl-oxidase activity in vitro by approximately 62%. Zinc ($5 \times 10^{-4} M$) in the incubation medium also inhibited the enzyme activity.

Miyahara and co-workers[69] found that cadmium solutions (2 mg/kg) added to organ cultures of chick embryonic femurs decreased the incorporation of 3-H-proline into collagenous-digestible protein, but there was no alteration in the hydroxylation of proline to hydroxyproline. This indicates that the inhibitory effect of cadmium on collagen synthesis is not caused by an inhibition of hydroxylation (the state of helix formation, Figure 2), but is due to an inhibition of collagenous peptide synthesis, which agrees with the decreases in lysyl-oxidase activity reported by Iguchi and Sano.[35]

Iguchi and Sano[36] gave rats cadmium (50 or 100 mg Cd per kilogram food) and zinc (50 mg Zn per kilogram food) in the diet for 6 to 8 weeks. The crude enzyme was isolated from femur, tibia, and humerus. Lysyl-oxidase activity was decreased 40% in a group exposed to 50 mg Cd per kilogram for 6 weeks and it was decreased 87% in a group exposed to 100 mg Cd per kilogram for 8 weeks (Table 5). There was no effect in vivo of zinc exposure via food. The cadmium concentration in bone and in enzyme increased significantly (Table 5), but the copper concentration was unaltered. In plasma, the copper concentration decreased remarkably (Table 5). Pathological study of the bones showed that the cadmium-treated rats had a narrower cortical layer than the controls and the width of the tibiae was reduced by 15%.[36] The resting cartilage cells of the epiphyseal plates proliferated abnormally and the cell arrangement was irregular (Figure 5). The cartilage cell columns became shortened and the extent of provisional calcification was reduced. There was no increase in the number of osteoblasts and osteoclasts in the bone tissue. Iguchi and Sano[36] considered that these findings in the epiphyseal plates resembled in part those of osteolathyrism.

Thus, in vitro and in vivo in animals cadmium can affect collagen metabolism and

Table 5
LYSYL-OXIDASE ACTIVITY OF THE BONE EXTRACTS PREPARED FROM CADMIUM-TREATED, ZINC-TREATED, AND CONTROL RATS, AND METAL CONTENTS IN THE CRUDE ENZYME PLASMA AND BONE

Animal group	Lysyl-oxidase activity (tritium released cpm/mg of protein of tissue extract)	Metal content in crude enzyme (mg/kg of protein)[b]		Plasma copper (mg/l)	Metal content in bone epiphyses (mg/kg dry weight)	
		Cd	Cu		Cd	Cu
Control	115 ± 23 (7[a])	3.1 ± 0.47 (4)	77 ± 15 (4)	1.41 ± 0.16 (4)	0.018 ± 0.015 (4)	0.78 ± 0.17 (4)
Cadmium treated						
50 mg/kg for 6 weeks	72 ± 10* (8)	13.7 ± 1.6** (4)	77 ± 8 (4)	0.22 ± 0.08*** (4)	2.30 ± 0.40* (4)	0.79 ± 0.18 (4)
100 mg/kg for 8 weeks	15	—	—	—	—	—
Zinc treated	115 ± 8(4)	—	—	—	—	—

Note: The crude enzyme solutions were prepared for cadmium-treated and zinc-treated and control rats and were dialyzed against 0.05 M tris-acetate buffer, pH 7.7, containing 0.15 MNaCl. The final concentration in an assay mixture was adjusted to 2.4 mg protein/ml.

Asterisks indicate significantly different ($p <$ * 0.05, ** 0.01, *** 0.001) from control t-test.

[a]   The number of animals are in parentheses.
[b]   The crude enzyme solution was prepared from the combined epi- and metaphyses of five rats.

From Iguchi, H. and Sano, S., *Toxicol. Appl. Pharmacol.*, 62(1), 126—136, 1982. With permission.

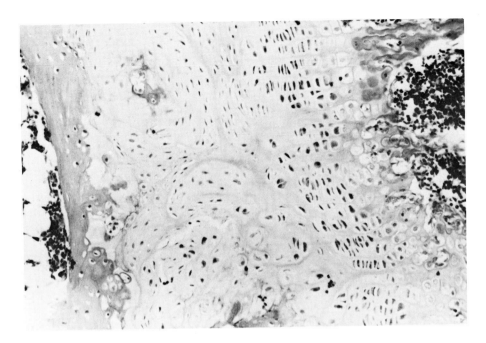

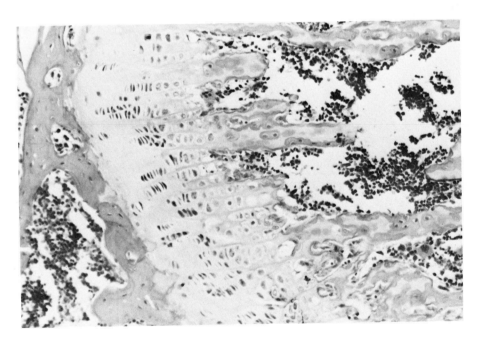

FIGURE 5. Epiphyseal cartilage plates of tibiae were stained with hematoxiline-eosin (Magnification × 200). Top, control; bottom, cadmium administered (100 mg/kg for 8 weeks). (From Iguchi, H. and Sano, S., *Toxicol. Appl. Pharmacol.*, 62(1), 126—136, 1982. With permission.)

the urinary findings of increased proline, and hydroxyproline in cadmium-exposed people indicated that the production of mature collagen fibers may be affected in humans also.

D. Histopathological Findings in Bone

In order to explain the mechanisms of cadmium toxicity on the bones, it is important to note whether or not renal effects occurred in the animals as well as the bone effects. Therefore, in this section observations of the kidney will be mentioned when these are available. Long-term exposure experiments will be mentioned first, as these are most relevant for the understanding of human cadmium poisoning.

Ishizaki and co-workers[38] first reported histopathological changes in bone induced by cadmium in male and female rats exposed to cadmium (300 mg Cd per liter in drinking water) and a low calcium diet for 24 weeks. Before the cadmium exposure started, the rats were fed a low calcium/low vitamin D diet for 1 month. The animals developed renal damage and all of them showed some degree of osteoporosis, and in addition osteomalacia was seen in two of five female rats.

Matsuda[65] exposed eight rats to cadmium in the diet (200 mg/kg for 10 months). Four of the rats developed bone changes: osteomalacia in one case, combined osteomalacia and osteoporosis in another case, and osteoporosis in the other two. When the rats were given combinations of low calcium and low vitamin D diets together with the cadmium, the occurrence of bone changes increased.[65] All 15 rats on a low calcium/low vitamin D diet with 100 to 200 mg/kg cadmium in the feed developed osteoporosis and 5 also developed osteomalacia. However, osteoporosis was also common in the rats with low calcium/low vitamin D diet without cadmium and osteomalacia occurred in one of four such rats. The rats exposed to cadmium had severe renal tubular damage, and Matsuda[65] concluded that renal osteomalacia occurred in some of these animals and that this osteomalacia as well as the renal damage itself was made worse by a low calcium diet.

Hirota[31] studied rabbits exposed to 200 mg Cd per liter in drinking water for 1 year (Section III.B). In one of eight rabbits histological examination revealed a picture of osteomalacia and osteoporosis. The rabbits had an increased average serum alkaline phosphatase level (Section III.B).

Kawai and co-workers[46] gave six groups of five male rats each, cadmium *ad libitum* in drinking water. One group (control group) received only tap water with no additional cadmium and four groups received 10, 50, 100, and 200 mg/ℓ cadmium, respectively, in tap water and were studied for 8½ months. The sixth group received the 200 mg/ℓ cadmium dose in tap water, but was studied for a longer period of 18 months. The results are summarized in Table 6 where it can be seen that the cadmium concentration in liver increases more or less linearly in relation to the total dose given. According to the authors, typical lesions of tubular atrophy with interstitial edema were found in the kidneys of the groups given 100 mg Cd per liter or more. Already at the 50 mg/ℓ dose level slight lesions appeared in some of the animals. This low dose gave an average kidney concentration of 38.4 mg/kg wet weight, which would correspond to a kidney cortex concentration of about 47 mg/kg. Decalcification and cortical atrophy of the femur occurred in the animals. These injuries are indicated in Table 6 by the decreasing cortical thickness in percent of the total femoral thickness. The calcium content also decreased with dose level. Histological findings in the bones occurred already in the group given 50 mg/ℓ in drinking water. Similar results, i.e., a decreased cortical thickness after cadmium exposure in mice for 4 months, were shown by Matsue and co-workers[66] and Muto and co-workers,[72] but fewer details were reported.

Bone changes were also found by Itokawa and co-workers who exposed four groups of five rats each to different combinations of calcium and cadmium in the diet: calcium-adequate and calcium-deficient diet, with or without 50 mg Cd per liter in drinking water.[39] After 120 days there was a considerable degeneration of the kidneys in the two groups exposed to cadmium. The greater part of the tubular epithelia was desquamated and vacuolized. Necrosis and partial hyalinization of the glomerular capillaries and adhesions between the Bowman capsule and glomerular capillaries were also observed. Thinning of the cortical osseous tissues was seen in the bones of the calcium-

Table 6

MAIN PATHOLOGICAL FINDINGS AFTER EXPERIMENTAL CHRONIC CADMIUM POISONING IN RATS[46]

Cd concentration in drinking water (mg/ℓ)	Exposure time (months)	Whole kidney Pathological finding	Whole kidney Cd (mg/kg)	Kidney cortex (Cd[a] mg/kg)	Atrophy	Bone Cortical thickness (%)	Bone Ca (mg/g)	Bone Cd (mg/kg)	Liver (Cd mg/kg)
0	8.5	—	N.D.	N.D.	—	28.7	142	0.90	N.D.
10	8.5	—	12.3	15	±	27.6	129	1.07	4.2
50	8.5	±	38.4	47	+	26.3	128	1.44	29.8
100	8.5	+	100.1	125	+	23.7	125	2.39	73.1
200	8.5	++	156.3	195	+	22.0	113	3.48	143.1
200	18	+++	136.8	171	++	19.8	114	2.87	234.7

Note: N.D. = no data.

[a] Recalculated based on a ratio of 1.25 between cortex concentration and whole kidney concentration (Volume I, Chapter 6, Section IV.B.2).

deficient rats with no cadmium exposure. In the cadmium-treated rats with calcium-adequate diet, fat deposition took place in the femoral spongiosa. In the group with both calcium deficiency and cadmium exposure, there were also some osteoid borders on the bone trabeculae and an increased number of osteocytes in the cortical tissue. Itokawa and co-workers[39] concluded that the bone findings were similar to the osteomalacia characterized by poor calcification as seen in humans.

In a review of the available data up to 1977, Nomiyama[77] disputes the conclusions drawn by Itokawa and co-workers, on the basis that bone histology was not carried out on all of the 20 rats and that extreme calcium deficiency alone can produce osteoid formation. However, the calcium-deficient rats of Itokawa and co-workers[39] did not develop osteoid, and in none of the other experiments on calcium-deficient animals referred to in this section was this reported. The diet used by Itokawa and co-workers has been described in detail.[40] Nomiyama[77] concludes that osteoporotic changes are the most common after long-term oral exposure to cadmium in animals and that osteomalatic changes may possibly appear under inadequate nutritional conditions. Most of the studies referred to above and also referred to by Nomiyama[77] report osteomalacia as well as osteoporosis.

More recent studies have confirmed that osteomalacia can be induced in animals by cadmium even when calcium and vitamin D intake are adequate. The cadmium-exposed rats without calcium deficiency studied by Kawamura and co-workers[47] (Section III.B) had severe renal damage as well as bone damage. In their femurs the histological changes were similar to osteomalacia.[47] The epiphyseal cartilage plate was remarkably thin and there was a decrease of cartilage cells of the proliferative and maturation zone. The provisional calcification zone was markedly thin, primary trabeculae formation had disappeared, and bone trabeculae were narrowed. Fat deposition in the femoral spongiosa was also evident and osteoid borders were observed on the bone trabeculae. In the group with calcium deficiency as well as cadmium exposure these changes were more prominent.

Takashima and co-workers[93] exposed rats to calcium-sufficient diets with additions of cadmium (10, 50, or 100 mg/kg) for 19 months. The cadmium concentrations in kidneys were not measured, but similar exposure duration to such high cadmium exposure levels have caused renal damage (Chapter 9, Section III.A.4). The cadmium level in femur increased from 1.07 mg/kg in the control group to 4.6 mg/kg in the 100 mg Cd per kilogram group. In the study by Kawai and co-workers[46] mentioned earlier in this section, rats were exposed to similar conditions and in that experiment a cadmium concentration in bone of 3.5 mg/kg corresponded approximately to a cadmium concentration in whole kidney of 156 mg/kg. Kawai and co-workers[46] found clear tubular pathological changes in the rats at such cadmium levels. Takashima and co-workers[93] reported degenerative changes both in the tubuli and glomeruli of the cadmium-exposed rats. There were significant decreases in the concentrations in femur of calcium (10%), zinc (15%), iron (33%), and sodium (3%) in the group with the highest cadmium exposure. The demineralization of the bone is in accordance with the effects on calcium metabolism (Section III.B).

Histological studies of the femurs showed that in the cadmium-exposed rats there were areas of osteoid in the cortical osseous tissues.[93] The occurrence of osteoid as well as irregularity of the thickness of cortical osseous tissues increased in the cadmium-exposed group in a dose-response manner.[93] As mentioned above, these bone effects occurred in the presence of renal damage. Other irregularities of the collagenous structure in the cartilage plates as well as fat tissue replacement of bone marrow were seen both in cadmium-exposed and control rats. The authors presumed that this was due to the influence of aging, but the irregularity of the collagen may very well have been associated with the cadmium exposure (Section III.C).

Further evidence for the development of osteomalacia in animals subject to long-term exposure via food has been reported by Nogawa and co-workers.[76] They exposed 9 groups of rats (about 10 rats in each group) for 500 days to various combinations of adequate, slightly deficient, and deficient vitamin D intake and cadmium exposure via the oral route (100 mg/kg in diet) or s.c. injection route (a total of 3.3 mg cadmium distributed over 22 injections). The accumulation of cadmium in tissues was similar for five of the groups with various combinations of oral and injected cadmium: in liver the concentrations were 290 to 330 mg/kg, in femur they were 3.4 to 7.5 mg/kg, and in whole kidney 124 to 146 mg/kg.

Renal lesions in the form of dilatation of tubuli, change of tubular epithelium, focal scarring, and focal lymphocytic interstitial infiltration were seen in the rats exposed to cadmium, with the more severe lesions occurring in the rats that had combinations of oral and injection exposure.[76] The calcium concentrations in liver, kidney, and femur were decreased in the groups with vitamin D deficiency and cadmium exposure further enhanced the decreased calcium concentrations.

Plasma calcium levels were also affected by vitamim D deficiency, but cadmium exposure did not influence the levels very much. On the other hand, urinary calcium excretion was increased in the group of rats with vitamin D-adequate diets and cadmium exposure, but unchanged or slightly decreased in the groups of rats with vitamin D-deficient diets with or without cadmium exposure. This indicates that the mechanism which controls the plasma calcium level can overcome cadmium-induced tendency towards an increase of urinary calcium excretion. Such an increase in excretion is therefore only likely to be seen in humans or animals with adequate vitamin D and calcium intake.

The histological investigation of the femurs revealed the same pathological changes in vitamin D-deficient control groups as those seen in the cadmium-treated groups (Figure 6A, B, C). In the vitamin D-adequate, cadmium-treated rats there was osteoid tissue formation, thickening of cortex, increase in numbers of osteoblasts and osteoclasts, abnormality of subchondral ossification, and fibrosis. These findings are in agreement with a diagnosis of osteomalacia.[9]

As mentioned earlier, there have been different opinions regarding the prominence of osteomalacia or osteoporosis in animals exposed to cadmium. There is some evidence that osteoporosis without osteomalacia is seen particularly after short-term high exposure to cadmium. Yoshiki and co-workers[107] exposed male rats to diets containing cadmium (0, 10, 30, 100, and 300 mg/kg) for 3 weeks. The diet was adequate in calcium. The group with the highest cadmium exposure was studied for a prolonged period of 12 weeks. They found osteoporotic lesions in tibias of the cadmium-exposed rats already after 3 weeks. Metaphysial trabeculae disappeared in varying degrees and the remaining trabeculae were narrowed as a result of cadmium exposure. There was no increase in the number of osteoclasts. There was retarded growth of the tibiae and the findings indicated inhibition of bone formation rather than acceleration of bone resorption. Very few or no osteoids were observed in the bone trabeculae even in the rats that were fed 300 mg Cd per kilogram.

A group of rats fed a rachitis-inducing diet for 3 weeks developed osteomalacia.[107] When the rachitis-inducing diet was combined with cadmium exposure for 3 weeks at 300 mg/kg food, the additional changes that occured were similar to the osteoporotic changes mentioned above.

Yoshiki and co-workers[107] point out that after 3 weeks there were no pathological kidney changes in the rats and the urinary excretion of cadmium had not increased. In the rats followed up for 12 weeks there was slight histological renal damage on termination of the experiments. The experiment by Yoshiki and co-workers[107] shows that subacute cadmium exposure affects calcium metabolism or bone metabolism in such a

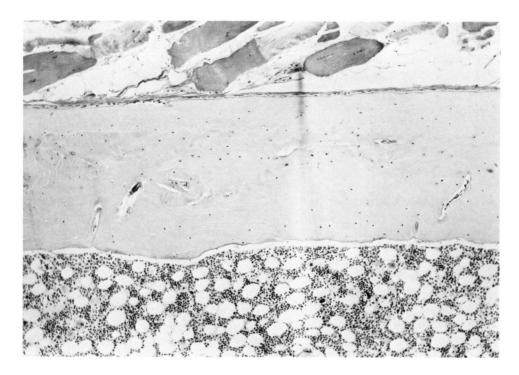

A

FIGURE 6.   (A) Cortical osseous tissue of femur: group 1 (vitamin D-adequate control rat). Cyanuric chloride, hematoxylin-eosin. (B) Group 2 (vitamin D-slightly-deficient control rat). (C) Group 6 (vitamin D-adequate, cadmium-treated rat). (From Nogawa, K., Kabayashi, E., and Konishi, F., *Environ. Res.*, 24, 233—249, 1981. With permission.)

way that osteoporosis develops and this occurs before histological renal changes have developed.

In conclusion, animal experiments show that both osteoporotic and osteomalacic lesions can be induced by oral and injection exposure to cadmium. Subacute exposure will induce osteoporotic changes exclusively and chronic exposure induces osteomalacic changes with or without concurrent osteoporotic changes. Depending on exposure conditions and types of bone changes looked for, entirely negative results may also emerge. In the experiments where osteomalacia was induced the animals also had severe renal damage. In some studies a histological picture similar to osteolathyrism was also reported. The data agree well with the findings that calcium absorption can be directly inhibited by oral cadmium exposure (Section III.B), that vitamin D hydroxylation in the kidney can be inhibited by the renal tubular damage induced by long-term cadmium accumulation in the kidney (Section III.A), and that cadmium inhibits lysyl-oxidase and the normal cross-linkage of collagen fibers (Section III.C).

E. The Japanese Monkey Study

The Japanese studies of rhesus monkeys referred to in Chapter 9, Section IV.A.1, have been extended in order to study the effects of cadmium exposure, vitamin D deficiency, and nutritional deficiency on bone metabolism and bone structure.[98] Forty healthy female rhesus monkeys were divided into eight groups according to Table 7. Group 8 contained more monkeys as it was intended to carry out vitamin D treatment experiments on some of them. The nutritional deficiency groups received low intakes

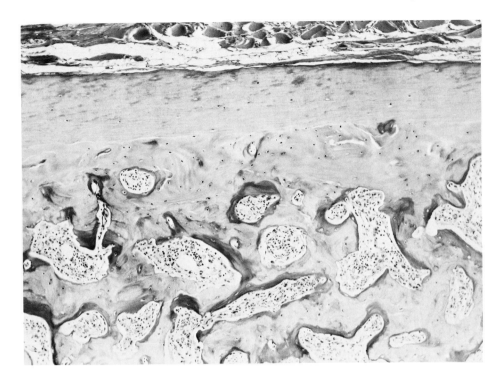

FIGURE 6B.

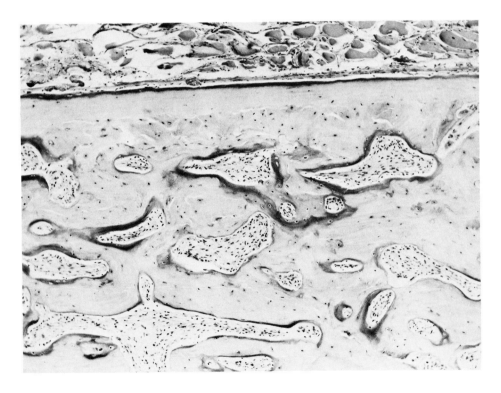

FIGURE 6C.

Table 7

TREATMENT OF EIGHT MONKEY GROUPS. AT THE END OF 1977, THE MONKEYS WERE ESTIMATED TO BE 4 TO 7 YEARS OLD. WHEN THE EXPERIMENT STARTED IN MAY 29, 1978, THEY WOULD HAVE BEEN ON THE AVERAGE ABOUT 6 YEARS OLD[79,80,98]

	Cd	Vitamin D deficiency	Low nutrition	No. of monkeys
Control Groups				
Group 1 (not treated)	−	−	−	5
Group 2 (low nutrition)	−	−	+	4, one was killed in 1983
Group 3 (Vitamin D deficiency)	−	+	−	4
Group 4 (Low nutrition and vitamin D deficiency)	−	+	+	4, all four were killed in 1983
Cadmium-treated groups				
Group 5 (Cd)	+	−	−	5, one died in 1979, one was killed in 1983
Group 6 (Cd and low nutrition)	+	−	+	4, one was killed in 1983
Group 7 (Cd and vitamin D deficiency)	+	+	−	4
Group 8 (Cd and low nutrition and vitamin D deficiency)	+	+	+	10, one died in 1979, three died in 1981, one died in 1983, the remaining five were killed in 1983

Note: Cd concentration added to food June 1978 to May 1979, about 3 mg/kg; from June 1979, about 30 mg/kg (still continuing for groups 5 to 7 in 1984). Groups 4 and 8 were treated with vitamin D from September 1981 to September 1983. Mainly oral treatment at dose levels (500 or 20,000 IU/kg), but some intramuscular treatment was also given.

Table 8

TYPICAL ANALYSIS RESULTS (% OF FRESH WEIGHT) OF THE
EXPERIMENTAL DIETS FOR THE MONKEYS[98]

Diet no.	Monkey group	Moisture	Crude protein	Crude fat	Crude fiber	Crude ash	Ca	P
IV	1	11.3	21.3	1.0	3.0	6.7	0.90	0.84
	2	8.7	14.2	1.4	2.5	4.1	0.31	0.31
	3	6.3	21.8	1.0	3.2	6.8	0.93	0.88
	4	7.1	14.4	1.4	2.6	4.2	0.36	0.36
	5	7.3	21.1	1.2	3.2	6.7	0.97	0.88
	6	5.7	14.3	1.4	2.7	4.1	0.36	0.34
	7	6.4	21.6	1.1	3.1	6.7	0.97	0.89
	8	6.1	14.7	1.4	2.3	4.2	0.34	0.33

of protein, phosphorus, and calcium. Cadmium chloride was added to the basic "chow" of half of the groups. Initially, in 1978 the exposure level was 3 mg Cd per kilogram chow, but this was increased after a year to 30 mg Cd per kilogram chow (Table 7). Analysis of the diets given[98] showed that the chow in the nonexposed groups contained on average about 0.2 to 0.7 mg Cd per kilogram; the 3 mg/kg chow actually contained only 1.6 to 2.9 mg Cd per kilogram; and the 30 mg/kg chow contained 19 to 31 mg Cd per kilogram, depending on the batch.

The chow was made up of casein, corn, wheat, flour, alfalfa, soybean cake, sucrose, vitamin mixture, and salt mixture. Some analytical results for the chow are shown in Table 8. In the low nutrition diets the protein content was lowered to 14% by giving a different mixture of the constituents. The calcium and phosphate content was also lowered, and as seen in Table 8, these diets contained less fiber. Vitamin D deficiency was created by excluding vitamin D3 from the vitamin mixture. The daily calcium intake from the low nutrition diet was estimated to be about 90 to 120 mg/kg body weight, which was slightly less than the assumed requirement of 150 mg/kg.

The report by the Tertiary Monkey Experiment Team[98] includes 2-year follow-up data on the effects on the monkeys and further follow-up data (up to 5.5 years) are presented in more recent reports by Nomura and co-workers[79,80] and other authors published in *Kankyo Hoken Report* by the Japanese Public Health Association.

Kidney biopsies were taken from two monkeys in each group after 12, 18, and 50 months. An additional 2 monkeys in groups 4 and 8 were biopsied at 50 months. The dry weight concentration of cadmium, zinc, and copper was measured and recalculated to wet weight values by dividing by four.[98] (In fact, it says in the report that the value is multiplied by four, but this is a misprint, according to Nomiyama, personal communication). No explanation for the factor 4 was given. It was also pointed out by the Tertiary Monkey Experiment Team[98] that "the amount of biopsy specimens was very small, and exact metal determinations were inevitably difficult in such specimens".

Table 9 shows the surprisingly high cadmium levels in the "control groups" (groups 1 to 4). On the other hand, the basic diet contained about 0.25 mg Cd per kilogram (calculated from data by the Tertiary Monkey Experiment Team),[98] and each day the monkeys ate about 110 g of food, which means that each year the monkeys would swallow about 10 mg Cd. At 5% absorption and one third of the body burden in the kidney (Volume I, Chapter 6, Sections II and IV), about 170 μg would be accumulated each year in the kidney. Ten percent of this amount may be excreted each year (Chapter 6, Section V) giving 610 μg total accumulation after 4 years. With a total kidney weight (both kidneys) of 30 g,[4] the average cadmium concentration can be expected to be about 20 mg/kg wet weight and in renal cortex the concentration may be 25% higher

Table 9

CADMIUM CONCENTRATION IN RENAL BIOPSY SPECIMENS AND
AUTOPSY SPECIMENS OF RENAL CORTEX OBTAINED FROM
EXPERIMENTAL MONKEYS (MG/KG WET WEIGHT). AUTOPSY
DATA FOR LIVER SPECIMENS IN BRACKETS[51-53]

Group no.		3 ppm Cd		30 ppm Cd			Liver (at time of autopsy)
		12	18	36	50	63	
I	101	17	56		46		
	102	6	11		33		
II	138	nd	1		28		
	140					[c 70 / m nd]	
	139	nd	12		19		3.2
III	106	nd	37		46		
	108	nd	45		48		
IV	110	61	105		81		
	111	nd	58		53		
	112				29		
	113				37		
V	114		[c 338 / m 109]				
	115					[c 1116 / m 365]	609
	117	33	173		1255		
	118	47	168		627		
VI	121	110				[c 381 / m 220]	1237
	122	128	486		911		
VII	125	29	228				
	126	97	547				
VIII	129						
	130			[751]			285
	131				693		
	132	279	766		850		
	133						
	134				1155		
	135	50	[c 529 / m 215]				
	136			[611]			
	119				1129		240
	124			[1016]			364

Note: nd = not detectable; c = cortex; m = medulla. Controls: 153-1   92
153-2   99

From Kimura, M., Watanabe, M., and Otaki, N., References 51—53, Japanese Public Health Association, Tokyo. With permission.

or 25 mg/kg (Chapter 6, Section IV.B.2). The measured average cadmium concentration at 50 months based on the data in Table 9 was 42 mg Cd per kilogram (SD = 16 mg/kg, n = 10). The average zinc concentration was 43 mg/kg (SD = 12 mg/kg) and the average copper concentration was 5.3 mg/kg (SD = 1.2 mg/kg).[51] These concentrations are similar to those found by Elinder[21] in horses.

The extremely high cadmium levels in the kidneys of the cadmium-exposed monkeys (Table 9) are unusual as in other animal experiments the levels have rarely exceeded

300 to 400 mg/kg, particularly when renal damage has developed (Chapter 9, Section V.B). The liver cadmium levels were also extremely high in the autopsy cases (Table 9). After 3 years of cadmium exposure they were about 300 mg/kg wet weight and after 5 years they were 600 mg/kg in one monkey on a "basic" diet and 1200 mg/kg in one monkey on a "nutrition-deficent" diet (Table 9). The higher level in the latter monkey of group 6 may be due to a higher gastrointestinal absorption at low calcium intake (Chapter 6, Section II.B.3.a.).

In spite of the high renal and liver cadmium levels, the indicators of renal damage used by Kimura and co-workers (total proteinuria, β_2-microglobulin in urine) showed significant changes after 39 months only in group 8[52] and in both groups 6 and 8 after 66 months.[53]

Another test of renal function is the PSP test. It was slightly decreased at 39 months in groups 6 and 8,[97] and this decrease progressed in group 6 until 60 months, whereas in group 8 it returned to normal during vitamin D treatment. The urinary β_2-microglobulin excretion of group 8 also decreased during treatment.[53] The mechanisms for these apparent changes in renal function during treatment are unclear. It was pointed out in Chapter 9, Section IV.A.1, that these monkeys must have been extremely resistant to cadmium compared to other experimental animals. Therefore, the rhesus monkey may not constitute an ideal animal model for studying systemic effects of cadmium. Another monkey species, the marmoset monkey, differs completely from other tested animal species in regard to the metabolism of arsenic.[102]

The plasma levels of vitamin D metabolites were measured. Before exposure started, the levels of 25 (OH)D3 were 10 to 20 $\mu g/\ell$, slightly lower than the average normal level in humans (22 μ/ℓ).[98] After 6 months administration of the vitamin D-fortified diet ("control diet") (275 IU vitamin D3 daily), the plasma levels in groups 1, 2, 5, and 6 had increased to 40 to 50 $\mu g/\ell$ and levels above 40 $\mu g/\ell$ were sustained until month 54.[90] In contrast, groups 3, 4, 7, and 8 (not receiving vitamin D supplement) had plasma 25(OH)D3 levels below 10 $\mu g/\ell$ after 6 months and these levels stayed between 5 and 10 $\mu g/\ell$ until month 54. There was no obvious difference between cadmium-treated and control groups. The pattern was the same for 24,25(OH)D3. The finding that vitamin D3-metabolites were found in plasma of groups with deficient vitamin D3 intake indicated, according to the Tertiary Monkey Experiment Team,[98] that some vitamin D was contained in the basic diet of the monkeys.

The 1-alpha-25(OH)D3-metabolite (Kalcitriol®) did not change significantly during the first 18 months of exposure. The preexposure level was 70 to 90 ng/ℓ, slightly higher than the normal human average of 43 ng/ℓ. In all groups the average at 6, 12, and 18 months' exposure was between 50 and 200 ng/ℓ.[98] At 24 months a decrease in groups 4, 7, and 8 appeared and this was sustained up to 54 months' exposure (Figure 7). In the first four sampling periods (6, 12, 18, and 24 months) the average plasma Kalcitriol® levels of the cadmium-treated monkeys were lower than the levels of the control groups in 14 comparisons out of 16. In the subsequent four sampling periods (36, 42, 48, and 54 months) for groups 1 to 3 and 5 to 7, the levels of the cadmium-exposed groups were lower than the controls in 10 comparisons out of 12.[90]

The probability of getting 24 lower values out of 28 (or a more extreme result) just by chance is 0.00009 (based on a binominal distribution calculation). Thus, a decrease in plasma Kalcitriol® is clearly associated with cadmium exposure. This effect appears to develop already after 6 months exposure. In group 7 there is a tendency for progressive decrease up to month 54. (We do not have information for groups 4 and 8 because their data were not included in the report.[90])

A vitamin D3 treatment experiment was also carried out.[79] In the 40th month of exposure some of the monkeys in groups 4 (low nutrition, low vitamin D, low cadmium) and 8 (low nutrition, low vitamin D, high cadmium) were given vitamin D3

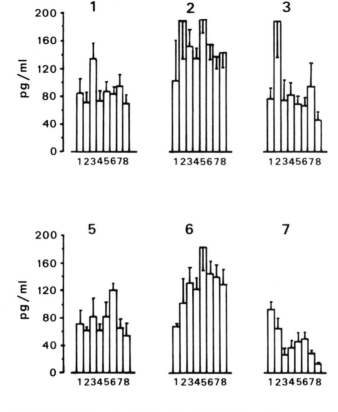

FIGURE 7.   Plasma levels of 1-alpha-25(OH)2-D at different exposure durations for monkeys. Vertical bars from left to right: 1 = 0 month; 2 = 12 months; 3 = 24 months; 4 = 30 months; 5 = 36 months; 6 = 42 months; 7 = 48 months; 8 = 54 months. (Modified from Suda, T., Shiina, Y., and Abe, E., *Kankyo Hoken Report No. 49,* Japanese Public Health Association, Tokyo, 1983, 64—67.)

treatment orally, either 500 IU/kg body weight per day or 20,000 IU/kg body weight per day (Table 7).

Two weeks after treatment started an increase of all three vitamin D metabolities was seen in plasma. The increase was much higher for the high vitamin D treatment group, but there was no obvious difference between the cadmium-exposed group 8 and the control group 4. The short-term effect on the plasma Kalcitriol® level of treatment with vitamin D may thus be less affected by cadmium than the long-term steady-state level.

The serum alkaline phosphatase decreased slowly with time in all monkey groups except groups 4 and 8. The two groups showed a dramatic increase during the first 24 months of exposure.[98] Serum inorganic phosphorus decreased to a plateau at half the level of the other groups already after 3 months. The serum Ca × P product was similarly decreased in groups 4 and 8.

X-ray measurement of the cross-sectional area of the femoral cortex was carried out at 12, 36, and 48 months of exposure (Figure 8). Only in groups 4 and 8 was there a significant decrease with time. The cortical cross-sectional area at 36, 48, and 60 months in the cadmium-exposed group 8 was consistently lower than the area of group 4.[108]

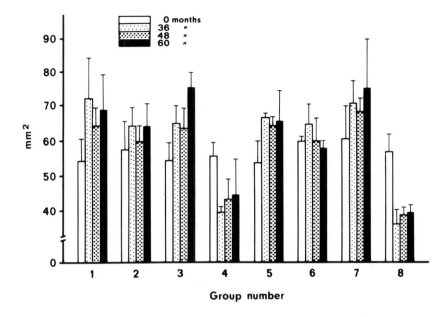

FIGURE 8. Average cross-sectional area of femoral cortex in the eight groups of monkeys at the start of exposure and after 36, 48, and 60 months of cadmium exposure. (From Yoshiki, S., Tachikawa, T., Yamaguchi, A., Yamazaki, T., and Yamana, H., *Kankyo Hoken Report,* in press.)

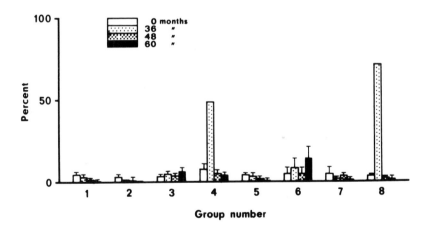

FIGURE 9. Average volume of osteriod seams (as a % of the trabecular surface area) in iliac bone biopsy of the eight groups of monkeys at the start of exposure and after 36, 48, and 60 months of cadmium exposure (From Yoshiki, S., Tachikawa, T., Yamaguchi, A., Yamazaki, T., and Yamana, H., *Kankyo Hoken Report,* in press.)

Three of the monkeys in groups 4 and 8 developed "Looser's zones", a sign of osteomalacia, already after 12 months. Another sign is the relative volume of osteoid tissue in the bones (Figure 9). Again groups 4 and 8 stand out and there is a tendency for group 8 (cadmium exposed) to have a higher relative volume of osteoid at 36 months of exposure. After that, both groups 4 and 8 were treated with vitamin D (Table 7) and the osteoid volume returned to normal (Figure 9). In group 6, which developed renal damage and did not receive vitamin D treatment, there is a tendency for an increased osteoid volume with time.

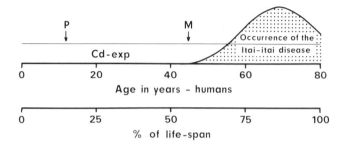

P : Puberty
M: Menopause

FIGURE 10.   Comparison of life span of humans and rhesus monkeys, with reference to the time relationship of cadmium exposure and occurrence of bone disease.

These extensive monkey experiments have shown osteomalacia in monkeys at a combination of low nutrition and deficient vitamin D intake even without any additional cadmium exposure. There is a possibility that the osteomalacia is more pronounced in group 8 which had cadmium exposure. Low nutrition or deficient vitamin D intake could not separately induce osteomalacia-like changes in group 6 that received low nutrition and additional cadmium. These changes only developed when renal functional damage had developed in group 6.

Neither group 5 nor group 7, in spite of the cadmium exposure, developed renal functional changes. As one main mechanism for the development of cadmium-induced osteomalacia is considered to be renal damage and its effect on vitamin D metabolism (Section III.A), there is no reason to expect such an osteomalacia in any of the monkeys of groups 5 and 7.

These monkeys were considered to be about 6 years old when the cadmium exposure started (Table 7). However, their weights were 3.9 to 6.4 kg,[98] which according to the *Biology Data Book*,[5] agrees better with an age of 2.5 to 4.5 years. Rhesus monkeys are sexually mature at about age 4 years, are fertile for 20 years, and live for about 30 years.[5] In terms of their total life span, the monkeys in the Japanese Tertiary Monkey Experiment Team started their exposure at "puberty" and still after 6 years of exposure have only passed about a third of their life span (Figure 10). Cadmium and bone metabolism are quite different at young age than at older age.[9] The data for this monkey study so far are, therefore, not comparable to the situation for the women in Japanese areas where Itai-itai disease has been found (Section V). None of the Itai-itai patients developed severe bone symptoms before menopause at about age 45, in spite of being exposed to cadmium virtually from birth (Figure 10). The monkeys need to be followed up for at least another 5 years while being exposed to cadmium before the occurrence of cadmium-induced osteomalcia may be expected to start occurring. If menopause is a necessary factor in the development of such osteomalacia, another 15 years of follow-up may be needed (Figure 10). Another factor of importance may be pregnancies. Most of the Itai-itai patients had several pregnancies (Section V). None of the monkeys had been pregnant.

Another difficulty in comparing the monkey studies with the human data is the small number of monkeys in each group. With four monkeys in a group, a response rate of 25% has to be reached in order to record one case of osteomalacia in a group. Only in the most severely polluted parts of Fuchu area (Appendix) did the prevalence rates of Itai-itai disease reach such high values.

However, the latest reports from these studies[53,108] indicate that in group 6, with low nutrition level and cadmium exposure, concomitant renal functional damage and pathological bone changes similar to osteomalacia are developing. With a longer follow-up period, the full-blown changes seen in other animals and humans may also be seen in the monkeys.

F. Other Effects

In 1980, Miyahara and co-workers studied the effect of cadmium on bone resorption in tissue cultures.[70] Tibias prelabeled with ^{45}Ca and ^{32}P from nine-day-old chick embryos were cultivated for 6 days and cadmium chloride (concentration range: 0.2 to 30 mg Cd per kilogram) was added to the culture medium. The ^{45}Ca released as a percentage of the total radioactivity was slightly (10%) increased in the group which received 1 mg Cd per kilogram ($p < 0.05$) and decreased to the same extent in the 30 mg Cd per kilogram group.[70] Also the phosphorus release (^{32}P) was increased while at the same time the phosphorus content of bone was decreased in the 1 mg Cd per kilogram group. This indicated that cadmium increased net resorption of bone and not simply the exchange of ^{45}Ca for the stable calcium in the medium. Bone lysosomes were incubated with cadmium at 0°C for 60 min and the free activity of acid phosphatase was the same.[70] It was found that at a cadmium concentration up to 50 mg Cd per kilogram, there was no effect on the free activity of acid phosphatase. On the other hand, release of acid phosphatase from embryonic chick tibias cultured for 2 days in the presence of different concentrations of cadmium was significantly decreased already at 1 mg Cd per kilogram.[70] This suggests that cadmium inhibited the synthesis of acid phosphatase rather than the enzyme activity.

Hydroxyproline is normally released into the medium from the embryonic chick femurs in the culture. Miyahara and co-workers[70] showed that there were significant decreases (to about one half of the control level) of the hydroxyproline recovered both in the media and in the bone cultures with 1 or 30 mg Cd per kilogram. This indicates that cadmium does not only have an effect on the mineralization and demineralization of the bone, but also on the absorption of the collagens in the bone.

Long-term exposure of rats to cadmium (1 year; 75 mg Cd per kilogram in food) causes decreased activity of alkaline phosphatase in the rats' femurs.[14] At 36 weeks of exposure, when the cadmium concentration in bone was 0.9 mg/kg, there was no difference from a control group, but at 1 year of exposure the alkaline phosphatase activity was decreased to half that found in the control group, and the femoral cadmium concentration was 0.1 mg/kg. By giving zinc (200 mg Zn per kilogram) simultaneously with the cadmium, the effects could be prevented and by giving this amount of zinc without cadmium, the alkaline phosphatase activity was significantly increased (15 to 25%). It should be mentioned in this context that a negative study based on the remarkable hypothesis that Itai-itai disease is ''zinc poisoning'' was accepted for publication.[6] They studied parameters of Haversian bone remodeling in dogs with excessive dietary zinc intake and found no effect except an increased zinc concentration in bone.

Finally, a subacute experiment of the compressive strength of rat femurs[81] showed that after a 4 week exposure to 10 mg Cd per liter in drinking water all compressive strength parameters were decreased.

IV. HUMANS SUBJECT TO OCCUPATIONAL EXPOSURE

Considering the large number of workers who have been exposed to cadmium in industry, the very high exposures they sustained in the past and the high prevalence of severe renal tubular damage (see Chapter 9, Section IV.B.1), the reported number of workers with bone effects is small.

The first report of bone effects among cadmium workers by Nicaud and co-workers in 1942[74] presented clinical information on 2 male and 4 female cases in a cadmium-nickel battery factory with a total of 20 workers. The diagnostic tools were crude, the diagnosis of bone effects mainly relying on X-ray investigations. These showed lines of pseudofractures, especially in the shoulder blades, pelvis, femur, and tibia. The workers complained of pain in the back and the extremities and of difficulties in walking.

Renal effects were only investigated with the urine boiling test, which is a very crude test for proteinuria. The results were negative, but it is quite possible that the workers had tubular proteinuria, as Friberg[23] studied the urine of other workers in the same factory with the nitric acid test, TCA test, and electrophoresis and found that they had proteinuria of the same type as that found in the Swedish battery workers.[23]

A total of 55 cases of various bone effects (mainly osteomalacia) following occupational cadmium exposure have been reported (Table 10). In one factory[74] the prevalence of osteomalacic changes was as high as 30% (Table 10). Even though none of the studies included reference groups, it is clear that such a high prevalence is likely to be higher than the prevalence of osteomalacia in the general population in that place and at that time.

The daily intake of calcium, vitamin D, and protein may modify the manifestations of the bone effects, and this could explain the geographical differences found. Friberg[23] looked for signs of ostemalacia in an X-ray study of the 11 most highly exposed workers out of a total of 43 participating in his study. They had been exposed to very high levels (more than 1 mg Cd per cubic meter) of cadmium oxide dust for 9 to 34 years. No bone changes were found, even though several of the workers had severe renal tubular damage. Most of these workers have since been under medical surveillance, but no development of bone disease has been reported. However, no systematic biochemical screening for changes in calcium metabolism has been carried out. In Sweden, the daily intake of calcium via milk is likely to have been high, even 40 years ago.

Valetas[103] studied 20 workers in a cadmium-nickel accumulator factory which had also been studied by Nicaud and co-workers.[74] Valetas reports some interesting clinical information about the cases. Of the 20 workers, 16 had various experiences of painful muscle contractions at night. In most of them neurological examination showed no abnormalities. In those cases where serum calcium was measured, it was normal. The reason for the contractions is unclear. Of the 20 workers, 6 had pathological bone findings: pain, decalcification, and pseudofractures (reported in 4 cases). The kyphosis and lumbar curvature of the spinal column were accentuated.

Eager to find a treatment for the disease, Valetas[103] reported that "the treatment had a remarkable effect and it had totally improved the prognosis". The workers were given i.v. injection of 10 mℓ, 10% calcium gluconate every second day, and "massive" doses of vitamin D (exact dose not stated). After 2 to 3 months, the symptoms had completely reversed, but the improvement in the X-ray findings took longer (not reported how much longer). Four years later they are still in "recovered state", but "it is true that they are still given every four months a cure of calcium and vitamin D".[103] Apparently, the treatment had to continue to be effective. Valetas[103] suggests that one could prevent cadmium poisoning by giving the workers a treatment of calcium and vitamin D once per month. He coined the work "Cadmiose" for the bone disease

Table 10

BONE EFFECTS REPORTED AMONG CADMIUM WORKERS[78]

Cadmium compounds	Cadmium in air (µg/m³)	Years of exposure	No. of workers examined	No. of abnormal workers	Abnormal findings of bone	Method of bone examination	Proteinuria	Ref.
CdO Dust	?	8—13	20	6	Osteomalacia	X-ray	?	74, 103
Fume	40—50	32	?	1	Decalcification	Autopsy	Tubular	13
Fume	?	8—30	?	7	Pseudofractures	X-ray	Common	27, 82
Dust	?	13	?	1		X-ray	Tubular (1 case)	
?	130—1,170	1—2	80	10	Osteoporosis	X-ray	7 cases	32
				3	Pseudofracture			
				13	Sclerotic foci			
Dust	500	?	38	1	Osteomalacia	X-ray, biopsy autopsy	Positive	1, 2
Dust and fume	?	?	12	1	Osteomalacia and osteoporosis	X-ray and clinical	tubular	48
Dust	?	36	?	1	Osteoporsis and osteomalacia	X-ray and clinical	Tubular	1, 12
Various Cd compounds	100—5,500	(16, 19)	285 (32)	1	Osteomalacia	X-ray	12 out of 54	64
	?	?	?	2	Pseudofracture	X-ray		95
				5	Enostosis	X-ray	?	96
				3	Periosteal proliferation and consolidation			

among cadmium workers and emphasized that this was a new occupational disease which should be eligible for workers' compensation.

In 1955 Bonnell found, at autopsy, severe decalcification in a cadmium worker exposed to cadmium oxide fume for 32 years.[13] The worker died from chronic renal failure. Before his death proteinuria had been found on repeated examinations. Electrophoresis analysis revealed that this proteinuria was of the type common in chronic cadmium poisoning.

Gervais and Delpech[27] found the same lines of pseudofractures as described by Nicaud and co-workers[74] in eight male workers (ages = 49 to 63; exposure times = 8 to 30 years) exposed to cadmium oxide fume (seven workers) and cadmium oxide dust (one worker). These workers had not been exposed for 10 to 20 years and the symptoms were diagnosed after exposure had ceased. In most cases there was proteinuria which the authors, probably erroneously, attributed to lead. Pujol and co-workers[82] have shown that in a worker with bone changes in the same factory, tubular proteinuria and other signs of tubular impairment were present. Adams and co-workers[2] found one case of osteomalacia in a group of workers exposed to cadmium oxide dust. These authors expressed little doubt that this osteomalacia was caused by renal disease since the worker had multiple tubular defects.

More recently, Kazantzis[48] presented detailed clinical data for six cadmium-exposed workers he had followed-up for more than 15 years. All of them had severe renal tubular damage and disturbances in calcium metabolism, mainly documented in the form of increased urinary calcium excretion and the occurrence of renal stones.

One of the workers developed severe pains when turning over in his sleep, 12 years after the initial examination.[48] Radiological examination of the bones showed a pseudofracture at the neck of the femur (Figure 11) and in the middle of the fibula. An iliac crest biopsy showed widened osteoid seams. Urinary abnormalities found were tubular proteinuria, renal glycosuria, generalized aminoaciduria, high hydroxyproline, low osmolality, and high pH, all of them indicating renal tubular damage (see Chapter 9, Section IV.B.2). A number of abnormalities were also found in the blood, including low creatinine clearance and high phosphate clearance as well as signs of hyperchloremic acidosis.

The osteomalacia responded to treatment with vitamin D, but only when bicarbonate and phosphate supplements had been added. Three months of vitamin D alone did not heal the lesions. Kazantzis[48] stated that the Fanconi syndrome first diagnosed in 1961 had progressed to a most serious consequence: osteomalacia.

Further details about the case reported by Adams and co-workers[2] were given by Adams.[1] This worker with osteomalacia became quite handicapped from bone deformities, which led to a need for bilateral hip replacements. He died in 1976 and the autopsy showed decalcified bones in general and in the undecalcified bone parts there was mild osteomalacia with thin osteoid seams. There was extensive new formation of subperiostal woven bone and nodules of packed lamellar bone in the cancellous bone, with wide osteoid seams.[1] The cadmium concentrations were 172 mg/kg in liver, 77 mg/kg in kidney, 8.3 mg/kg in spine, 3.2 mg/kg in ribs, and 2.1 mg/kg in femur.

Adams[1] and Blainey and co-workers[12] also reported another case of osteomalacia, who had been a cadmium battery worker for 36 years until 1964. He had had a partial gastrectomy for duodenal ulcer in 1956, when he had no proteinuria. In 1958 proteinuria was first diagnosed. When he retired in 1964 he was in "apparent good health" (Adams did not say whether his proteinuria had deteriorated), but his serum alkaline phosphatase was 41 King-Armstrong units ("normal" range 3 to 13). In 1972 he was seen again, frail, with a "waddling gait" and with limb pains. His serum alkaline phosphatase was 47 King-Armstrong units and X-rays showed generalized osteoporosis. Low molecular weight proteinuria was found even though his general renal func-

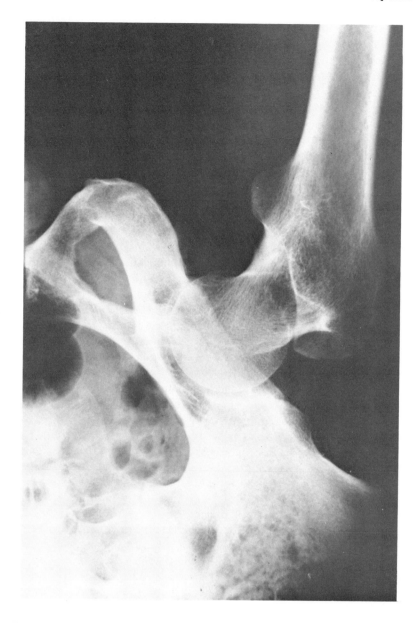

FIGURE 11.   Radiograph of right femur of subject no. 1 showing pseudofracture.
(From Kazantzis, G., *Environ. Health Perspect.*, 28, 155—159, 1979. With permission.)

tion was "excellent". A bone biopsy showed that undecalcified sections had trabeculae
lined by osteoid seams, indicating osteomalacia. Treatment with 50,000 units vitamin
D per day and additional calcium for a month relieved the pain, and after 3 months
his waddling had improved. After a year he could walk well and the vitamin D treatment was discontinued. He stayed reasonably well but developed heavy glycosuria after
a few years. In 1977 he died of carcinoma of the gall bladder and no autopsy was
carried out.

Another recent case of osteomalacia in a cadmium worker[64] had a similar history of
tubular proteinuria developing about 10 years before the osteomalacia. Symmetrical
pseudofractures in the femurs were seen and they responded to treatment with calcium

and vitamin D. However, the renal tubular damage progressed to chronic renal insufficiency.

In other countries such as Japan, Belgium, and the U.S.A. some attempts to find bone effects in cadmium workers have been made, but no systematic studies have been published. Bone effects among workers in Poland[32] and the Soviet Union[95,96] have been reported (Table 10), but not as much detail is given as in the other studies.

In the absence of exactly comparable diagnostic methods and criteria for kidney damage or bone damage in these studies, no definite conclusions can be drawn from apparent variations in the prevalence of bone effects in the different factories studied. Different exposure levels may play a role as may differences in nutritional habits. The intake of fat-soluble vitamins such as vitamin D and of calcium has been much higher in, e.g., Sweden and U.S.A. than in Japan at the time Itai-itai disease started to occur (Section V.A). Furthermore, in France the intake of milk is likely to have been lower than in Sweden or in the U.K. The poor nutritional conditions during and just after World War II in central Europe may have played an important role. Of interest also is that some of the French cases and almost all of the Itai-itai disease patients (Section V.D) were women and thus, more susceptible to calcium and vitamin D deficiencies.[9] In addition, most women living in the Itai-itai area, and particularly the Itai-itai patients, had a large number of pregnancies (Section V and Appendix).

Some studies have shown an increased urinary calcium excretion and renal stone formation among cadmium workers (Chapter 9, Section IV.B.1.c). An increase in the renal stone occurrence among cadmium workers was first reported by Friberg.[23] This was not a finding among Itai-itai disease patients (Section V.B), so it is likely that it only occurs when calcium intake is higher and before serious bone damage has developed. For example, in 1976 Scott and co-workers reported various abnormalities in calcium metabolism in coppersmiths exposed to cadmium.[85] Of 27 workers, 22 had increased urinary calcium excretion (the authors did not give exact data); 13 had lowered inorganic phosphorus in plasma; 3 had raised alkaline phosphatase values; and 5 (18.5%) had a history of renal stone disease.

Further studies of the coppersmiths and other cadmium-exposed workers[86] showed that calcium levels in urine up to three times the normal maximum (6 mmol/ℓ) occurred. Many of the workers had a high cadmium exposure as their blood cadmium values were in the range 10 to 30 μg/ℓ. Normal levels are less than 1 μg/ℓ (Chapter 5). In a subgroup of the workers whole body in vivo neutron activation analysis[87] showed a significant decrease in calcium content of bone as compared to nitrogen content. The authors suggest that this finding may be a forerunner of later skeletal changes.

In conclusion, a number of studies of cadmium-exposed workers have shown an adverse effect of cadmium on calcium metabolism and from case studies there is a strong indication that osteoporosis and/or osteomalacia can be produced by long-term cadmium exposure under certain circumstances.

V. HUMANS SUBJECT TO EXPOSURE FROM THE GENERAL ENVIRONMENT

Cases of bone effects among people exposed to cadmium in the general environment have only been reported from Japan. The exposure levels through food in some cadmium-polluted areas in Japan are the highest reported in the world (Chapter 3, Section III.D.2)

When the first cases were noted in Japan in the 1940s, the etiology was unclear and the name "Itai-itai disease" (ouch-ouch disease) was used to identify the syndrome because of the severe pain complained of by all the patients. During recent years, there has been a lot of controversy regarding the definition and diagnosis of Itai-itai disease, partly due to the legal implications of "compensation for pollution disease" which follows if a case is diagnosed as Itai-itai disease.

In this section we will briefly review the main features of Itai-itai disease and other manifestations of cadmium-induced bone effects and present the more recent data on cases occurring outside the original "endemic" area. A more detailed description of earlier studies of Itai-itai disease and a comprehensive list of references are given in the Appendix.

A. Brief History of Itai-Itai Disease

The first case of this severe bone disease may have occurred in the 1920s when there was a report about "rheumatic disease" in the Fuchu area of Toyama Prefecture (Appendix). After the war, the number of cases became so high that a local general practitioner, Dr. Noboru Hagino, reported at a local medical society meeting about an unusually high occurrence of osteomalacia in the area.[28] Treatment with vitamin D improved the symptoms in most cases, which further supported the diagnosis of osteomalacia. However, very large doses had to be used and Hagino was not satisfied that nutritional deficiency only was the cause of the disease.

Contacts between Dr. Hagino, Dr. K. Yoshioka, an agricultural economist, and Dr. J. Kobayashi, an agricultural chemist specializing in trace element research, became instrumental in the discovery of the link with cadmium. Yoshioka had been involved with the local farmers in claims for compensation of crop damage caused by discharge water from a large mine upstream from the polluted area. He and Hagino proposed the idea that some pollutant in the mining discharge water had contributed to the causation of the disease. Yoshioka asked Kobayashi to help analyze samples from the area and it was eventually found that both the environment and autopsy tissues had unusually high concentrations of cadmium. In 1961 it was postulated in a report to a scientific meeting[29] that cadmium played a role in the etiology of Itai-itai disease.

The local health authorities carried out case studies and environmental studies in the early 1960s. Based on these results, the Japanese Ministry of Health and Welfare in 1968[41] concluded that "Itai-itai disease is caused by chronic cadmium poisoning, on condition of the existence of such inducing factors as pregnancy, lactation, imbalance in internal secretion, aging, and deficiency of calcium".

The research on the disease has since been intensified as there was concern in Japan that the disease may occur in other polluted areas. Three-stage health surveys were carried out by Japanese Government authorities in a number of cadmium-polluted areas (Appendix), and a large number of people with renal tubular damage were found (see Chapter 9, Section IV.B.2), but none of the few cases with bone disease was declared as being true Itai-itai disease. Some other reports conclude that such cases have indeed occurred (Section V.E), but the opinions of the Japanese researchers in this field are very divided.

A series of major reviews of the disease and its etiology have been published.[24,25,99,100,106] These authors present different points of view on the causation of the kidney and bone effects seen in Itai-itai disease, but the detailed etiological mechanism and the possible role of nutritional factors are still debated. The accumulating experimental data (Section II) assist in explaining how cadmium exposure may give rise to Itai-itai disease.

B. Clinical Features of Itai-Itai Disease

The patients present with severe pain from the bones. Even light pressure causes pain and the patients have difficulties sleeping or even breathing. Usually there are 1 or several (up to 75) fractures. Apparently these occur spontaneously. Compression fractures in the spine may occur and these cause a shortening of the stature. The patients

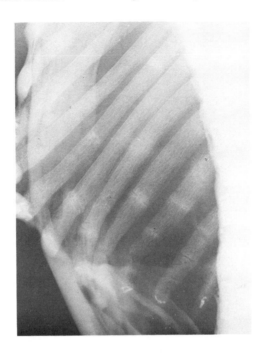

FIGURE 12. X-ray picture of a typical case with "pseudofractures", characteristic of osteomalacia. (Murata, personal communication.)

also develop a duck-like gait and progressive difficulties in walking. Eventually they become confined to bed and without treatment they deteriorate and die of the disease.

Most of the patients in the Fuchu area were postmenopausal women who had had several pregnancies. No hereditary factors were implicated. X-ray investigations show severely decalcified bones and in many cases fractures; some of these may be partly or fully developed "pseudofractures" (Figure 12). Pathological investigations at autopsy have found various degrees of osteoporosis and osteomalacia in almost all Itai-itai disease patients studied (Chapter 9, Table 6 and Appendix). Clinical chemistry of the blood showed a very high level of alkaline phosphatase and slightly lowered levels of serum calcium and phosphorus. This is in agreement with a diagnosis of osteomalacia. Hypochromic anemia and low serum iron were found in many of the early patients. This may have increased their uptake of cadmium from the intestines (Chapter 6, Section II.B.3.e) and thereby increased their susceptibility to severe cadmium poisoning.

The vast majority of patients also had signs of renal tubular damage (Chapter 9, Table 6 and Appendix). Proteinuria, glucosuria, and aminoaciduria are common findings. In the earlier studies rather crude diagnostic methods were used, which may explain the lack of positive urine findings in some patients. During recent years tests for low molecular weight proteinuria have been used and a high excretion of such proteins has been found in all recently tested patients.

As discussed in Chapter 9, Section IV.B.2.c, confirmation of the renal tubular damage is often found in pathological studies after autopsy (Chapter 9, Table 6). A renal tubular function test (the PSP test) was also affected in a majority of patients.[73]

Large doses of vitamin D in combination with anabolic steroids will alleviate the pain after some weeks treatment and healing of the pseudofractures may occur after several months treatment. The recovery is much slower than what would be expected if the osteomalacia was due purely to a nutritional deficiency of vitamin D (Appendix).

C. Cadmium Exposure of the Patients

As is discussed in the Appendix, the daily intake of cadmium via food may have been as high as 600 μg with an additional intake from river water, which was used for cooking and drinking. It has been suggested by Ishizaki (personal communication) that in all the areas where bone effects of cadmium have occurred, the people of the area have consumed cadmium-polluted river water as well as cadmium-polluted rice.

Analysis of tissue cadmium at autopsy also showed very high body burdens. The liver concentrations were in the range 60 to 150 mg/kg (Chapter 9, Tables 5 and 6), which is in the same range as the industrial workers with chronic cadmium poisoning. At "normal" Japanese daily cadmium intakes of about 40 μg/day, the liver cadmium would be on average 2 mg/kg. The levels in Itai-itai disease patients may thus indicate daily intakes of more than 1000 μg/day.

Because of the severe kidney damage (Chapter 9, Section IV.B.2.c), the renal cortex concentrations were relatively low (Chapter 9, Tables 5 and 6).

D. Epidemiological Studies

The first epidemiological studies of Itai-itai disease in 1962 to 1965 (Appendix) were limited to the area along the Jintsu river (where patients had been found) and one control area. A total of 1100 people (mainly women) over age 40 were studied in the exposed area and 508 women over age 40 in the control area. Not one single patient or suspected patient was found in the control area.

In 1967 a somewhat larger area was studied and all men and women over age 30 were included in the target group. A three-stage screening process was carried out on 6717 people (Appendix). The first stage involved a questionnaire about diseases or symptoms in the family and an analysis of protein and glucose in urine. The combined results were assessed (Appendix) and those with suspected Itai-itai disease were selected for the second screening.

This second screening included a medical interview, clinical examination, and X-ray of the upper arm and shoulder. Out of 1911 people selected for these tests, 1400 participated and of them 451 were selected for a third screening. This selection was mainly based on alterations in bone. The last stage screening included further X-rays, orthopedic examination, and more detailed urine and blood analysis (calcium, phosphorus, alkaline phosphatase, etc.). Fifty cases of Itai-itai disease were diagnosed using a set of criteria, which included renal tubular damage and bone changes similiar to osteomalacia (Appendix).

Detailed prevalence or incidence rates in relation to age or dose have not been calculated, but no cases were found under age 45 and only one patient was male. It appears from the descriptive data (Table 11) that the prevalence increased with age. The prevalence among women over age 50 was above 20% in certain hamlets within the afflicted area (Appendix, Figure 6).

Analysis of cadmium in rice field soil (Appendix, Figure 7) showed that hamlets with the highest prevalence of Itai-itai disease also had the highest cadmium contamination levels.

A modified three-stage screening method for Itai-itai disease was used in ten other cadmium-polluted areas. The details of the method are given in the Appendix. The final stage of the screening procedure involves a Committee decision as to whether an individual case can be diagnosed as Itai-itai disease (Japanese Environment Agency Committee for Differential Diagnosis of Itai-itai Disease and Cadmium Poisoning).

A number of cases with bone and kidney effects living in other polluted areas than the Fuchu area have been considered by the Committee during the years, but none of them has been officially diagnosed as Itai-itai disease. There was one autopsied case

Table 11

NUMBER OF ITAI-ITAI PATIENTS REGISTERED BY TOYAMA
PREFECTURE HEALTH AUTHORITIES, SEPTEMBER 1968[59]

Age	I M	I F	i M	i F	(i) M	(i) F	O_{ob} M	O_{ob} F	Total M	Total F	Total (both sexes)
30—39											
40—49		2				1		2		5	5
50—59		6		2			3	14	3	22	25
60—69		31		14		5	12	40	12	90	102
70—79		12		5		6	14	30	14	53	67
80—89		2	1	6		1	3	5	4	14	18
Total		53	1	27		13	32	91	33	184	217

Note: I = definitely Itai-itai patients (typical bone manifestations on X-ray); i = people deeply suspected for Itai-itai disease (bone signs but not typical); (i) = people suspected for Itai-itai disease (slight bone signs on X-ray); O_{ob} = people needing follow-up (urinary and/ or blood signs only).

reported by Takebayashi[94] which clearly had osteomalacia as well as all the other signs of Itai-itai disease. The report stated that as this was the first autopsy case with such findings, "more detailed studies of several autopsy studies are required" before any link to cadmium can be proven. It was later found (Tsuchiya, personal communication) that a bone biopsy 2 years earlier failed to show osteomalacia. The Committee concluded that this proved that the case was not Itai-itai disease (Tsuchiya, personal communication).

E. Recent Cases of Bone Effects

In Toyama Prefecture, where the Itai-itai disease was first detected, the local prefectural health office has continued screening studies in the Fuchu area. Since 1967 about 200 cases have been diagnosed using the original diagnostic criteria (Appendix). A few cases are still diagnosed each year (Table 12). No control groups are studied though, so it is not known whether some of the diagnosed cases are unrelated to cadmium exposure. The sudden increase in "observation cases" in 1972 was the result of changes in diagnostic criteria and specific surveys that year. Since 1979, due to the establishment of a medical school in Toyama Prefecture, renewed efforts have been made to carry out autopsies on deceased Itai-itai disease patients. Twenty-seven cases in the age range 68 to 94 years were autopsied from 1979 to 1982.[54] All had some degree of osteomalacia. Ten had very severe findings. All 27 also had severe to very severe osteoporosis and all except 3 had clear histopathological signs of renal tubular damage. Most cases had been treated with large doses of vitamin D including the three without clear renal tubular damage. They had received 9 to 32 million IU of vitamin D. Four cases had received no vitamin D treatment, but all cases had renal tubular damage. Thus, this damage can not have been caused by the vitamin D treatment as suggested by some.[100] The autopsy findings in these recent Itai-itai disease patients from Fuchu area Toyama Prefecture are similar to those reported in the earlier cases (Chapter 9, Tables 5 and 6).

Five case reports of cadmium-exposed people with bone effects in the Ikuno area of Hyogo Prefecture have been published.[75] They were all living close to the Ichi river, which was polluted from the Ikuno mine. All five cases complained about severe bone pains similar to those in patients from Fuchu area.

They also had very low density bones as seen on X-rays[75] and fractures or dense

Table 12

NUMBER OF ITAI-ITAI DISEASE PATIENTS DIAGNOSED OR DECEASED IN DIFFERENT YEARS[45]

Year	Recognized Itai-itai disease			"Observation cases"		
	Diagnosed	Died	Remaining	Diagnosed	Deleted	Remaining
1967	73	3	70	155	2	153
1968	44	12	102	33	50	136
1969	3	8	97	1	113	24
1970	4	4	97	2	23	3
1971	1	5	93	1	1	3
1972	0	11	82	138	5	136
1973	1	2	81	21	26	131
1974	3	12	72	7	17	121
1975	0	7	65	4	24	101
1976	0	6	59	5	5	101
1977	0	9	50	4	8	97
1978	1	1	50	4	13	88
1979	0	5	45	1	6	83
1980	1	5	41	3	20	66
1981	1	3	39	1	7	60
1982	0	4	35	2	6	56
Total 1967—1982	132	97		382	326	

areas in the cortical bone, which according to the authors were similar to the "pseudo-fractures" of osteomalacia (Figure 13). All five cases also had high urinary excretion of low molecular weight protein (RBP) (Table 13) and a very high serum alkaline phosphatase. The urine data indicated that they had renal tubular damage and the blood data that these bone findings may be in agreement with a diagnosis of osteomalacia. The urinary cadmium excretion in all cases was as high as in industrial workers with chronic cadmium poisoning. The cases were referred to the Committee for Differential Diagnosis of the Environment Agency and it was concluded that none of them had Itai-itai disease.[100] The five cases were treated with large doses of vitamin D in combination with anabolic steroids. It took 2 to 3 months of treatment before the pain disappeared and in some cases treatment for 6 months or more made it possible for the patients to stand up and walk unaided.

One of the five cases (Mrs. A) was studied by autopsy.[75] The cadmium concentration in liver (75 mg/kg) (Chapter 9, Table 5, case 1N) was as high as for Itai-itai disease patients in Fuchu. The renal cortex level (53 mg/kg) was low, as expected, because this patient had severe renal tubular damage. Autopsy material from the same patient was tested by Miyasaki.[71] The bones showed osteoporosis, but almost no osteoid tissue. The latter would have been a sign of current osteomalacia. However, as she had been treated with large doses of vitamin D (Shibata, personal communication), most of the osteoid tissue is likely to have disappeared as a result of the treatment. The tubular epithelium in the kidneys showed "atrophy, desquamation, cell dissociation, vacuolization, and hydropic swelling". Miyasaki still concluded that "the present case had only slight interstitial fibrosis". Cadmium analysis of the tissues showed similar results to those above. In 1982 only one of the five cases was still alive (Shibata, personal communication). It was not possible to carry out autopsies on any of the other cases.

There have also been some disputed cases of Itai-itai disease in Tsushima area of Nagasaki Prefecture. One case several years ago was never investigated in detail (Kobayashi, personal communication), but recently another case was autopsied.[94]

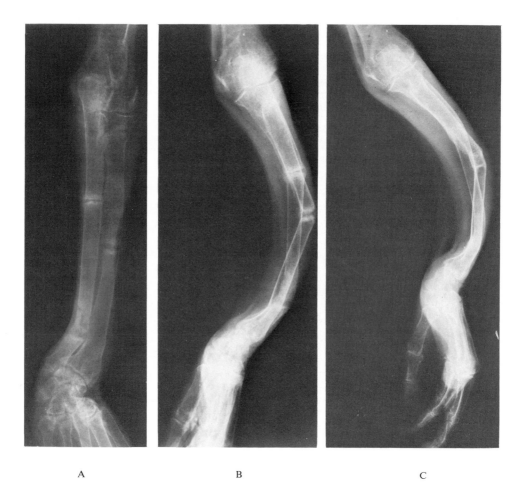

A   B   C

FIGURE 13.   X-rays of the left radius and ulna of Mrs. A. before (A, B) and after (C) treatment for 360 days. (From Nogawa, K., Ishizaki, A., Fukushima, M., Shibata, I., and Hagino, N., *Environ. Res.*, 10, 280—307, 1975. With permission.)

The case was a woman, aged 75 at autopsy, who had been screened for cadmium exposure since 1962. Since that time her urine cadmium concentration decreased progressively from 11.5 to about 3 to 4 $\mu g/\ell$ in 1970 to 1974 (Appendix, Table 5). She had persistent proteinuria and glycosuria and also very high excretion of low molecular weight proteins (Appendix, Table 5). The plasma alkaline phosphatase level increased with time (from 7 to 14 KA units in 1967 to 1971 up to 26.5 KA units in 1974), while at the same time plasma calcium was relatively low (8 to 9.5 mg/100 mℓ). While there was a decreased concentration of most amino acids in urine between 1970 and 1973, the concentration of hydroxyproline and proline increased (Appendix, Table 6). The autopsy showed severe renal tubular damage and a mixture of osteoporosis and osteomalacia in bone.

There is one more area where suspected cases have been found (Kakehashi river basin of Ishikawa Prefecture). Of 31 people studied in detail with blood and urine tests, bone X-ray, and clinical examinations, 11 had osteoporosis and 2 had osteomalacia-like "Looser's zones" as seen by X-ray.[37] All the 31 cases had elevated low molecular weight proteinuria (20 to 120 mg β_2-microglobulin per liter urine), and all except 1 case had a tubular reabsorption of inorganic phosphorus (TRP) below the normal minimum

Table 13

BIOCHEMICAL DATA OF FIVE CASES WITH BONE DISEASE IN THE
POLLUTED IKUNO AREA. DATA FROM THE FIRST SAMPLES
COLLECTED FROM WHICH ALL MEASUREMENTS WERE AVAILABLE[75]

Case no.	1	2	3	4	5	"Normal" range
Serum						
Calcium (mg/ℓ)	83	85	105	100	101	90—110
Phosphorus (mg/ℓ)	28	23	26	30	28	30—40
Alkaline phosphatase (King-Armstrong units)	41	44	33	35	47	3—13
Urine						
Calcium (mg/ℓ)	113	38	25	55	203	
Phosphorus (mg/ℓ)	211	223	225	277	581	
RBP (scale)	+++	+++	+++	+++	+++	—
Cadmium (μg/g creatinine)	38	46	25	22	26	0.1—2.0
Creatinine (mg/ℓ)	180	330	720	270	690	

of 83%. The two lowest TRP values (4.7 and 45%) were recorded for the cases with osteomalacia-like "Looser's zones". Serum alkaline phosphatase was increased in several cases. The highest value was found in one of the cases with "Looser's zones".

These studies of the cadmium-polluted areas in Japan have shown that in a small number of cases with severe renal tubular damage, osteomalacia and/or osteoporosis is also found. The general nutritional and environmental conditions have improved in Japan since the time of the outbreak of Itai-itai disease in Toyama Prefecture. It is therefore to be expected that the number of cases of this disease would be very low if it now occurred in other cadmium-polluted areas. Unfortunately, none of the studies to date has analyzed the data on bone findings in both the polluted and control areas in such a way that a quantitative assessment of the difference in prevalence could be carried out.

VI. MECHANISMS FOR DEVELOPMENT OF CADMIUM-INDUCED BONE EFFECTS

The preceding sections have shown that cadmium exposure in animals and humans may affect calcium and vitamin D metabolism and in animals effects on the bone collagen have also been shown. The points where cadmium could exert its effects are shown in Figure 1. Some of these effects are secondary to the renal damage induced by cadmium (Chapter 9).

The exact biochemical basis for the effects of cadmium is not known. Several of the cadmium effects will decrease the calcium available for mineralization of bones (Section III.B). In addition, the cadmium-induced inhibition of hydroxylation of vitamin D into Kalcitriol® in the kidney (Section III.A) will decrease the availability of a "hormone" which is necessary for the proper mineralization process. The former effect is likely to manifest itself as osteoporosis and the latter as osteomalacia. In addition, cadmium may affect collagen via lysyl-oxidase (Section III.C).

Lysyl-oxidase is a copper-dependent enzyme (see Section II.D) and it may be that cadmium replaces copper in the enzyme and inhibits its activity (Iguchi and Sano, personal communication). It has repeatedly been shown (Section III.D) that a combination of osteoporosis and osteomalacia can be produced in animals by cadmium ingestion. The findings are dose dependent, though. The higher the dose, and the shorter the time the animal is exposed, the more likely it is that the main bone finding will be osteoporosis. After long-term low level exposure, osteomalacia will develop, particularly if the animal's vitamin D intake is marginal. The recent studies on monkeys have not been carried on long enough to give conclusive results (Section III.E), but in one group with low calcium intake and cadmium exposure there were emerging renal effects and signs of osteomalacia.

No clinical findings that can be directly related to the effect of cadmium on collagen metabolism (Section III.C) have been reported so far. However, the studies to date have not been looking specifically for such findings.

In human beings, as in animals, different effects are likely to be produced depending on the cadmium exposure conditions and the exposure to calcium, vitamin D, and sunlight, as well as the general nutritional status of those exposed.

If cadmium exposure comes via food, the high cadmium content in the intestines will interfere with the calcium absorption via the intestinal mucosa cells (see Section III.B). This will decrease the calcium available to the body, but it is likely to lead to clinical effects only if the calcium intake is low or marginal. In industrial workers exposed to cadmium via inhalation, there may also be a high gastrointestinal intake of cadmium due to swallowing of cadmium dust cleared from the lungs and contamination of hands and food.[3]

As cadmium accumulates in the renal cortex, it will interfere with calcium metabolism in three different ways. Adenylcyclase will be inhibited, which causes PTH to be less effective in stimulating vitamin D hydroxylation. The vitamin D hydroxylase will be inhibited which further decreases the production of Kalcitriol®, a hormone necessary for cadmium absorption via the intestines and for the proper mineralization of bone. Further, the renal reabsorption of calcium will decrease and urinary excretion increase causing losses of calcium from the body.

Calcium is an essential element for the function of muscles and nerves. The bone contains a sizeable reserve of calcium to maintain the serum calcium level above a minimum. It is therefore natural that the effects of cadmium on calcium storage in the bones will take some time to manifest themselves as clinical bone disease. If the exposed population has a high calcium intake via food, clinical bone effects may not develop at all. Instead the victim will develop renal stones (Chapter 9, Section IV.B.1.c).

In susceptible populations, clinical osteoporosis and/or osteomalacia have developed after long-term cadmium exposure (Sections IV and V). The close epidemiological association between cadmium exposure and bone effects in the Fuchu area in Japan (Section V) indicates that cadmium exposure is a prerequisite for the development of the disease. It is possible that suboptimal dietary conditions were also necessary to produce the bone effects. In Japan in the 1940s many people must have been subject to these conditions. Still, among the number of cadmium-polluted areas in Japan, only Fuchu had such a high occurrence of the severe bone disease. Possible reasons for this observation could be the following:

1. Fuchu had the most severely polluted area and one of the largest exposed populations.[100]
2. Fuchu had the longest high exposure duration (Chapter 9, Table 12).
3. People were using polluted river water in the household as drinking water.
4. Special attention was paid to the diagnosis of this disease in this area.

VII. DOSE-RESPONSE RELATIONSHIPS

As the effects of cadmium on bones are mainly indirect and due to effects of cadmium on the kidneys or the intestines, the relationship between cadmium concentrations in bone and the occurrence of the effects may be misleading. The daily intake of cadmium (determining the cadmium concentration in the intestines) and the body burden of cadmium (determining the cadmium concentration in the kidney) are likely to be of greater interest.

In the Fuchu area, where most of the Itai-itai disease patients lived, the daily cadmium intake of the patients has been estimated to be in the range 600 to 2000 μg (in 1967), but earlier exposure may have been higher or lower. No exact data are available. It appears (Chapter 9, Table 12) that none of the other polluted areas had as high exposure as did Fuchu. Thus, the low number of Itai-itai cases in the other polluted areas indicates that the intake may have to exceed about 1 mg/day for an extended period (decades) before cadmium-induced bone disease develops.

The body burdens of Itai-itai disease patients and osteopenia cases in industry were similar. The kidney data do not reflect body burden because they decrease when the renal damage occurs (Chapter 9, Section IV.B.2.b), but the liver data are not likely to decrease much if the exposure is sustained. Due to a change in eating habits in the terminal period before autopsy, the daily cadmium intake may be reduced and the liver concentrations may decrease at a half-time of about 6 years (Chapter 7, Section IV.C). Therefore, in a year there may be a 10% decrease.

The reported cadmium concentrations in liver of Itai-itai patients and osteopenia cases in industry vary between 60 and 150 mg/kg (see Sections IV and V.C). Due to terminal changes, the range 100 to 200 mg/kg may better represent the peak liver concentration. At middle age, in Japanese nonpolluted areas, the average cadmium concentration in liver is about 5 mg/kg[55] when the daily intake is about 50 μg. The liver cadmium concentrations in the cases with cadmium-induced bone disease would therefore correspond to cadmium intakes via food in the range 1000 to 2000 μg Cd per day.

Due to lack of more detailed data on the prevalence and/or incidence of bone effects in industry as well as in the general environment, a quantitative organ dose-response relationship cannot be established.

VIII. SUMMARY AND CONCLUSIONS

In animals, cadmium has been shown to affect vitamin D metabolism, calcium metabolism, and collagen metabolism. These effects manifest themselves as osteomalacia and osteoporosis, and have been produced experimentally in several species. The combined effect of cadmium, poor nutrition, and low vitamin D intake may be particularly harmful.

These findings in animal experiments are supported by the epidemiological data. About 50 industrial workers have suffered osteomalacia or osteoporosis after cadmium exposure. The incidence appears to have peaked 30 to 40 years ago when the dietary conditions may have been deficient in the countries with reported cases. Similarly, the highest incidence of bone effects in the general environment appears to have occurred about 30 to 40 years ago in Japan. A total of about 150 definite cases has been found there. The etiology of the bone disease in many of these cases is still debated vigorously. We conclude that cadmium was a necessary factor for the development of the high incidence of osteomalacia and osteoporosis in one polluted area. Poor nutrition may also have been a necessary factor, but the epidemiological association between cadmium exposure and the bone disease shows that poor nutrition alone cannot explain the observations. Sporadic cases have been seen in other polluted areas where the likely exposure levels were lower during the period 30 to 40 years ago when dietary deficiencies were likely to be common.

The quantitative organ dose-response relationships for the bone effects are not known, and the cadmium levels in the intestines, the kidney, and the serum may be of greater importance for the occurrence of effects than the levels in the bone. It appears that a long-term daily intake via food or water of more than 1 mg cadmium may be needed to cause the severe bone disease.

REFERENCES

1. Adams, R. G., Osteopathy associated with tubular nephropathy in employees in an alkaline battery factory, in *Cadmium-Induced Osteopathy,* Shigematsu, I. and Nomiyama, K., Eds., Japanese Public Health Association, Tokyo, 1980, 66—73.
2. Adams, R. G. Harrison, J. F., and Scott, P., The development of cadmium-induced proteinuria, impaired renal function, and osteomalacia in alkaline battery workers, *Q. J. Med.,* 38, 425—443, 1969.
3. Adamsson, E., Piscator, M., and Nogawa, K., Pulmonary and gastrointestinal exposure to cadmium oxide dust in battery workers, *Environ. Health Perspect.,* 28, 219—222, 1979.
4. Akahori, F. and Masaoka, T., Further studies on effects of dietary cadmium on rhesus monkeys. II. General health check, in *Recent Studies on Health Effects of Cadmium in Japan,* Environment Agency, Tokyo, 1981, 10—27.
5. Altman, P. L. and Dittmer, D. S., Eds., *Biology Data Book,* Federation of American Societies for Experimental Biology, Bethesda, Md., 1972.
6. Anderson, C. and Danylchuk, K. D., The effect of chronic excess zinc administration on the Haversian bone remodelling system and its possible relationship to "Itai-itai" disease, *Environ. Res.,* 20, 351—357, 1979.
7. Ando, M., Shimizu, M., Sayato, Y., Taniumura, A., and Tobe, M., The inhibition of vitamin D-stimulated intestinal calcium transport in rats after continuous oral administration of cadmium, *Toxicol. Appl. Pharmacol.,* 61, 297—301, 1981.
8. Avioli, L. V., Osteoporosis: pathogenesis and therapy, in *Metabolic Bone Disease,* Vol. 1, Avioli, L. V. and Krane, S. M., Eds., Academic Press, New York, 1977, 307—386.
9. Avioli, L. V. and Krane, S. M., Eds., *Metabolic Bone Disease,* Vol. 1, Academic Press, New York, 1977.
10. Barrow, M. V., Simpson, C. F., and Miller, E. J., Lathyrism: a review, *Q. Rev. Biol.,* 49, 101—128, 1974.
11. Bengtsson, G.-E., Carlin, CH., Larsson, A., and Svanberg, O., Vertebral damage in minnows, *Phoxinus phoxinus* L., exposed to cadmium, *Ambio,* 4, 166—168, 1975.
12. Blainey, J. D., Adams, R. G., Brewer, D. B., and Harvey, T. C., Cadmium-induced osteomalacia, *Br. J. Ind. Med.,* 37, 278—284, 1980.
13. Bonnell, J. A., Emphysema and proteinuria in men casting copper-cadmium alloys, *Br. J. Ind. Med.,* 12, 181—197, 1955.
14. Bonner, F. W., King, L. J., and Parke, D. V., Cadmium-induced reduction of bone alkaline phosphatase and its prevention by zinc, *Chem. Biol. Interact.,* 29, 369—372, 1980.
15. Broadus, A. E., Kaminsky, N. I., Hardman, J. G., Sutherland, E. W., and Liddle, G. W., Kinetic parameters and renal clearances of plasma adenosine 3′,5′-monophosphate and guanosine 3′,5′-monophosphate in man, *J. Clin. Invest.,* 49, 2220—2236, 1970.
16. Ceresa, C., An experimental study of cadmium intoxication, *Med. Lav.,* 36, 71—88, 1945 (in Italian).
17. Coburn, J. W., Brickman, A. S., Sherrard, D. J., Singer, F. R., Baylink, D. J., Wong, E. C. G., Massry, S. G., and Norman, A., Clinical efficacy of 1,25-dihydroxy-vitamin D3 in renal osteodystrophy, in *Vitamin D: Biochemical, Chemical and Clinical Aspects Related to Calcium Metabolism,* Normna, A. W., Schaefer, K., Coburn, J. W., DeLuca, H. F., Fraser, D., Grigoleit, H., and Herrath, D., Eds., Walter DeGruyter, Berlin, 1977, 657—666.
18. Cundy, T., Kanis, J. A., Heynen, G., Earnshaw, H., Clemens, T. L., O'Riordan, J. L. H., Merritt, A. L., and Compston, J. E., Lack of direct effect of 1,25-dihydroxy vitamin D3 on mineralisation of bone and secretion of parathyroid hormone, *Clin. Sci.,* 59, 15p, 1980.
19. De Luca, H. F., *Vitamin D. Metabolism and Function,* Springer-Verlag, Basel, 1979.
20. Dent, C. E. and Stamp, T. C. B., Vitamin D, rickets and osteomalacia, in *Metabolic Bone Disease,* Avioli, L. V. and Krane, S. M., Eds., Academic Press, New York, 1977, 237—306.

21. Elinder, C.-G., Early Effects of Cadmium Accumulation in Kidney Cortex, with Special Reference to Cadmium and Zinc Interactions, Doctoral thesis, Department of Environmental Hygiene, Karolinska Institute, Stockholm, Sweden, 1979.

22. Feldman, S. L. and Cousins, R. J., Influence of cadmium on the metabolism of 25-hydroxycholecalciferol in chicks, *Nutr. Rep. Int.,* 8, 251—259, 1973.

23. Friberg, L., Health hazards in the manufacture of alkaline accumulators with special reference to chronic cadmium poisoning. Doctoral thesis, *Acta Med. Scand.,* 138(Suppl. 240), 1—124, 1950.

24. Friberg, L., Piscator, M., and Nordberg, G., *Cadmium in the Environment,* Chemical Rubber Co., Cleveland, 1971.

25. Friberg, L., Piscator, M., Nordberg, G. F., and Kjellström, T., *Cadmium in the Environment,* 2nd ed., CRC Press, Boca Raton, Fla., 1974.

26. Fullmer, C. S., Oku, T., and Wasserman, R. H., Effect of cadmium administration on intestinal calcium absorption and vitamin D-dependent calcium-binding protein, *Environ. Res.,* 22, 386—399, 1980.

27. Gervais, J. and Delpech, P., Cadmium intoxication, *Arch. Mal. Prof. Med. Trav. Secur. Soc.,* 24, 803—816, 1963 (in French).

28. Hagino, N., About investigations on Itai-itai disease, *J. Toyama Med. Assoc.,* Dec. 21, 7, 1957 (in Japanese).

29. Hagino, N. and Yoshioka, K., A study on the cause of Itai-itai disease, *J. Jpn. Orthoped. Assoc.,* 35, 812—815, 1961 (in Japanese).

30. Haussler, M. R. and McCain, T. A., Basic and clinical concepts related to vitamin D metabolism and action, *New Engl. J. Med.,* 297, 974—983, 1041—1050, 1977.

31. Hirota, M., Influences of cadmium on the bone salt metabolism, *Acta Sci. Med. Univ. Gifu,* 19, 82—115, 1971 (in Japanese).

32. Horstowa, H., Sikorski, M., and Tyborski, H., Chronic cadmium poisoning in the clinical and radiological picture, *Med. Pracy,* 17, 13—15, 1966 (in Czech).

33. Howard, G. A. and Baylink, D. J., Matrix formation and osteoid maturation in vitamin D deficient rats made normocalcaemic by dietary means, *Miner. Electrolyte Metab.,* 3, 44—50, 1980.

34. Howard, G. A., Turner, R. T., Sherrard, D. J., and Baylink, D. J., Human bone cells in culture metabolize 25-hydroxy-vitamin D_3 to 1,25-dihydroxy vitamin D_3 and 24,25-dihydroxy vitamin D_3, *J. Biol. Chem.,* 256(5), 7738—7740, 1981.

35. Iguchi, H. and Sano, H., Effect of cadmium on bone lysyloxidase, *Jpn. J. Hyg.,* 35, 224, 1980 (in Japanese).

36. Iguchi, H. and Sano, S., Effect of cadmium on the bone collagen metabolism of rat, *Toxicol. Appl. Pharmacol.,* 62(1), 126—136, 1982.

37. Ishikawa Prefecture, *White Paper on the Environment of Ishikawa Prefecture,* Prefectural Authority, Kanazawa, 1982, 189—192 (in Japanese).

38. Ishizaki, A., Tanabe, S., Matsuda, S., and Sakamoto, M., *Jpn. J. Hyg.,* 20, 398—404, 1966 (in Japanese).

39. Itokawa, Y., Abe, T., Tabei, R., and Tanaka, S., Renal and skeletal lesions in experimental cadmium poisoning, *Arch. Environ. Health,* 28, 149—154, 1974.

40. Itokawa, Y., Yagi, N., Kaito, H., Kamohara, K., and Fujiwara, K., Influence of diet on the induction of hepatic ceroid pigment in rats by polychlorinated biphenyls, *Toxicol. Appl. Pharmacol.,* 36, 131—141, 1976.

41. Japanese Ministry of Health and Welfare, Opinion of the Welfare Ministry with regard to "ouch-ouch" disease and its causes, May 9, 1968 (in Japanese).

42. Jones, H. S. and Fowler, B. A., Biological interactions of cadmium with calcium, *Ann. N.Y. Acad. Sci.,* 355, 309—318, 1980.

43. Kacew, S., Merali, Z., and Singhal, R. L., Comparison of the subacute effects of cadmium exposure upon nucleic acid, cyclic adenosine 3′,5′-monophosphate and polyamine metabolism in lung and kidney cortex, *Toxicol. Appl. Pharmacol.,* 38, 145—156, 1976.

44. Kanis, J. A., Brown, C. B., Cameron, E. C., Cundy, T., Platts, M. M., Paterson, A., and Russell, R. G. G., The role of vitamin D metabolites in the osteomalacia of renal disease, *Curr. Med. Res. Opin.,* 7, 294—315, 1981.

45. Kato, T., Overall analysis of the occurrence of Itai-itai disease patients and their mortality, in *Kankyo Hoken Report No. 49,* Japanese Public Health Association, Tokyo, 1983, 121—124 (in Japanese).

46. Kawai, K., Fukuda, K., and Kimura, M., Morphological alterations in experimental cadmium exposure with special reference to the onset of renal lesion, in *Effects and Dose-Response Relationships of Toxic Metals,* Nordberg, G. F., Ed., Elsevier, Amsterdam, 1976, 343—370.

47. Kawamura, J., Yoshida, O., Nishino, K., and Itokawa, Y., Disturbances in kidney functions and calcium and phosphate metabolism in cadmium-poisoned rats, *Nephron,* 20, 101—110, 1978.

48. Kazantzis, G., Some long-term effects of cadmium on the human kidney, in *Cadmium 77, Proc. 1st Int. Cadmium Conf. San Francisco 1977,* Metal Bulletin Ltd., London, 1978, 194—198.

49. Kazantzis, G., Renal tubular dysfunction and abnormalities of calcium metabolism in cadmium workers, *Environ. Health Perspect.*, 28, 155—159, 1979.

50. Kimura, M., Otaki, N., Yoshiki, S., Suzuki, M., Horiuchi, N., and Suda, T., The isolation of metallothionein and its protective role in cadmium poisoning, *Arch. Biochem. Biophys.*, 165, 340—348, 1974.

51. Kimura, M., Watanabe, M., and Otaki, N., Effects of nutritional factors and cadmium exposure in monkeys. VII. Analysis of heavy metals, in *Kankyo Hoken Report No. 49,* Japanese Public Health Association, Tokyo, 1983a, 76—83 (in Japanese).

52. Kimura, M., Watanabe, M., and Otaki, N., Determination of heavy metals and beta-2-microglobulin, in *The Influence of Nutritional Factors on Cadmium Administration in Monkeys,* Tertiary Monkey Experiment Team, Japanese Public Health Association, Tokyo, March 1983b, 66—82.

53. Kimura, M., Watanabe, M., and Otaki, N., Effects of nutritional factors and cadmium exposure on monkeys. VII. Analysis of heavy metals. Proc. of the 1984 Cadmium Research Conference organised by the Japanese Public Health Association, *Kankyo Hoken Report,* in press (in Japanese).

54. Kitagawa, S., Mitane, Y., and Murai, E., Twenty-seven autopsy cases of Itai-itai disease in Toyama Prefecture, in *Kankyo Hoken Report No. 49.* Japanese Public Health Association, Tokyo, 1983, 140—145, (in Japanese).

55. Kjellström, T., Exposure and accumulation of cadmium in populations from Japan, the United States and Sweden, *Environ. Health Perspect.*, 28, 169—197, 1979.

56. Kjellström, T., Lööw, H., and Sjöberg, H.-E., Renal adenylcyclase activity and urinary excretion of cyclic AMP in cadmium-exposed rats, in press.

57. Kodicek, E., Recent advances in vitamin-D metabolism, in *Clinics in Endocrinology and Metabolism, I.* MacIntyre, I., Ed., W. B. Saunders, London, 1972, 305—323.

58. Kobayashi, J., Effects of cadmium on calcium metabolism of rats, in *Trace Substances in Environmental Health — VII,* Hemphill, D. D., Ed., University of Missouri, Columbia, 1974, 263—266.

59. Kubota, J., Tsuchiya, K., Koizumi, A., and Inoue, T., Environmental Health in Japan (Exhibition, Tokyo), The Organizing Committee of the 16th International Congress on Occupational Health, obtainable from the "Occupational Health Service Center", Japanese Industrial Safety Association, Tokyo, September 22—27, 1969.

60. Larsson, S.-E. and Piscator, M., Effect of cadmium on skeletal tissue in normal and calcium-deficient rats, *Isr. J. Med. Sci.*, 7, 495—497, 1971.

61. Lawson, D. E. M., *Vitamin D,* Academic Press, London, 1978.

62. Lorentzon, R. and Larsson, S.-E., Vitamin D metabolism in adult rats at low and normal calcium intake and the effect of cadmium exposure, *Clin. Sci. Mol. Med.*, 30, 559—562, 1977.

63. Maehara, T., Studies on the calcium metabolism of rat bone after long-term administration of heavy metal (cadmium), *J. Jpn. Orthoped. Assoc.*, 42, 287—300, 1968.

64. Marouby, J., Osteopathy in cadmium workers in France, in *Cadmium-Induced Osteopathy,* Shigematsu, I. and Nomiyama, K., Eds., Japanese Public Health Association, Tokyo, 1980, 34—77.

65. Matsuda, S., *J. Juzen Med. Soc.,* 76, 239, 1968 (in Japanese with English abstract).

66. Matsue, R., Fukuyama, Y., Ishimoto, M., Kubota, K., and Takayanagi, N., Bone damage in experimental cadmium poisoning by oral administration in the mouse, *Med. Biol.,* 81, 33—36, 1970.

67. Memmos, D. E., Eastwood, J. B., Harris, E., O'Grady, A., and De Wardener, H. E., Response of uremic osteoid to vitamin D., *Kidney Int.,* 21(Suppl. 11), 50—54, 1982.

68. Merali, Z. and Singhal, R. L., Influence of chronic exposure to cadmium on hepatic and renal cyclic AMP-protein kinase system, *Toxicology,* 4, 207—214, 1975.

69. Miyahara, T., Komurasaki, T., and Kozuka, H., Inhibitory effect of cadmium on collagenous peptide synthesis of embryonic chick bone in tissue culture, *Toxicol. Lett.,* 6(3), 137—139, 1980a.

70. Miyahara, T., Miyakoshi, M., Saito, Y., and Kozuka, H., Influence of poisonous metals on bone metabolism. III. The effect of cadmium on bone resorption in tissue culture, *Toxicol. Appl. Pharmacol.,* 55, 477—483, 1980b.

71. Miyasaki, K., Autopsy findings and discussion of the effects of cadmium on the human body, in *Cadmium Studies in Japan — A Review,* Tsuchiya, K., Ed., Elsevier, Amsterdam, 1978, 309—318.

72. Muto, Y., Shidoji, Y., and Suzuki, S., Cadmium induced bone changes, *Nutr. Food,* 28, 1—5, 1975 (in Japanese).

73. Nakagawa, S., A study of osteomalacia in Toyama Prefecture (so-called Itai-itai disease), *J. Radiol. Phys. Ther. Univ. Kanizawa,* 56, 1—51, 1960 (in Japanese with English summary).

74. Nicaud, P., Lafitte, A., and Gros, A., Symptoms of chronic cadmium intoxication, *Arch. Mal. Prof. Med. Trav. Secur. Soc.,* 4, 192—202, 1942 (in French).

75. Nogawa, K., Ishizaki, A., Fukushima, M., Shibata, I., and Hagino, N., Studies on the women with acquired Fanconi syndrome observed in the Ichi River basin polluted by cadmium, *Environ. Res.,* 10, 280—307, 1975.

76. Nogawa, K., Kobayashi, E., and Konishi, F., Comparison of bone lesions in chronic cadmium intoxication and vitamin D deficiency, *Environ. Res.,* 24, 233—249, 1981.

77. Nomiyama, K., Experimental studies on animals. *In vivo* experiments, in *Cadmium Studies in Japan: A Review,* Tsuchiya, K., Ed., Elsevier, Amsterdam, 1978. 47—128.

78. Nomiyama, K., Literature review on cadmium-induced osteopathy among cadmium workers in Japan and other countries, in *Cadmium-Induced Osteopathy,* Shigematsu, I. and Nomiyama, K., Eds., Japanese Public Health Association, Tokyo, 1980, 57—59.

79. Nomura, T., Kimura, M., Tanioka, Y., Tanimoto, Y., Tazaki, H., Suda, T., Yoshiki, S., Shima, E., and Mizaki, J., Effect of nutritional cadmium exposure on monkeys. I. Introduction, in *Kankyo Hoken Report No. 49,* Japanese Public Health Association, Tokyo, 1983, 39—45 (in Japanese).

80. Nomura, T., Kimura, M., Tanioka, Y., Tanimoto, Y., Tazaki, H., Suda, T., Yoshiki, S., Hata, M., and Deguchi, N., Effects of nutritional factors and cadmium exposure in monkeys. I. Introduction, Proc. of the 1984 Cadmium Research Conference organised by the Japanese Public Health Association, *Kankyo Hoken Report,* in press (in Japanese).

81. Ogoshi, K., Aoki, Y., Kurumatani, N., Moriyama, R., and Nanzai, Y., The changes in mechanical properties of rat bones under the low dose level of cadmium. I. The compressive properties, *Jpn. J. Hyg.,* 36, 584—595, 1981 (in Japanese).

82. Pujol, M., Arlet, J., Bollinelli, R., and Charles, P., Renal tubular disorders in chronic cadmium poisoning, *Arch. Mal. Prof. Med. Trav. Secur. Soc.,* 31, 637—648, 1970 (in French).

83. Raisz, L. G., Bone metabolism and calcium regulation, in *Metabolic Bone Disease,* Vol. 1, Avioli, L. V. and Krane, S. M., Eds., Academic Press, New York, 1977, 1—49.

84. Sano, H. and Iguchi, H., Discussion of urinary amino acid analysis in relation to changes in proline metabolism and Itai-itai disease, in *Kankyo Hoken Report No. 31,* Japanese Public Health Association, Tokyo, 1974, 95—96 (in Japanese).

85. Scott, R., Mills, E. A., Fell, G. S., Hussain, F. E. R., Yates, A. J., Paterson, P. J., McKirdy, A., Ottoway, J. M., Fitzgerald-Finch, O. P., Lamont, A., and Roxburgh, S., Clinical and biochemical abnormalities in coppersmiths exposed to cadmium, *Lancet,* August 21, 396—398, 1976.

86. Scott, R., Patterson, P. J., Burns, R., Ottoway, J. M., Hussain, F. E. R., Fell, G. S., Dumbuya, S., and Iqbal, M., Hypercalciuria related to cadmium exposure, *Urology,* 11, 462—465, 1978.

87. Scott, R., Haywood, J. K., Boddy, K., Williams, E. D., Harvey, I., and Paterson, P. J., Whole body calcium deficit with cadmium-exposed workers in hypercalciuria, *Urology,* 15, 356—359, 1980.

88. Slorik, D. M., Adams, J. S., Neer, R. M., Horlick, M. F., and Potts, J. T., Jr., Deficient production of 1,25-dehydroxy-vitamin D in elderly osteoporotic patients, *New Engl. J. Med.,* 305, 372—374, 1981.

89. Suda, T., Horiuchi, N., Ogata, E., Ezawa, I., Otaki, N., and Kimura, M., Prevention by metallothionein of cadmium-induced inhibition of vitamin D activation reaction in kidney, *FEBS Lett.,* 42, 23—26, 1974.

90. Suda, T., Shiina, Y., and Abe, E., Effects of nutritional factors and cadmium exposure on monkeys. V. Plasma levels of vitamin D metabolites, in *Kankyo Hoken Report No. 49,* Japanese Public Health Association, Tokyo, 1983, 64—67 (in Japanese).

91. Sugawara, Ch. and Sugawara, N., Cadmium toxicity for rat intestine specially on the absorption of calcium and phosphorus, *Jpn. J. Hyg.,* 6, 511—516, 1974.

92. Sumino, K., Hayakawa, K., Shibata, T., and Kitamura, S., Heavy metals in normal Japanese tissues, *Arch. Environ. Health,* 30, 487—494, 1975.

93. Takashima, M., Moriwaki, S., and Itokawa, Y., Osteomalacic changes induced by long-term administration of cadmium to rats, *Toxicol. Appl. Pharmacol.,* 54, 223—228, 1980.

94. Takebayashi, S., First autopsy case, suspicious of cadmium intoxication, from the cadmium-polluted area in Tsushima, Nagasaki Prefecture, in *Cadmium-Induced Osteopathy,* Shigematsu, I. and Nomiyama, K., Eds., Japanese Public Health Association, Tokyo, 1980, 124—138.

95. Tarasenko, N. J. and Vorobjeva, R. S., Hygienic problems connected with the cadmium use, *Vestnik, Akad. Med. Nauk SSR* 28, 37—43, 1973 (in Russian with English summary).

96. Tarasenko, N. J., Vorobjeva, R. S., Shabalina, L. P., and Cvetkova, R. P., Cadmium in the environment, its effects on calcium metabolism, *Gig. Sanit.,* 9, 22—25, 1975 (in Russian).

97. Tazaki, H., Hata, M. Deguchi, N., and Fujioka, T., Effects of nutritional factors and cadmium exposure on monkeys. IV. Urinary and renal findings, Proc. of the 1984 Cadmium Research Conference organised by the Japanese Public Health Association, *Kankyo Hoken Report,* in press (in Japanese).

98. Tertiary Monkey Experiment Team, The influence of nutritional factors on cadmium administration in monkeys, in *Recent Studies on Health Effects of Cadmium in Japan,* Environment Agency, Tokyo, 1981, 211—300.

99. Tsuchiya, K., Causation of ouch-ouch disease, an introductory review. I. Nature of the disease, *Keio J. Med.,* 18, 181—194, 1969.

100. Tsuchiya, K., Ed., *Cadmium Studies in Japan — A Review,* Elsevier, Amsterdam, 1978.

101. Tsuruki, F., Otawara, Y., Wang, H. L., Moriuchi, S., and Hosoga, N., Inhibitory effect of cadmium on vitamin D stimulated calcium transport in rat duodenum *in vitro, J. Nutr. Sci. Vitaminol.,* 24, 237—242, 1978.

102. Vahter, M., Marafante, E., Lindgren, A., and Dencker, L., Tissue distribution and subcellular binding of arsenic in Marmoset monkeys after injection of [74]As-arsenite, *Arch. Toxicol.,* 51, 65—77, 1982.

103. Valetas, P., Cadmiose or Cadmium Intoxication, Doctoral Report, School of Medicine, University of Paris, 1946 (in French).

104. Washko, P. W. and Cousins, R. J., Effect of low dietary calcium on chronic cadmium toxicity in rats, *Nutr. Rep. Int.,* 11, 113—127, 1975.

105. Wasserman, R. H., Corradino, R. A., Taylor, A. N., and Morrissey, R. L., in *Cellular Mechanisms for Calcium Transfer and Homeostatis,* Nichols, G. and Wasserman, R. H., Eds., Academic Press, New York, 1971, 293—312.

106. Yamagata, N. and Shigematsu, I., Cadmium pollution in perspective, *Bull. Inst. Public Health (Tokyo),* 19, 1—27, 1970.

107. Yoshiki, S., Tanagisawa, T., Kimura, M., Otaki, N., Suzuki, M., and Suda, T., Bone and kidney lesions in experimental cadmium intoxication, *Arch. Environ. Health,* 30, 559—562, 1975.

108. Yoshiki, S., Tachikawa, T., Yamaguchi, A., Yamazaki, T., and Yamana, H., Effects of nutritional factors and cadmium exposure on monkeys. VI. Bone tissue observations, Proc. of the 1984 Cadmium Research Conference organised by the Japanese Public Health Association, *Kankyo Hoken Report,* in press (in Japanese).

Chapter 11

OTHER TOXIC EFFECTS

Carl-Gustaf Elinder

TABLE OF CONTENTS

I. INTRODUCTION

The major health risks associated with long-term occupational or environmental exposure to cadmium are found in connection with the effects which take place in the lungs, the kidneys, and the skeleton. These effects have been discussed in preceding chapters (see Chapters 8 to 10).

Other types of toxic effects have been observed in humans as well as in experimentally exposed animals and these will be discussed in this chapter.

Additional toxic effects, mainly acute ones, have been observed in experimentally exposed animals, but not in humans. It is often difficult to infer the possible hazards involved for humans from such experimental results. Nevertheless, animal studies are of value for a general understanding of the toxicology of cadmium. For example, it has been shown repeatedly that a single administration of cadmium salts to most species of animals gives rise to massive hemorrhagic necrosis of the testes (Section VI.A). Similar effects have not been reported for humans exposed to cadmium either orally or by inhalation. However, this phenomenon, i.e., cadmium-induced testicular necrosis, has resulted in basic knowledge about the interactions taking place between various trace elements and the influence of protein binding, etc. on the type of toxic manifestations observed.

The literature dealing with the different aspects of the toxicity of cadmium administered to animals has become very extensive and we will not attempt to review all studies. A good overview and summary of data obtained in various animal experiments has been given by Fielder and Dale.[79] In this chapter attention is given to adverse effects in humans and supportive evidence from experimentally exposed animals. Cellular, subcellular, and biochemical effects of cadmium are not discussed here. The cadmium-binding protein, metallothionein, is reviewed in Chapter 4. Biochemical effects of cadmium, especially with regard to enzymes, are discussed in an extensive review by Vallee and Ulmer.[274] For information about the cellular and subcellular effects of cadmium, reference is made to, e.g., Morselt et al.[173]

II. EFFECTS ON THE GASTROINTESTINAL TRACT AND DOSE-RESPONSE RELATIONSHIPS

A. Acute Effects

Ingestion of food or beverages contaminated with significant amounts of cadmium gives rise to acute symptoms of food poisoning. Contamination may be caused by cadmium-plated containers, cans, etc. Typical symptoms are nausea, vomiting, abdominal pains, and diarrhea.[52,160,272] Although outbreaks of food poisoning due to cadmium decreased considerably after the Second World War, when the use of cadmium was prohibited in cooking utensils and food containers, more recent reports indicate that this problem still exists.[19,185,228] The emetic threshold dose for cadmium has been estimated to be in the order of 3 to 90 mg.[44] This corresponds to a concentration, exceeding 15 mg/ℓ, of soluble cadmium salts in solutions.

As rodents do not usually regurgitate stomach contents, exposure to higher doses is possible. The acute oral LD_{50} for soluble compounds, such as cadmium acetate, cadmium chloride, and cadmium sulfate, is in the order of 50 to 400 mg/kg body weight. Insoluble compounds, such as cadmium selenide and cadmium sulfide, have been reported to have oral LD_{50} values exceeding 5 g/kg body weight.[44]

B. Chronic Effects

Data on the long-term effects of cadmium on the gastrointestinal tract are scarce. Richardson and Fox[230] fed Japanese quail a diet containing a very high content of cadmium, 75 mg/kg, from the time of hatching to 4 weeks of age. The villi of the proximal small intestines were short and thick. The microscopical findings were similar to those occurring in human malabsorption syndromes. Sugawara and Sugawara[259] reported histopathological changes in duodenal mucosa of rats fed 50 mg Cd per liter drinking water for 9 months (see Chapter 10, Section II.B).

Murata et al.,[174] treating 60 Itai-itai patients in Toyama, Japan, mentioned in their report that several of these patients had atrophic gastritis, enteropathy, and reduced absorption of fat. These authors suggested that cadmium-induced enteropathy was of importance for the development of the Itai-itai disease, especially in relation to the absorption of calcium which in the case of fatty stools will decrease.

III. EFFECTS ON THE LIVER AND DOSE-RESPONSE RELATIONSHIPS

A. Acute Effects

1. Animals

After high peroral, parenteral, or inhalative exposure to most cadmium compounds, high concentrations of the metal will be found in liver (Volume I, Chapter 6). Shortly after exposure, cadmium in liver is bound to high molecular weight proteins, but already after about 8 hr most of it is bound to metallothionein, a low molecular weight protein (Chapter 4, Section VI; Chapter 6, Section IV.B.3). The overt signs of acute toxicity in the liver which are seen shortly after parenteral administration of cadmium are most likely related to nonmetallothioein-bound cadmium. When metallothionein has been produced in the liver, this protein sequesters cadmium and thereby decreases its toxicity.

Andreuzzi and Odescalchi[8] injected rabbits with single i.v. doses of cadmium, ranging from 1.25 to 3 mg Cd per kilogram; thereafter, they determined the aspartate aminotransferase (ASAT) activity in the serum at different intervals over a period of 72 hr. In the rabbits given 3 mg/kg, the ASAT activity increased considerably after 17 hr, and 60% of the animals died within 24 hr. In rabbits given 2.5 mg/kg, ASAT activity increased more than 10-fold after 24 hr, and 40% of the rabbits died within 48 hr.

Rabbits given 2 and 1.25 mg/kg, respectively, survived for more than 72 hr. After 24 hr of exposure the increase in ASAT activity was approximately 4-fold, but within normal range again 72 hr. later. Dudley et al.[70] performed similar studies on rats. Injections of 3.9 mg cadmium (as $CdCl_2$) per kilogram body weight brought about severe hepatic damage within hours, including a dramatic increase in plasma enzymes released from the liver, as well as pronounced morphological changes. Hoffmann et al.[113] gave rats a single dose of 6 mg cadmium acetate per kilogram body weight, corresponding to about 4 mg Cd per kilogram, intravenously, and performed light and electron microscopic examinations of the liver 4, 8, and 16 hr later. Changes were more profound in the parenchymous cells than in the Kupffer cells. The most prominent changes were, e.g., parenchymal cell necrosis, deterioration of rough endoplasmatic reticulum, proliferation of smooth endoplasmatic reticulum, and mitochondrial degenerative changes.

Hadley et al.[108] and Pence et al.[200] examined the hepatic metabolism of hexobarbital in rats, 3 days after a single i.p. dose of 0.84 mg Cd per kilogram body weight in the form of cadmium acetate. Male rats, but not female rats, exhibited a significantly decreased hepatic metabolism, confirming the assumption that cadmium at these dose levels may influence the hepatic microsomal metabolism. In accordance with these studies, Teare et al.[265] reported markedly decreased activity in two drug metabolizing enzymes in rat liver homogenates (aniline hydroxylase and nitroreductase). In addition to this, there was a decrease in the content of microsomal cytochrome P-450 in liver homogenates obtained from rats given a single i.p. injection of 1.5 mg Cd per kilogram body weight in the form of cadmium chloride 6 days earlier. Confirmatory data on the inhibitory effects of cadmium injections on the microsomal drug metabolizing enzymes were also reported by Ando[7] and Lui et al.[161]

Chelating agents may influence distribution and effects from cadmium (Chapter 6, Section VII.B). Cadmium chloride given to mice (3.2 mg Cd per kilogram body weight) was much more toxic when given together with nitrilotriacetic acid (NTA) or sodium tripolyphosphate (STPP).[72] Mice given cadmium together with one of the chelating agents (NTA/STPP) developed severe, often fatal, liver necrosis within 24 hr after the administration, whereas no such effects were seen in animals that received either cadmium or one of the chelating agents.[72] In spite of the increased liver toxicity seen in mice given cadmium simultaneous with the chelating agent, the liver concentration of cadmium in these animals was only about 50% of that in mice given cadmium alone. As discussed in Chapter 6, Section VII.B, the high acute toxicity of cadmium when given together with NTA or STPP can possibly be related to a rapid uptake in the liver cells of nonmetallothionein-bound cadmium.

Glutathione (GSH), a nonprotein thiol which has a key function in the liver cells detoxification of various xenobiotics, appears to be involved in the mechanism behind acute liver effects seen shortly after injections of cadmium.[69] Rats that were pretreated with agents (phorone or diethyl maleate), which deplete the glutathione concentration in liver cells, were much more susceptible to a subsequent challenge with 1.3 or 2.0 mg Cd per kilogram body weight given intravenously. Glutathione-depleted animals had high mortality (40 to 80%) and revealed profound morphological and biochemical signs of liver toxicity. Control rats, given cadmium alone, had no mortality and only minor morphological and biochemical changes. On the other hand, pretreatment with cysteine, which increases the liver glutathione levels, markedly decreased the liver toxicity of 3.9 mg Cd per kilogram also given intravenously.[69] Thus, glutathione appears to have a protective function in relation to acute effects from cadmium on the liver.

2. Humans

A 23-year-old man who drank 5 g CdI$_2$ dissolved in water in order to commit suicide, in addition to kidney damage (Chapter 9, Section III.B.), developed signs of pronounced liver damage and metabolic acidosis before he died.[291] Apart from this case report, data on the acute effects of cadmium on liver in humans are not available.

B. Chronic Effects

1. Animals

Exposure to high doses of cadmium via inhalation, perorally or parenterally, can give rise to morphologic changes in the liver.[87,227,258] Prodan in 1932 decribed swollen cells and polymorphonuclear leukocyte infiltration in livers obtained from cats exposed to 2 to 16 mg Cd per cubic meter via inhalation. Friberg[87] observed "cirrhotic changes" in livers obtained from rabbits which had been given cadmium sulfate injections, 0.65 mg Cd per kilogram, body weight, 6 days/week over a period of 2 to 4 months. Stowe et al.[258] exposed rabbits to cadmium in drinking water, 160 mg/ℓ, constantly for 6 months. Light microscopy disclosed that, in contrast to controls, the cadmium-exposed rabbits showed depletion of glycogen. Inflammatory cell infiltrates were frequent in the portal regions and biliary hyperplasia was often present. Electron microscopy showed that the most striking changes took place in the endoplasmic reticulum, which increased in exposed animals. The mean cadmium concentration in liver on termination of exposure was 188 mg/kg wet weight.

Compared to morphological examination, measurement of liver enzymes in plasma and serum appears to be a less sensitive effect parameter for evaluating the long-term toxic effects of cadmium on liver. Axelsson and Piscator[16] determined the ASAT activity in serum from rabbits given injections of cadmium chloride (0.25 mg Cd per kilogram body weight) 5 days/week for 11 to 29 weeks. Compared to controls, no difference was apparent after 11 weeks of exposure, whereas significant increases in activity occurred after 17 weeks. The concentration of cadmium in liver at this time was about 450 mg/kg wet weight. In the experiment discussed above, Stowe et al.[258] were unable to show any significant increase in serum ASAT, alanine aminotransferase (ALAT), lactate dehydrogenase (LDH), alkaline phosphatase, or any evidence of impaired liver function as evaluated by BSP test and blood coagulation tests. This was true despite the fact that marked morphological changes were observed in the liver. Increased activity of liver enzymes in serum or plasma has, however, been noticed as a result of a more extensive exposure. Faeder et al.[74] gave rats s.c. injections of cadmium, 0.25, 0.5, and 0.75 mg Cd per kilogram body weight as CdCl$_2$, 3 days/week for 8 weeks. After 6 weeks the plasma glutamyl transpeptidase (GT) and ASAT activities increased in the two groups exposed to 0.5 and 0.75 mg/kg, respectively, and were accompanied by ultrastructural changes in hepatocytes. By this time, cadmium concentrations in liver were in the order of 90 to 120 mg/kg wet weight. Similar results were obtained by Chapatwala et al.[46] They gave female rats daily i.p. injections of cadmium chloride (doses ranging from 0.25 to 1.25 Cd mg/kg body weight) and noticed increased activity of ASAT and ALAT in serum already after 2 weeks of exposure.

Assay of enzymes from liver homogenates enables identification of the early signs of impaired liver function. Sporn et al.[256] exposed rats perorally to cadmium, 1 mg/ℓ

drinking water constantly for 1 year, and 1 to 25 mg/kg in food for 42 to 84 days. Rats given cadmium in drinking water showed an increase in the activity of phosphorylase and a decrease in aldolase activity, indicating that cadmium may interfere with carbohydrate metabolism in liver. Rats given larger amounts of cadmium in food over shorter periods of time exhibited an influence of cadmium on the oxidative phosphorylation in the liver mitochondria. Further evidence that long-term cadmium exposure may affect carbohydrate metabolism in the rat has been provided by Merali et al.[168] In rats given daily i.p. injections of cadmium chloride, 0.25 and 1 mg (0.15 and 0.6 mg Cd)/kg body weight, significant increase was seen in several different types of gluconeogenic enzymes following 45 and 21 days of exposure to the lower and higher doses, respectively. Consistent with increased activity of gluconeogenic enzymes, liver glyco- gen levels decreased to about 50% of the level in controls after 21 days of exposure in both groups. At the same time, blood glucose levels increased. Increased activity of liver adenyl cyclase, the enzyme forming cyclic AMP, as well as an increase in the total content of cyclic AMP were also observed. Furthermore, it was found that the metabolic effects brought about by cadmium exposure persisted for 21 days after cessation of exposure. Unfortunately, tissue levels of cadmium were not reported. In a later experiment, Merali and Singhal[166] showed that simultaneous exposure to selenium, in conjunction with cadmium, prevented cadmium-induced increases in hepatic gluconeogenic enzymes. An increased activity of gluconeogenic enzymes was also noticed in the experiment conducted by Chapatwala et al.[46] discussed above.

Other biochemical studies on the effect of long-term exposure of liver enzymes to cadmium are also available. Copius Peereboom-Stegeman et al.[54] gave rats s.c. injections of 0.5 mg cadmium chloride (0.3 mg Cd) per kilogram three times a week. After 12 to 13 weeks, liver homogenates displayed an increase in the activity of alkaline phosphatase and a decrease in the glycogen content.

A summary of the toxic effects of cadmium in liver as observed in animal experiments is given in Table 1.

2. Humans

Reports on the effect of cadmium on human liver function are rare. Friberg[87] mentioned that he found nonspecific signs of liver disease in serum samples of 4 of 19 workers with severe cadmium poisoning (Chapter 9, Section IV.B.1). One of the subjects studied showed a positive Takata reaction, two had positive thymol reactions and one had a positive Takata together with a positive thymol reaction. Increases in serum gammaglobulin levels were found in several workers. In most other investigations liver function has not been extensively studied but it is evident from reports submitted by Bonnell,[36] Kazantzis et al.,[132] and Adams et al.[1] that, compared to the pronounced changes in renal function, major changes in liver function are seldom found in cadmium-exposed workers. Piscator et al.[219] examined 48 women exposed to cadmium dust in a Swedish alkaline battery plant for less than 1 year and up to 30 years. Twenty-six constituents of serum were analyzed, several of which would have shown signs of change in the presence of liver disease. No evidence of liver disease or dysfunction was found in any of the female workers examined. From Japan, slight elevation in plasma ASAT such as signs of liver involvement have been reported as occurring in Itai-itai patients as well as in persons under observation for suspected Itai-itai disease.[178] Plasma ASAT was 38 ± 5 units/ℓ in 31 inhabitants of nonpolluted areas, in contrast to 53 ± 12 units/ℓ for 37 Itai-itai patients and 54 ± 15 units/ℓ for 53 patients under observation.

Table 1

SUMMARY OF EFFECTS IN LIVER FROM CADMIUM IN ANIMALS

Species	Exposure	Dose	Period	Concentration in liver (mg/kg wet weight)	Effects	Ref.
Cat	Inhalation	2—16 mg Cd/m²	24 hr	6.1	Morphological changes	227
Rat	Peroral	10 mg Cd	Daily for 2 months	56	Morphological changes	227
Rat	Peroral	250—500 mg CdCl₂/kg	In diet for 100 days	N.A.	Morphological changes	290
	s.c. injections	0.65 mg Cd/kg	6 days × 9—18 weeks	N.A.	Morphological changes / Signs of cirrhosis	87
	s.c. injections	0.25 mg Cd/kg	6 days × 11—29 weeks	After 17 weeks about 450 mg/kg	After 17 weeks, serum ASAT increases significantly	17
Rat	Peroral	1—25 mg Cd/kg in food; 1 mg Cd/l in water	42—84 days; 335 days	N.A.	Changed activities observed in liver enzymes, e.g., decreased activity of aldolase	256
Rabbit	Peroral	160 mg Cd/l in water	1—6 months	After 6 months, 188 mg/kg	After 6 months, morphological changes; depletion of glycogen deposits; electron microscopy revealed striking changes of the endoplasmic reticulum; no changes in serum ASAT, ALAT, LDH isoenzymes, or alkaline phosphatase	258
Rat	s.c. injections	0.25—0.75 mg Cd/kg	3 days × 8 weeks	After 4 weeks, 25—70 mg/kg; after 6 weeks, 80—120 mg/kg; and after 8 weeks, 120—200 mg/kg	After 6 weeks of exposure, increase in plasma ALAT and γ-GT	74
	i.p.	0.25—1.0 mg CdCl₂/kg (0.15—0.6 mg Cd/kg)	Daily for 21 and 45 days	N.A.	After 21 days, increase is four gluconeogenic enzymes in rats given 1 mg/kg body weight; after 45 days, a significant increase also in rats given	168

Table 1 (continued)
SUMMARY OF EFFECTS IN LIVER FROM CADMIUM IN ANIMALS

Species	Exposure	Dose	Period	Concentration in liver (mg/kg wet weight)	Effects	Ref.
					0.25 mg/kg; significant decrease in liver glycogen content in both groups after 21 days; increase in adenyl cyclase and cAMP content after 21 days	
	s.c. injections	0.5 mg $CdCl_2$/kg (0.3 mg Cd/kg)	3 days × 28 weeks	N.A.	After 12—13 weeks, a marked increase in liver alkaline phosphatase; electron microscopic examination showed that liver cells from cadmium-exposed animals often were deprived of their glycogen depots	55
	i.p.	0.25-1.25 mg Cd/kg	Daily for 2 to 4 weeks	N.A.	After 2 weeks, increase in serum ASAT and ALAT; after 2 to 4 weeks, increase in liver gluconeogenic enzymes	46

Note: N.A., not available; ASAT, aspartate aminotransferase (= GOT glutamic oxaloacetic transaminase); ALAT, alanine aminotransferase (= GPT glutamic pyruvic transaminase); LDG, lactate dehydrogenase; γ-GT, γ-glutamyltranspeptidase.

IV. EFFECTS ON THE HEMATOPOIETIC SYSTEM AND DOSE-RESPONSE RELATIONSHIPS

A. Acute Effects

An acute hemolytic effect following a single injection of cadmium sulfate (about 25 mg Cd per kilogram was observed in dog by Athanasiu and Langlois as early as in 1896.[15] Elevated hemoglobin concentration and increased packed cell volume (PCV) have been seen in animals[227] as well as in humans[33] a couple of days after inhalation of high concentrations of cadmium. This elevation is probably related to the hemoconcentration secondary to edema of the lungs.

B. Chronic Effects

1. Animals

Anemia is a common finding in animals after both peroral and parenteral exposure to cadmium. In connection with peroral exposure, decreased hemoglobin concentration and decreased PCV are among the early signs of cadmium toxicity. These effects are probably due to the influence of cadmium on the absorption and availability of certain essential metals, such as zinc and iron.

When cadmium is administered parenterally, high doses are necessary in order to produce any effect at all on the hematopoietic system. Hemolysis is probably one of the mechanisms in the pathogenesis underlying hematological effects which result from parenteral administration of cadmium.

a. Peroral Exposure

Wilson et al.[290] fed rats food containing 31 to 500 mg Cd per kilogram ($CdCl_2$) and found that the lowest dose produced anemia after a couple of months. Following cessation of exposure, there was a tendency towards normalization in hemoglobin levels. These authors also noted increases in reticulocytes and eosinophils in a group given 125 mg Cd per kilogram for up to 7 months. Examination of the bone marrow showed that it was hyperplastic. Decker et al.[64] exposed rats to cadmium via drinking water, 0.5 to 50 mg/ℓ (as $CdCl_2$), constantly for 1 year. In rats given 50 mg/ℓ hemoglobin levels decreased considerably after 2 weeks only. Rats given 0.5 to 10 mg/ℓ did not develop anemia. Similarly, Prigge et al.[225] observed significant decreases in hemoglobin concentration and PCV in male rats given 25 to 100 mg Cd per liter in drinking water for 7 or 8 weeks. Female rats appeared to be less sensitive. In a later experiment on female rats exposed to identical levels of cadmium in drinking water, 25 to 100 mg/ℓ, the same author[223] found no significant change in either hemoglobin concentration or PCV. A significant decrease in serum-iron was noted, indicating some effect from cadmium on the metabolism of iron.

In agreement with these experiments, Cousins et al.[58] showed that growing swine responded with a marked decrease in the hematocrit when given a basic diet supplemented with more than 50 mg Cd per kilogram as $CdCl_2$.

In contrast to the above-mentioned experiments, Kotsonis and Klaassen[147] found no significant changes in hemoglobin concentration or PCV in rats given cadmium, 10 to 100 mg Cd per liter, in drinking water for periods of up to 24 months. Possibly, the discrepancy in the results obtained under similar exposure conditions can be assigned to differing contents of other trace elements, such as zinc and iron as well as vitamins, in the experimental diets.

Fox and Fry[84] and Fox et al.[85] studied Japanese quail fed a diet containing 75 mg/kg of cadmium. In addition to iron, the presence of ascorbic acid prevented anemia. According to these authors, ascorbic acid constituted a good prophylactic due to its ability to bring about an increase in the absorption of iron from the intestinal tract.

Pond and Walker[220] showed that a single prophylactic injection of iron can prevent occurrence of cadmium anemia in growing rats exposed for 4 weeks to 100 mg/kg of cadmium in the diet. Addition of iron to the food had the same preventive effect.

Another way to investigate possible effects on the hematopoietic system is to measure heme precursors. Mahaffey et al.[163] determined heme precursors (ALA, coproporphyrin, and uroporphyrin) in urine from rats on a diet containing cadmium (50 mg/kg), lead (200 mg/kg), arsenic (50 mg/kg), and combinations of these elements. Cadmium exposure alone did not result in any major changes in excretion of heme precursors. However, there was an increase in excretion of coproporphyrin and uroporphyrin in rats with combined exposure to lead and cadmium, compared to rats exposed to lead only. In vitro studies of the activity of the enzyme pyrimidine 5′-nucleotidase in red blood cells have shown that cadmium ions can inhibit its activity, and that this inhibition could be counteracted by metallothionein.[172]

b. Exposure Due to Inhalation

Friberg[87] found that rabbits exposed to cadmium oxide dust in air developed slight anemia. No anemia was observed in control rabbits. A prominent finding was eosinophilia. Of the white cells in exposed rabbits, 25% were eosinophils compared with 3% in a control group. There was no eosinophilia in rabbits exposed to nickel containing dust, despite the fact that these rabbits also showed certain lung changes. Unfortunately, this effect of cadmium has not to our knowledge been examined in subsequent studies on the experimental toxicity of cadmium.

Prigge et al.[225] demonstrated no significant change in either hemoglobin concentration or in PCV in rats exposed to 0.2 mg Cd per cubic meter as cadmium chloride aerosols for 66 days. In a later experiment, Prigge[224] was again unable to show any depression of the hematopoietic system in rats exposed to cadmium oxide in air, using concentrations ranging from 25 to 100 μg Cd per cubic meter administered constantly for periods of up to 3 months. On the contrary, results indicated that inhalation of cadmium oxide evoked a significant dose-dependent increase in the hemoglobin concentration. This increase was probably due to the lower oxygen tension as a result of lung damage which was observed in the exposed animals. Prigge[224] suggested that the anemia seen in rabbits exposed to cadmium oxide fumes as reported by Friberg [87]could have been due to clearance of cadmium particles from the tracheobronchial system to the gastrointestinal tract, thus giving rise to gastrointestinal exposure. This gastrointestinal exposure of cadmium may in turn, as discussed in Section IV.B.1.a, produce a decrease in the hemoglobin concentration possibly via influence on other, essential metals.

c. Parenteral Exposure

Parenteral administration of cadmium has been shown repeatedly to give rise to anemia. In rabbits injected with cadmium sulfate, 0.65 mg Cd per kilogram body weight, 6 days/week, pronounced anemia was evident after 2 months of exposure.[87] It was found that simultaneous administration of iron had a beneficial effect on the anemia, indicating that the anemia was partly due to iron deficiency.[88] Berlin and Friberg[28] gave ^{59}Fe to cadmium-exposed rabbits [1 mg Cd (as $CdCl_2$) per kilogram body weight, 6 days/week, s.c. injection]. One group received the isotope before exposure, a second group after 2 months of exposure, and a third group comprised controls. This study demonstrated that an increased destruction of erythrocytes took place in cadmium-exposed animals. No definite difference between controls and exposed animals was observed in regard to utilization of iron for hemoglobin synthesis. Beneficial effects from parenterally administered iron were seen. This indicates that hemoglobin synthesis was not blocked by cadmium but that cadmium possibly caused decreased uptake of iron from the intestines. Berlin et al.[31] found that in rabbits given cadmium, deposits of iron increased in bone marrow, but no decrease in erythropoietic activity occurred. Other changes in the bone marrow were observed, but these changes were regarded as nonspecific and as being produced by general toxic effects. The rabbits were rather severely intoxicated after having received s.c. injections of 1 mg Cd per kilogram body weight daily for 7 months. Berlin and Piscator[29] studied blood and plasma volumes in cadmium-poisoned rabbits and concluded that the low hemoglobin concentration in blood could be due partly to increases in plasma volume.

Axelsson and Piscator[17] gave rabbits cadmium chloride 0.25 mg Cd per kilogram body weight, subcutaneously, 5 days/week for 11 to 29 weeks. After 11 weeks of exposure they found no detectable amounts of haptoglobin in plasma, indicating hemolysis. The hemolytic anemia persisted in some animals for up to 6 months after cessation of exposure. These animals also continued to excrete large amounts of cadmium in urine.[217] In a group of rabbits observed after cessation of exposure to 0.25 mg Cd per kilogram body weight, 5 days/week for 24 weeks, hemoglobin levels rose again, but 30 weeks later were still significantly lower than those in a control group.

Nordberg et al.[184] found that cadmium in erythrocytes taken from exposed mice is mainly stored in a low molecular weight protein, probably metallothionein. If it is assumed that the tubular reabsorptive capacity is exceeded, the high urinary excretion of cadmium in rabbits with hemolytic anemia can be explained partly by the release of cadmium-containing low molecular weight proteins from the erythrocytes.

2. Humans

Somewhat lowered hemoglobin concentrations and decreased PCV have been observed repeatedly among cadmium-exposed workers when compared to controls.[32,36,87,269] Bernard et al.[32] examined 42 male workers employed in cadmium production and a control group of 77 nonexposed. Exposure periods ranged from 2.3 to 47 years, average 24.5 years. Exposure levels varied from 3 to 67 μg Cd per cubic meter. The average blood hemoglobin concentration in the exposed group was 141 g/ℓ and PCV was 43.2%. In the control group corresponding figures were 149 g/ℓ and 45.0%. These differences were, however, not statistically ($p > 0.05$) significant. In a group of 16 workers exposed to cadmium for from 5 to 30 years, a significant correlation was found between high cadmium levels in blood and low hemoglobin levels.[216] In this group of workers the mean haptoglobin level was below normal. Moreover, in several cases the haptoglobin concentration was at the lower normal limit, indicating that slight hemolysis might have occurred. In a group of previously exposed workers who had not been exposed to cadmium for at least 10 years, haptoglobin concentrations were generally normal or somewhat elevated and did not differ from levels in a group of workers who had never been exposed to cadmium. This indicates that hemolysis is an effect which occurs during, and not after, exposure.

In heavily exposed workers the number of white cells is generally normal, but an increase in the number of eosinophilic cells has been observed.[87,175,226] No explanation of this phenomenon has yet been reported, but it is nevertheless of interest that it has also been observed in heavily exposed rabbits (Section IV.B.1.b). Bone marrow from 19 cadmium-exposed workers was examined microscopically by Friberg.[87] No pathological changes were found.

V. EFFECTS ON BLOOD PRESSURE AND THE CARDIOVASCULAR SYSTEM

A. Acute Effects

Parental administration of high doses of cadmium, several milligrams per kilogram body weight, gives rise to severe endothelial cell damage in the small vessels of several organs 15 to 30 min after exposure.[93] These vascular lesions are most evident in the peripheral nervous system[89] and in the testes.[191] These studies suggest that the immediate cause of death in animals given excessive doses of cadmium may be brought about by its dramatic effects on the endothelial cells, which secondarily cause impairment of microcirculation and cellular respiration in several well-perfused organs. Other evidence (Section III.A.1) indicates that severe liver damage is often the cause of death in acute exposure studies of animals. This damage could be secondary to vascular effects in the liver. Further discussion on endothelial effects is to be found in the section devoted to effects of cadmium on the testes and on the nervous system (see Sections VI.A. and VIII.A.2)

Single or a small number of parenteral administrations of cadmium salts to animals can give rise to either a decrease or an increase in blood pressure. Acute response depends on administered dose as well as on the species. In general, large doses will result in a fall in blood pressure, whereas relatively low doses may give rise to increased diastolic and/or systolic blood pressure. Thus, Dalhamn and Friberg[61] found that an i.v. injection of cadmium sulfate (0.3 to 0.5 mg Cd per kilogram body weight) in the cat and the rabbit caused a rapid fall in blood pressure within 15 sec. Perry and Yunice[206] found that an intraarterial injection of 0.1 to 0.4 mg Cd per kilogram body weight caused increased diastolic blood pressure in rats, whereas larger doses (0.8 to 3 mg/kg body weight) caused decreased pressure. Increased blood pressure was produced in rats by i.v. injection of cadmium in doses of 0.02 to 2 mg/kg body weight[207] and by i.p. injections of 0.2 to 2.4 mg/kg body weight.[201] The latter authors also found considerable increases in renin activity in blood of rats given an intraperitoneal dose of about 1 mg Cd per kilogram body weight. Maximum activity was observed 8 hr after injection, but a significant increase was seen up to 1 month after injection.[202]

B. Chronic Effects
1. Animals
a. Hypertensive Effects Detected
i. Oral Exposure

Hypertension was repeatedly produced in female rats of the Long Evans strain by Schroeder and co-workers who administered cadmium in drinking water for long periods of time.[125,240,245] Control rats were given a special cadmium-free diet and the exposed animals were given 5 mg/ℓ of cadmium (as chloride) in double deionized water together with essential elements. In rats receiving cadmium in drinking water, hypertension was usually manifested after 1 year. Analysis of cadmium in tissues revealed concentrations in the liver and kidneys of the same magnitude as that in American adults, i.e., on an average around 6 and 40 mg/kg wet weight in liver and kidney, respectively. When the cadmium to zinc ratio was above 0.8, the animals were always hypertensive. Histological examination[125] disclosed renal arterial, and arteriolar lesions. According to these authors, the changes were indistinguishable from those accompanying benign hypertension from other causes.

The findings of Schroeder and associates were extended and confirmed by Perry and co-workers in several experiments.[203,210-214] Their extensive work on cadmium-induced hypertension in rats has been summarized by Perry et al.[213,214] and Perry and Kopp.[205] In 6 separate experiments, a total of 144 female Long Evans rats were exposed to 5 mg Cd per liter in drinking water. Blood pressure was measured and compared to that of 267 control animals after 6 and 12 months of exposure. A significant increase in systolic blood pressure was seen in all six series. The average cadmium-induced increase in systolic blood pressure for the exposed group compared to controls ranged from 6 to 23 mmHg. The normal systolic blood pressure for rats is between 100 and 110

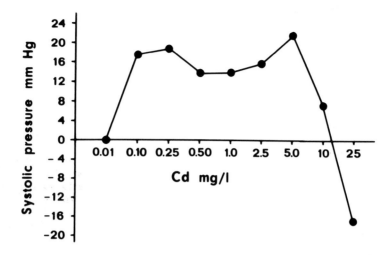

FIGURE 1.   Average change in systolic blood pressure induced by exposure to different cadmium levels (0.01 to 25 mg/ℓ) in drinking water for 18 months. (From Perry, H. M., Jr., Erlanger, M., and Perry, E. F., *Environ. Health Perspect.*, 28, 251—260, 1979. With permission).

mmHg.[213] The overall average increase for all exposed animals compared to controls following 6 and 12 months of exposure was 12 and 17 mmHg, respectively. The average renal cadmium concentration after 6 and 12 months of exposure to 5 mg Cd per liter in drinking water was about 14 and 29 mg/kg wet weight, respectively.

Perry and co-workers have also examined the cadmium-induced increase in blood pressure found in female Long Evans rats as a function of length of exposure and dose levels in water. Exposure periods have varied from 3 to 24 months and cadmium levels in drinking water from 0.1 to 10 mg/ℓ. Even the lowest cadmium concentration in water, 0.1 mg/ℓ, after 3 months of exposure induced an average increase in systolic blood pressure of 5 mmHg. The same concentration at double the length of exposure, 6 months, induced an increase of 10 mmHg. The most marked increase in blood pressure, 24 mmHg, was seen in rats exposed to 1 mg/ℓ for 24 months. Exposure to 1 mg/ℓ for 24 months results in an average renal cadmium concentration of approximately 12 mg/kg wet weight.[212] In the maximum exposure groups, i.e., those given 10 mg/ℓ in drinking water for 24 months, a significant decrease in blood pressure was observed. The average renal concentration in these rats can be estimated to be around 65 mg/kg.[212] Figure 1 shows the relationship between average change in systolic blood pressure in rats exposed for 18 months to nine different levels of cadmium in drinking water, from 0.1 to 25 mg/ℓ. The data show that cadmium-induced increase in blood pressure constitutes an effect seen at relatively low levels of exposure only. This results in renal cadmium concentrations ranging from less than 3 mg/kg to no higher than 60 mg/kg wet weight. The most conspicuous effects on blood pressure have been observed at cadmium levels in kidney of 5 to 30 mg/kg. At higher exposure levels, resulting in renal cadmium concentrations exceeding 50 mg/kg, a decrease in blood pressure is seen.

In this context, it should be pointed out that it is by no means firmly established that the hypertensive effect of cadmium in rats is caused by changes in renal function. At present most researchers tend to assume that this is the case. It is possible that the critical organ for elicitation of hypertension in rats is some organ other than the kidney. By taking into account data on cadmium in kidneys it is, nevertheless, possible to form some idea of the degree of exposure and body burden.

Most experiments showing cadmium-induced hypertension have been carried out on female Long Evans rats. In one study Perry et al.[210] exposed male Long Evans rats, female Sprague Dawley rats, and a group of female Long Evans rats to 5 mg Cd per liter in drinking water. Similar hypertensive responses were observed in all these groups, indicating that the hypertensive effect of cadmium is relevant not only for one strain or sex of rats. However, results obtained by other investigators indicate that sex of rats might be of importance, Petering et al.[215] exposed male and female Sprague Dawley rats to 4.3, 8.6, and 17.2 mg Cd per liter in the form of $CdCl_2$ in drinking water for 39 weeks. A dose-dependent increase in blood pressure was seen among the male rats, but not among the females. Ohanian and Iwai[186] examined the effect on blood pressure of low level peroral exposure to cadmium in the form of cadmium acetate in drinking water, 1 mg Cd per liter, in female Dahl hypertensive resistant (R) and hypertensive sensitive (S) rats given a diet containing either a low or high salt content. The effect on blood pressure of high salt intake was much more pronounced than was that of cadmium. After 12 weeks of exposure, cadmium-exposed S rats exhibited significantly higher blood pressure than that seen in S controls. It was also found that S rats had significantly higher cadmium concentrations in kidney, about 12 mg/kg wet weight; when compared to R rats, the latter had concentrations of approximately 6 mg/kg wet weight.

In rats perorally exposed to cadmium, the hypertensive response is influenced by the content of other trace elements in the diet. Schroeder et al.[247] noticed that when cadmium was given to rats in hard water, i.e., water rich in calcium and/or magnesium, less hypertension occurred. Perry et al.[209,210] showed that the pressure effect from 2.5 and 10 mg Cd per liter in drinking water was inhibited by the simultaneous addition to water of either 3.5 mg/ℓ of selenium, 20 mg/ℓ of copper, or 100 to 200 mg/ℓ of zinc. The addition of 1 mg/ℓ of lead, however, enhanced the hypertensive response.

The specially prepared rye-based diet with a very low cadmium concentration, used by Schroeder, and Perry and Erlanger, appears to be crucial for the development of hypertension in rats.[204,248]

According to Schroeder et al.[248] the cadmium-induced hypertension in rats is gradually normalized when exposure is discontinued. Rats that had developed hypertension after exposure to 5 mg Cd per liter in drinking water for 513 days, became normotensive 200 days after cessation of exposure.

Schroeder and Buckman[243] reported that hypertension induced by cadmium could be reversed by administering the chelating agent cyclohexane-1,2-diamine-*NNN′N′*-tetraacetic acid (CDTA), which binds cadmium more firmly than zinc. When an i.p. injection of a disodium zinc chelate was given to nine rats previously administered cadmium in drinking water, blood pressure returned to normal in all the animals within 2 weeks. After treatment there was a decrease in cadmium levels in the kidneys, and the authors concluded that the lowering of cadmium to zinc ratio could have caused the decrease in blood pressure.

ii. Parenteral Exposure

Hypertension has also been induced by parenteral administration of cadmium salts. Schroeder et al.[246] and Schroeder and Buckman[243] found that hypertension developed in rats after one or two i.p. injections of cadmium acetate or citrate (1 to 2 mg Cd per kilogram body weight). Over a period of 1 month, the induced level of hypertension became equal to that seen in rats on which partial constriction of one renal artery had been performed. Similar results have been reported by Chiappino and Baroni[48] for Sprague Dawley rats. These authors also found a hyperplasia of the juxtaglomerular apparatus and of the glomerular zone of adrenal cortex. They concluded that stimulation of the renin-aldosterone system had probably taken place.

Results obtained by Ohanian and Iwai[187] indicate that sex and strain differences may also exist in regard to hypertensive responses in rats given cadmium parenterally. After repeated injections of cadmium acetate, 1 or 2 mg Cd per kilogram (total dose 4 mg Cd per kilogram) hypertension was observed in female Dahl hypertensive sensitive rats, but not in the male-sensitive or female- and male-resistant rats.

There are also a number of reports indicating that cadmium may affect renal management of sodium. Vander[275] showed that a single i.v. dose of a cadmium-cysteine complex enhanced sodium reabsorption in the renal tubules. Perry et al.[208] produced permanent sodium retention in female rats with four injections of cadmium, 1 mg/kg body weight, administered at 1-month intervals. Lener and Musil[157] using cadmium acetate exposed rats for 16 months to 5 mg Cd per liter in drinking water. The animals were also exposed to an increased concentration of sodium chloride in drinking water. At the end of the exposure period, a reduced excretion of sodium was found after a sodium chloride load. This was interpreted as being due to the fact that cadmium enhanced proximal tubular reabsorption of sodium. Renal concentrations of cadmium were on an average 4.8 and 1.3 mg/kg wet weight in the exposed and control groups, respectively. Taking into account the cadmium dose given in drinking water and the long exposure time, values reported for the exposed group are surprisingly low. In experiments on rats conducted by Decker et al.[64] Piscator and Larsson,[218] and Perry et al.,[210] much higher concentrations were found after 1 year at the same exposure level.

Retention of sodium may increase susceptibility to cadmium as indicated by the findings by Lener and Bibr.[156] A group of Long Evans rats was given 2% NaCl for 50 days before receiving two doses of cadmium citrate, 2 mg Cd per kilogram body weight with an interval of 84 days. There was a significant increase in systolic blood pressure in the cadmium-exposed animals on high sodium intake, whereas cadmium alone did not cause hypertension.

Hypertension has also been induced in dogs. Thind et al.[267] and Thind and Fischer[266] gave female mongrel dogs weekly i.p. injections of 2 mg cadmium acetate per kilogram body weight, for periods exceeding 18 weeks. There was a slight, but significant ($p <$ 0.05), increase in systolic and diastolic blood pressure. In spite of the rather high concentration of cadmium in kidney and liver, about 160 and 450 mg/kg wet weight, 4 to 6 weeks after termination of exposure, the authors reported that there were no overt signs of cadmium toxicity. Kidney morphology and function were not studied in any detail, though.

b. No Hypertensive Effects Detected

There are several studies in which no hypertensive effects from cadmium have been reported. Lener[154] and Lener and Bibr[155] reported that hypertension did not appear when Wistar female rats were given cadmium in drinking water, 5 mg/ℓ, and observed for 16 months, or when Long Evans rats were put on 2% NaCl for 12 days and then given three i.p. doses of cadmium citrate, 1.1 and 2 mg per kilogram body weight, at intervals of 1 and 2 weeks. Repeated measurements of blood pressure revealed no hypertension up to 4 weeks after the final injection. Castenfors and Piscator[42] using female Sprague Dawley rats, could not induce hypertension by injection of cadmium chloride, 0.5 mg Cd per kilogram, 3 days/week for 6 months. In another experiment the rats were given cadmium in drinking water, 5 mg/ℓ, for 1 year, and monthly determinations of blood pressure revealed no difference compared to controls. Porter et al.[221] gave different groups of female and male Sprague Dawley and Long Evans rats repeated i.p. doses of cadmium for 80 days. Accumulated doses ranged from 1.7 to 6.7 mg Cd per kilogram body weight. No significant effect on blood pressure was demonstrated. In a later experiment, the same authors[222] investigated the vascular responsiveness, in vivo and in vitro, of aortic smooth muscle obtained from rats given

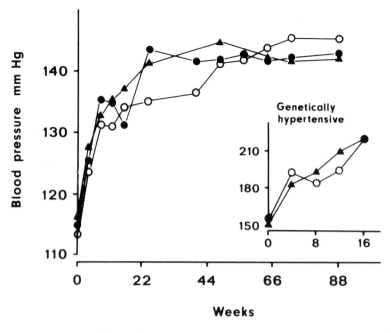

FIGURE 2.   Influence of cadmium in drinking water on blood pressure of OSU brown rats and (inserted in figure) genetically hypertensive rats. (From Whanger, P. D., *Environ. Health Perspect.*, 28, 115—121, 1979. With permission.)

two i.v. doses totalling 3.0 mg Cd per kilogram body weight, in the form of acetate. Both in vivo and in vitro cadmium-exposed rats responded less than control rats to drugs which normally cause an increase in blood pressure.

Doyle et al.[68] gave female and male Long Evans rats cadmium in drinking water, 5 mg/ℓ, for 320 days. No significant effects on blood pressure were disclosed. There was, however, a significantly higher retention of ^{24}Na given intraperitoneally to cadmium-exposed rats compared to controls. This could tentatively be judged to be an early sign of developing hypertension. Kotsonis and Klaassen[146] measured the systolic blood pressure in Sprague Dawley rats daily for 14 days following a single oral dose of differing amounts of cadmium, 25 to 150 mg/kg body weight. They found no significant change in any of the groups. In a later experiment rats were exposed to cadmium in drinking water, 10 to 100 mg Cd (as CdCl$_2$) per liter for 25 weeks.[147] A slight but insignificant increase in blood pressure was seen when exposure terminated. Whanger[287] gave OSU brown rats and genetically hypertensive rats (sex not stated) 10 and 20 mg Cd per liter in drinking water for periods of up to 88 weeks, and found no hypertensive effect (Figure 2).

Whanger[287] drew attention to the fact that researchers finding hypertensive effects in rats, caused by cadmium, have generally used a rye flour-based diet, whereas researchers finding no increase in blood pressure used other types of diets. This hypothesis has been confirmed by Perry and Erlanger.[204] By giving the rats a commercial stock diet, and not the rye-based diet normally used by Perry and co-workers, the hypertensive effect from cadmium in drinking water was largely diminished and after longer periods of exposure totally eliminated.

Neither hypertensive effect nor increase in plasma renin levels was found by Eakin et al.[71] upon examination of rats given two different doses of cadmium in drinking water, 10 and 20 mg/l, and observed for a period of 88 weeks. Another failure to produce hypertension in rats by giving cadmium in drinking water was reported by Fingerle et al.[80] The authors continuously exposed Sprague Dawley rats to cadmium in drinking water at concentrations of 0.5, 12.6, and 31.5 mg/l. Blood pressure was measured repeatedly during the experiment. There was no evidence of any hypertensive effect whatsoever. It was suggested in the report that the type of anesthesia used during the measurements of blood pressure might be crucial for the outcome of experiments of this type.

c. Effects on the Vascular System and the Myocardium

Fowler et al.[83] exposed rats to cadmium concentrations in drinking water ranging from 0.2 to 200 mg/l for 12 weeks and noticed morphological changes in the vascular system of the kidneys in all the exposed rats after 6 weeks of exposure. The walls in the small and median arteries were thickened and the larger arteria appeared more dilated in exposed rats when compared to controls. Blood pressure was not measured but this finding could possibly be associated with hypertension. However, the vascular changes observed were not more predominant in the most highly exposed group of rats compared to the less exposed. The cadmium concentration in kidney after 6 weeks of exposure ranged from 0.6 to 28 mg/kg wet weight in the different groups.

In a series of experiments on rats, Kopp and co-workers[140-143] were able to demonstrate certain effects of cadmium on the myocardium in addition to hypertension. In rats given 5 mg Cd per liter in drinking water for 24 weeks, ECG recordings differed significantly from controls in several characteristics; e.g., the PR interval was lengthened and the ventricular depolarization time (QS interval) was prolonged in cadmium-exposed rats. Electrograms of "His's bundle" revealed a 30% increase in the A-H interval. In heart tissue homogenates from exposed animals, the content of ATP was lowered compared to controls.[140] In later similar experiments[141,142] a marked decrease in the content of glycerophosphorylcholine was also found in exposed rats. The data indicate a depressive effect of cadmium on the heart. The cadmium content in heart muscles from exposed rats has been in the order of 0.1 mg/kg compared to about 0.01 mg/kg in controls. Corresponding concentrations of cadmium in kidney were in the order of 30 to 40 mg/kg in exposed rats compared to less than 0.1 mg/kg in controls.[142] In 1983, Kopp et al.[144] reported similar cardiovascular changes in rats given a lower dose of cadmium, 1 mg/l in drinking water.

Morphological changes in the myocardium of rats due to long-term inhalative exposure to cadmium oxide fumes at concentrations of 0.16 and 1 mg Cd per cubic meter have been reported.[135] According to the authors, the ultrastructure of the intercalated discs of the cardiac papillary muscles of exposed rats was different from that of controls — it became wider as a result of cadmium exposure. The meaning of this observation is as yet unknown. Unfortunately, the data are not presented in any detail. For example, no information is given on the number of animals in control and exposed groups.

d. Conclusions

Reported effects of cadmium on blood pressure of experimental animals are contradictory. Some research groups report marked increases in blood pressure, whereas this effect has not been reported by others using similar experimental conditions. Several factors appear to influence the degree of hypertensive response. Experiments on rats have shown that, apart from species and/or strain, sex and diet are also of importance. Hypertension has mainly been seen when rats have been given a rye-based diet, but not when fed other types of chow.[204] Furthermore, exposure levels are of great importance. Hypertension is an early effect of cadmium exposure, seen predominantly at comparably low cadmium levels. Excessive exposure does not lead to an increase in blood pressure. Nevertheless, it is clear that long-term exposure to cadmium, under certain circumstances, might induce an increase in blood pressure in rats.[205]

In addition to high hypertension, experimental work on rats[140-142,144] has shown that cadmium in drinking water, 1 and 5 mg/ℓ, has a depressive effect on the myocardium.

2. Humans
a. Exposed Workers

Despite the abundance of data showing that cadmium under certain environmental and nutritional exposure conditions will induce hypertension in animals, there are very few results available from studies of cadmium-exposed workers. Friberg[87] examined 43 workers with a mean exposure time to cadmium oxide dust of 20 years and 15 workers with a mean exposure time of 2 years. The study included physical and roentgenological examinations of the heart; electrocardiographic examination at rest and after exercise; and measurement of blood pressure. No increased prevalence of hypertension or cardiac disease was found. Electrocardiographic changes occurred to the same extent as those found in a group of sawmill workers who were not exposed to cadmium. The majority of the subjects had perfectly normal blood pressure. There was nothing to support the assumption that higher frequency of hypertension occurred in the cadmium workers than would normally be expected in workers in similar age groups. As Friberg did not examine a control group with regard to the prevalence of hypertension, it is impossible to use his data for a more precise evaluation. In 1974 a review of the medical records of employees at the same factory (Johansson and Kjellström, unpublished data) showed no indications of an increased prevalence of hypertension.

Chest examination and blood pressure measurements have also been reported in other studies[36,37,114,132] None of the resulting reports contained findings of cardiac disease or hypertension due to cadmium exposure. Hammer et al.[109] found no relationship between exposure to cadmium and blood pressure in superphosphate workers.

One study on cadmium-exposed workers in a battery factory has emerged from the U.S.S.R.[279] A total of 92 workers, 72 females and 20 males exposed to cadmium oxide dust at concentrations ranging from 0.04 to 0.5 mg/m³, were examined. Blood pressure was measured and electrocardiogram taken. No control group was examined. The authors reported increased prevalence of hypertension and increased absence from work due to hypertensive and ischemic heart disease among the exposed workers compared to what was considered normal. Furthermore, several types of abnormalities in the ECG of the exposed workers were observed. Of the exposed workers, 39% had tachycardia, between 11 and 13% were regarded as normal, and 26% had changes in the "R" spike compared to the normal 7 to 9%. Increased QRS period was observed in 45% of the workers compared to the normal values of 14 to 16%. The data presented in this Russian report are especially interesting in view of the recent findings in rats (Section V.B.1.c) which provide experimental evidence for myocardial effects from cadmium exposure. These results of the study are, however, presented in very condensed form excluding details and it is, therefore, difficult to draw clear-cut conclusions.

There are no indications of excess mortality due to cardiovascular or heart disease in cadmium-exposed workers. On the contrary, lower than expected mortality in cardiovascular diseases has been noticed in most of the cohort studies.[6,11,115,116,153,255] These observations strongly speak against the hypothesis that cadmium may cause hypertension as a result of occupational exposure, at least at those exposure levels that were common in the past.

Further research on the cardiovascular effects of cadmium in exposed workers would be of great value, especially at those dose levels commonly encountered by workers today.

b. General Population

Based on the observation that subjects who died from hypertensive and/or cardiovascular diseases have higher cadmium concentrations in tissues, especially liver and kidney, compared to persons dying from other causes, it has been suggested that cadmium is a causative factor in the development of these diseases.[241,242] Schroeder[242] suggested that the low zinc to cadmium ratio in kidney could be a determining factor for cardiovascular disease. In agreement with this hypothesis, Voors et al.[277] found a decreased zinc to cadmium ratio in kidney and aorta obtained from patients with cardiovascular disease.

Smoking is, however, an important confounding factor. This was not considered in the early studies. Smokers, apart from being more susceptible to cardiovascular disease, are also exposed to higher levels of cadmium through inhalation of tobacco smoke (Chapter 3). Obviously, then, cadmium in itself is not necessarily a causative factor in the development of cardiovascular disease. Indeed Østergaard[189] found a lower average cadmium concentration in kidney samples from nonsmoking hypertensive patients aged 45 to 65 (742 mg/kg ash weight in whole kidney samples) when compared to nonsmoking normotensive subjects of similar age (1260 mg/kg ash weight). This corresponds to kidney cortex values on a wet weight basis of about 12 and 20 mg Cd per kilogram, respectively (Chapter 5). Similar results were obtained by Cummins et al.[59] when they measured cadmium content in normo- and hypertensive patients using in vivo neutron activation. The cadmium content in kidneys of hypertensive nonsmoking patients was 2.4 mg compared to 3.9 for nonsmoking control persons. This corresponds to kidney cortex values of about 20 and 32 mg/kg wet weight (Chapter 5).

Several attempts have been made to relate cadmium in blood and urine to blood pressure. Measurements of cadmium in hypertensive patients have been compared to the levels found in normotensive patients of similar age. Unfortunately, the analytical validity usually has not been considered seriously and, therefore, most of the published results are of no value. For example, Bierenbaum et al.,[34] Glauser et al.,[98] Khera et al.,[133] and Revis and Zinsmeister[229] present data indicating that hypertensive patients have higher cadmium concentrations in blood or serum compared to normotensive. The published results on cadmium in blood and serum are, however, far too high and clearly erroneous (see Chapters 2 and 5); therefore, the authors' conclusions must be disregarded.

Smoking habits must be considered because smokers have higher blood cadmium levels when compared to nonsmokers (Chapter 5). In two studies in which published data on cadmium in whole blood lie within reasonable limits and in which blood cadmium levels in hypertensive nonsmoking patients were compared to normotensive nonsmoking patients, no significant differences in cadmium concentrations were disclosed.[26,27,282] In a third study on blood cadmium levels in nonsmokers, with and without hypertension, Tulley and Lehmann[271] found a slight and significant difference: hypertensive patients had 0.9 µg/ℓ and normotensive controls 0.6 µg/ℓ. However, since former smoking habits were not considered, the data are not convincing.

Comparison has also been made of urinary excretion of cadmium in hypertensive and normotensive patients,[164,169,257] but the results on urinary cadmium in the early studies were one order of magnitude higher than what is normally found in humans nowadays (Chapter 5) and thus, the reports cannot be evaluated. In one recent study,[257] the 24-hr urinary excretion of cadmium of 57 adult males and 59 adult females was related to systolic and diastolic blood pressure. When confounding factors such as age, smoking habits, and sex were considered there was no evidence of a positive correlation between urinary cadmium excretion and increasing blood pressure. On the contrary, there were significant negative correlations. The meaning of these negative correlations is unknown, but in any case the data do not support cadmium exposure as a cause of hypertension.

Different types of macroepidemiological approaches have also been considered. For example, Carroll[41] and Hickey et al.[110] found a correlation between cadmium concentration in the air of American cities and mortality due to hypertension and heart disease. Such epidemiological evidence is not very conclusive since many confounding factors are involved, e.g., population density, smoking habits, ethnic background of inhabitants in the cities, etc. Furthermore, as a rule, cadmium in air constitutes only a minor source of cadmium exposure compared to intake via food (Chapter 3).

The hardness of drinking water (the hardness of water is equal to the sum of its calcium and magnesium content) has repeatedly been reported to be associated with mortality in cardiovascular disease.[39,43] In soft water areas, the mortality in cardiovascular disease is higher than in hard water areas. Many researchers have speculated as to which substance or substances present in soft water might cause this increase in cardiovascular disease. Cadmium has sometimes been mentioned as one of the toxic metals that might occur in soft water and that may possibly give rise to cardiovascular disease.[82,242] The daily intake of cadmium from soft water is, however, generally more than ten times less than the intake from food (Chapter 3).

In Japan, the general population is more exposed to cadmium than in any other country. However, hypertension has not been found in patients suffering fom Itai-itai disease,[179] nor in people living close to the district where the endemic disease is found. This is also the case for people living in other cadmium-polluted areas of Japan.[251,270] In spite of these clinical observations, Nogawa et al.[180] observed an increased mortality in cerebrovascular and heart disease in a follow-up study on the causes of death in Japanese farmers with cadmium-induced proteinuria. In a cohort comprising 81 male farmers with proteinuria, 13 fatalities in cerebrovascular and heart disease had occurred between 1975 and 1979, compared to an expected figure of 5.9. These figures are based on regional statistics. In a similar female cohort, comprising 124 persons, the observed mortality in cardiovascular and heart disease over the same period of time was 15 compared to an expected figure of 9.8.

Suggestive evidence for an association between cadmium exposure and increased mortality has also been reported from the U.K.[119] Among 501 residents in the cadmium-polluted village of Shipham (see Chapter 3), there was a slight excess in death caused by diseases of the circulatory system (SMR = 113 for males and females together) and a significantly increased mortality in cerebrovascular disease (SMR = 140). No excess in these causes of death was found in a nearby control village. There were, however, no exposure data on an individual basis and one must therefore be careful before drawing general conclusions.

A large cohort study on mortality among inhabitants of four cadmium-polluted areas compared to mortality among inhabitants of four nonpolluted areas in the same prefectures has been reported by Shigematsu et al.[251] The total cohort was comprised of 333,000 people and mortality rates were investigated for 6 to 30 years in the four areas. There was no excess mortality in cerebrovascular disease, heart disease, or hypertensive disease among the inhabitants of the polluted areas when compared to the control areas. On the contrary, for several of the comparisons, there was a lower age adjusted mortality ratio among inhabitants of cadmium-polluted areas when compared to nonpolluted areas. One problem with this study is the fact that exposure was classified in a rather crude way. It appears that the "exposed population" also included a large number of persons not actually exposed to cadmium. Anyway, the results do not support the hypothesis that cadmium plays a role in the development of human hypertension. This study by Shigematsu et al.[251] is further discussed in Section I.B.2 of Chapter 12.

C. Summary and Conclusions

Experiments have been carried out mainly on female rats of the Long Evans strain. It is clear from these animal experiments that long-term exposure to relatively low levels of cadmium will induce an increase in systolic blood pressure when the animals are fed a rye-based diet containing a very low cadmium concentration. There are also several studies in which no hypertensive effects were recorded. Concentrations of cadmium in drinking water, which were shown to give rise to hypertension in rats, were in the order of 1 to 5 mg Cd per liter. The renal cadmium concentrations in rats exposed to these levels for periods from 6 to 12 months are around 10 to 40 mg/kg. It has not been clearly demonstrated whether this hypertensive effect in rats is the result of ongoing exposure or if it is the result of cadmium accumulation in the body, especially in the kidney. This is probably a crucial question since the daily intake of rats exposed to 5 mg Cd per liter in drinking water would be around 5 µg/kg body weight. This level is considerably higher than the daily intake of cadmium for environmentally exposed humans which would be in the order of 0.3 µg/kg body weight (20 µg/day, 70 kg man). However, the levels of cadmium in kidneys obtained from hypertensive rats are in the same range, about 10 to 40 mg/kg, as those found in middle-aged humans.

Results obtained on examination of the cardiovascular system of cadmium-exposed workers are sparse. Friberg,[87] examining workers in a Swedish accumulator plant, discovered no abnormalities. However, a report from the U.S.S.R., also based on workers in a battery plant, indicates that cardiovascular effects might occur.[279] This is of special interest since somewhat similar effects have also been recorded in exposed rats.[143,144] According to the studies on rats, the hypertensive effects occur at relatively low levels of cadmium in the kidneys. This makes it difficult to interpret cross-sectional studies on workers who had experienced heavy exposure to cadmium. Possibly many of the workers will have renal cadmium concentrations which are higher than those seen in hypertensive rats. Thus, the hypertensive effect, if it also exists in humans, might have disappeared. However, no excess of death due to cardiovascular diseases has been reported from cohort studies on causes of death among cadmium-exposed workers. Carefully conducted studies on possible cardiovascular effects in occupationally exposed workers are needed, especially the possibility that hypertension may develop at relatively low levels of exposure should be examined.

Avaiable data on the cardiovascular effects of cadmium in the general population are not conclusive. There are various uncertainties in connection with some of the early reports, e.g., confounding factors such as smoking have not always been taken into account and sometimes the analytical results published are in obvious error. More recent studies do not support a causal relationship between cadmium exposure and hypertension.

VI. EFFECTS ON THE REPRODUCTIVE ORGANS, ESPECIALLY THE TESTES, AND DOSE-RESPONSE RELATIONSHIPS

Parenteral administration of sufficiently high doses of cadmium salts to male mammals gives rise to testicular necrosis within 24 to 48 hr after administration. This phenomenon was first described by Parizek and Zahor[195] and Parizek.[191] As early as 1919, however, brief mention was made of a "bluish discoloration of the testicles" of experimental animals in a report on the pharmacology of cadmium,[4,249] but this information

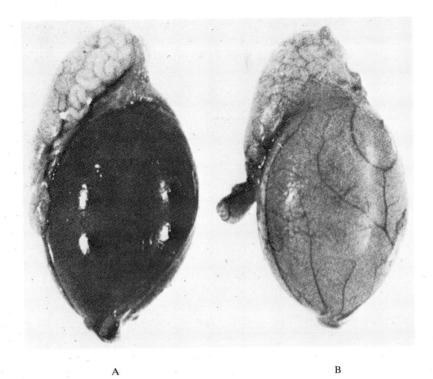

A   B

FIGURE 3.   Rat testis after cadmium chloride injection. (A) Severely congested testis
as a result of cadmium. (B) Normal testis. (From Chang, L. W., Reuhl, K. R., and
Wade, P. R., *Cadmium in the Environment, Part 2,* Nriagu, J. O., Ed., John Wiley &
Sons, New York, 1982, 783—839. With permission.)

seems to have escaped attention at the time. A vast literature on the subject has been
accumulated, especially during the 1960's. The literature was reviewed and discussed
by Gunn and Gould in 1970[99] and by Barlow and Sullivan in 1982.[22] Acute and chronic
effects of cadmium in male animals are discussed in Sections VI.A and VI.B.

Female reproductive organs may also be damaged by cadmium, and this is discussed
in Sections VI.C and VI.D. Effects on the placenta are reviewed in Section VII.A.1.b.
Only very few data which relate cadmium to effects on the reproductive organs of
humans are available, and these are presented in Section VI.E.

Additional information is available in the extensive conference proceeding, *Reproductive and Developmental Toxicity of Metals,* including cadmium, which was published in 1983.[50]

A. Acute Effects in Male Animals
1. Cadmium-Induced Testicular Necrosis

According to the description provided by Parizek and Zahor[195] and Parizek,[191] a s.c.
injection of 0.02 mmol of cadmium chloride or lactate per kilogram (2.2 mg Cd per
kilogram body weight) to rats caused microscopic changes in the testicles within the
first few hours after injection. The organs first became swollen and dark red or purple
(Figure 3). Weight then decreased rapidly and the testicles became small, hard, and
yellowish. At the same time, the weight of the seminal vesicles and the prostate decreased as a result of decreased endocrine activity in the testicles. Histologically, a
capillary stasis and edema of the interstitium were observed 2 to 4 hr after injection,

followed by extensive hemorrhages. Regressive changes of the seminiferous epithelium were seen 4 to 6 hr after injection, progressing to total necrosis within 24 to 48 hr.

At a given time following acute necrosis (about 1 month in the rat), a revascularization of testicles[145,176] and a regeneration of Leydig cells occurred.[3] Simultaneously, the endocrine activity of the testes returned.[3,191] When the injected dose was large enough, the germinal epithelium did not regenerate and the testicles thus functioned as endocrine organs only.[3,101,126] Occasionally, approximately 1 year after injection, interstitial cell tumors developed[101,231] (see also Chapter 12).

Simultaneous to the alteration in the testicles, morphological changes in the spermatozoa of the ductus deferens and proximal parts of the epididymis occur, but spermatozoa in distal parts of the epididymis sometimes remain unaltered. In some cases animals become permanently sterile as early as 24 hr after injection,[127] but if androgens are administered fertility can be maintained up to 9 days after injection of cadmium.[107] Cadmium is also extremely toxic for sperm cells in vitro.[288] The normal sperm motility is inhibited at cadmium concentrations exceeding $1.6~\mu M (= 180~\mu g/\ell)$.[239]

Laskey et al.[150] examined the testicles and other reproductive organs (seminal vesicles, epididymis, and sperm count) of male rats given cadmium chloride at doses from 0.18 to 17 mg Cd per kilogram. A 10 and 60% mortality occurred in the two highest dose groups. Fourteen days after the treatment, the weight of the testes, seminal vesicles, and epididymis of rats that had been treated with cadmium in doses of 1.8 mg Cd per kilogram or higher were significantly reduced. In these groups the human chorionic gonadotrophin (HCG)-induced serum testosterone concentration was reduced to less than 5% of that in control rats. Of particular interest in this study was that even groups of rats that had been given comparably low doses of cadmium (0.18, 0.34, and 0.83 mg Cd per kilogram) showed a reduced response when challenged with HCG, compared to control rats, although this effect was much less pronounced than in the high dose groups (1.8 or 3.7 mg Cd per kilogram).

As discussed in Section VI.B, the acute testicular damage caused by cadmium usually produced a long-term sterility.

2. Susceptibility of Different Animals to Testicular Necrosis at Different Dose Levels

The observations of testicular necrosis found by Parizek and Zahor[195] and Parizek[191] concerned mice and rats. Extensive later studies have confirmed these observations.[3,100-102,126,165] Similar changes have been shown to occur after systemic administration of cadmium salts to other experimental animals such as rabbits,[40,192] monkeys[96,97] guinea pigs,[124,192] and golden hamsters,[192] as well as domestic animals, e.g., calves.[199] However, some animal species such as frog, pigeon, rooster, armadillo and opossum[49] and domestic fowl[73,123] did not develop testicular necrosis, despite doses of 10 to 20 mg CdCl$_2$ per kilogram body weight (corresponding to 6.2 and 12.4 mg Cd per kilogram body weight). Chiquoine[49] suggested that cadmium necrosis is common to species possessing scrotal testes and is absent from those possessing abdominal testes. Opossum were an exception to this generalization.

Sensitivity to cadmium varies for different species, and also for different types of strains within the same species. The lowest dose effective for the elicitation of acute testicular changes in calves is 0.2 mg CdCl$_2$ per kilogram body weight (0.12 mg Cd per kilogram), given intravenously.[199] Subcutaneous injections of cadmium salts at a dose of 0.01 mmol/kg (1.1 mg/kg body weight) can produce total testicular necrosis in some animals, e.g., certain strains of mice.[105] For other animals, e.g., rabbits, higher doses are required. Doses in the order of 2.2 to 5.6 mg/kg body weight are usually used in rats.[100-102,126,191] Kotsonis and Klaassen[146] were also able to induce testicular necrosis after a single oral dose of cadmium, when it exceeded 50 mg Cd per kilogram body weight.

Some animal species, or certain strains of a species, are so resistant to the acute necrotizing effects of cadmium that even doses in the lethal range to not cause alterations.[103] Furthermore, the dose interval between no effect and total testicular necrosis is usually very narrow for a given type of animal.

3. Mechanism of Testicular Necrosis

The exact mechanism of cadmium-induced testicular necrosis is still unknown. It is clear, however, that the vascular bed and the blood flow of the testicles are affected very rapidly following injection of cadmium. This has been shown employing histological,[102,126] electron microscopical,[49,51] angiographical,[176] and functional techniques.[51,124,250,281] The endothelial cells in the small vessels of the testis are affected as early as 15 to 30 min after i.v. administration of cadmium.[93]

Aoki and Hoffer[9] showed that i.p. injection of 3.4 mg Cd per kilogram body weight in rats brought about destruction of the smaller vessels in the testis with leakage of intravenously administered carbon particles (Trypan blue) into the interstitium. This provided evidence of endothelial damage. An interstitial edema developed a couple of hours after the cadmium injection. This in turn decreased the capillary blood flow, giving rise to ischemia and testicular cell necrosis. A similar technique was used by Francavilla et al.[86] providing additional evidence for a vascular mechanism underlying cadmium-induced testicular necrosis. Using acriflavine, a UV-fluorescent dye, given simultaneously with cadmium to rats, the same authors were able to show that cadmium gave rise to increased permeability of the testicular capillary blood system. Capillary damage gives rise to massive vascular escape of fluids and blood substances into the interstitium which subsequently causes edema and circulatory stasis.

Decreased activity in certain enzymes of the testes has been reported to occur shortly after exposure. Hodgen et al.[112] detected an isoenzyme of carbonic anhydrase in rat testis not found in rat kidneys or erythrocytes. They suggested that this organ-specific carbonic anhydrase was the primary target of cadmium toxicity in the testicles. Hodgen and colleagues[111,112] also showed that the activity of this isoenzyme decreased 30 min after cadmium injection and subsequently ceased entirely. Tsang et al.[268] injected male rats with 1 mg Cd per kilogram body weight as cadmium chloride (i.p. injection). The cadmium treatment caused no changes in the levels of cyclic AMP and phosphodiesterase activity in the testis, but there was an increase in the testicular adenylcyclase (57%) and a decrease in the cyclic AMP-dependent protein kinase (40%) without any changes in its binding activity. In the prostate, on the other hand, the cyclic AMP levels and adenylcyclase levels were decreased 43%, protein kinase activity increased 25%, and cyclic AMP binding activity increased 115%. It is likely that cadmium has no specific effect on the cyclic AMP activity, but produced indirect effects due to changes in other metabolic activity. Omaye and Tappel[188] reported decreased activity of testicular glutathione peroxidase 36 hr after the administration of 2.8 mg Cd per kilogram body weight of cadmium chloride to rats.

4. Protective Measures

The necrotizing action of cadmium may be prevented by means of several specific treatments. Administration of zinc, selenium, or cobalt before or simultaneously with cadmium can counteract the necrotizing effect of cadmium. The most effective of these preventive metals is selenium. Twice the molar dose of selenium, 0.04 mmol/kg body weight, has been shown to protect rat testicles against the necrotizing effects of cadmium. This was seen in rats given 0.02 mmol $CdCl_2$ per kilogram body weight corresponding to about 2.2 mg/kg body weight.[128] The preventive effect of selenium has also been confirmed in mice.[106]

In all, 80 to 200 times the molar equivalent of zinc acetate given in 3 injections (5 hr before, simultaneously, and 19 hr after cadmium was administered) afforded complete protection of the testicles of the rat from the effects of 0.04 mmol CdCl₂ per kilogram (about 4.5 mg/kg) body weight.[190-192] These observations have been confirmed by other investigators, e.g., Gunn et al.,[100] who used a total of 3 mmol/kg of zinc acetate to prevent testicular damage as a result of 0.03 mmol CdCl₂ per kilogram in Wistar rats. Cobalt chloride given to rats at a dose of 30 mg CdCl₂ per kilogram body weight, 17 hr prior to administration of 6.60 mg CdCl₂ per kilogram, also protected the testicles from damage.[90] Complexing agents such as cysteine and dimercaptopropanol (BAL) have also been shown to prevent cadmium-induced necrosis of the testicles.[104]

It is of special interest to note that relatively small doses of cadmium, which do not normally cause necrosis of the testis, may serve as a protective measure. This is probably related to the induction of metallothionein synthesis (Chapter 4). Ito and Sawauchi[121] reported that injection of one half to one fourth of the testis-destructive dose of cadmium chloride, 0.1 mℓ of a 0.1% CdCl₂ solution per mouse, 2 days prior to the injection of a testis-destructive dose, prevented testicle destruction in more than half of the animals. This protective effect of pretreatment with cadmium was later confirmed by other investigators.[99,181]

B. Chronic Effects in Male Animals

The testicular damage caused by cadmium usually results in complete sterility.[22] Saksena et al.[235] observed a very reduced fertility when male rats previously given 5 mg CdCl₂ (3 mg Cd) per kilogram body weight were mated with nontreated females. Likewise Laskey et al.[150] found no viable sperms in *vas deferens* of rats treated with cadmium chloride at doses exceeding 1.8 mg/kg 14 days earlier. Kotsonis and Klaassen[146] observed a decreased fertility as well as testicular necrosis and reduced spermatogenesis 2 weeks after oral administration of 100 and 150 mg Cd per kilogram body weight.

According to most of the reports reviewed below, the effects of cadmium on the male reproductive organs, after long-term exposure to comparably low levels of cadmium, are mild and usually totally absent. One report from the U.S.S.R., however, presents contradictory data.[148] According to this report the index of spermatogenesis fell and motility of spermatozoa decreased in rats already at a daily oral exposure to 0.5 to 5 μg Cd per kilogram body weight. Other signs of toxicity, decreased weight and increase in serum alkaline phosphatase activity, are also mentioned in the report. The dose levels related to toxicity in this report are considerably lower than in any other report discussed in this chapter. We believe that there is some major methodological or printing error involved.

Piscator and Axelsson[217] reported on histological and electron microscopic examinations of testicles from rabbits exposed to repeated s.c. (0.25 mg Cd per kilogram, body weight, 5 days/week) injections of cadmium for as long as 24 weeks followed by an interim of 30 weeks before killing. The investigators observed no pathological changes in the testicles in spite of the presence of kidney damage. Nordberg[181] performed studies on mice of the CBA strain, which are very sensitive to cadmium-induced testicular necrosis. By means of repeated injections, considerable amounts of cadmium were accumulated in all the organs, including the testicles, but no histologically evident changes were found in this organ. The author suggested that the cadmium accumulated from long-term exposure is mainly bound to metallothionein and that this protein has a protective effect. Injections of cadmium bound to metallothionein did not produce testicular necrosis at doses which were effective when cadmium only was injected.[181]

The previously mentioned observations by Ito and Sawauchi[121] concerning the protective action of small doses of cadmium are consistent with the finding that no testicular necrosis has occurred in animals repeatedly exposed to cadmium. In agreement with these findings in mice, Sutherland et al.[260] found no evidence of testicular necrosis in rats given daily i.p. injections of cadmium chloride, 1 mg/kg body weight for 45 days. Long-term parenteral exposure to cadmium, however, gives rise to reduced testicular growth as well as increased activity in cyclic AMP.[260] Zenick et al.[293] conducted another long-term study in order to investigate the effect of peroral exposure of cadmium on the male reproductive organs. Male Long Evans hooded rats were exposed to 0, 17.2, 34.4, and 68.8 mg Cd per liter in drinking water over a period of 70 days. Testis weight, sperm count, number of abnormal sperms, and testis morphology were carefully examined. No effect of cadmium exposure was revealed in any of the exposed groups. Furthermore, reproductive outcome was completely normal. Similar negative results with regard to effects from cadmium in the male rat reproductive organs were reported by Battersby et al.[25] They gave rats cadmium in drinking water at a concentration of 5 and 50 mg Cd per liter and saw no morphological changes in the prostate. The testicles were not examined in this study.

Another observation which might be relevant when discussing long-term effects on the testicles is the decrease in "normal" proteinuria demonstrated in mice repeatedly injected with $CdCl_2$ over a period of several months.[182,183] Male mice and male rats normally excrete a high concentration of protein in urine. The major urinary protein is synthetized in the liver under the influence of testosterone.[81,232,233] The observed decrease in proteinuria during cadmium exposure is, therefore, an indication of the action of cadmium on the production of testosterone in the interstitial cells of the testicles. Nordberg[182] also observed lower secretory activity of the epithelium in the seminal vesicles after morphological examination of seminal vesicles from male CBA mice exposed to s.c. injections of cadmium chloride 5 days/week for 6 months. Nordberg's findings were consistent with the theory implying a decreased testosterone activity in cadmium-treated animals. This hypothesis, of lowered plasma testosterone levels following cadmium-induced testicular damage, was later confirmed by Zylber-Haran et al.[295] and is further discussed in Section VIII.A.4.

C. Acute Effects in Female Animals

The ovaries may also be affected by cadmium injections, although the damage usually appears to be less severe compared to what is seen in testes of male animals. Kar et al.[128] noticed that ovaries of prepubertal rats underwent morphological changes after injection of 10 mg $CdCl_2$ (6 mg Cd) per kilogram body weight. After 1 week recovery was complete. Parizek et al.[197] reported that massive ovarian hemorrhages could be brought about by injection of cadmium acetate or chloride at doses of 2.3 to 4.6 mg Cd per kilogram body weight in rats in persistent estrous. Watanabe et al.[283] found similar effects after s.c. injections of 3 and 6 mg $CdCl_2$ (1.8 and 3.6 mg Cd) per kilogram body weight in mice. In prepubertal female rats, injection of 5 mg $CdCl_2$ per kilogram body weight produced endothelial damage and other morphological changes in the blood vessels in the uterus and ovaries.[54] In hamsters, injections of cadmium chloride in doses exceeding 3 mg Cd per kilogram body weight inhibit the normal ovulation, but does not produce long-term sterility.[234]

D. Chronic Effects in Female Animals

To our knowledge only one report provides data on long-term effects of cadmium in the female reproductive organs. Copius Peereboom-Stegeman et al.[56] examined the small arteries in the myometrium of the uterus in female rats that had been injected with 0.036 and 0.18 mg $CdCl_2$ per kilogram body weight daily for 8 to 60 weeks. It appeared that cadmium exposure increased the thickness of the basal lamina in the blood vessels in a dose-related way, but there were also indications of an increased thickness of the basal lamina with age. Possibly cadmium exposure was accelerating the age-related changes in these blood vessels.

E. Humans

There are no data on testicular necrosis in humans. As many mammalian animal species, including monkeys, are susceptible to the action of cadmium on the testicles, it is highly probable that a similar effect would occur in human beings also. However, as cadmium is not used as a drug, humans are not likely to be exposed to injection. If the absorbed peroral dose necessary for elicitation of testicular changes is proportionally equal to the lowest s.c. dose in animals, i.e., approximately 1 mg Cd per kilogram, this would imply that 70 mg of cadmium must be absorbed within a short time by a 70-kg man to produce a similar effect. Such an exposure level is unrealistic and would probably induce vomiting.

Smith et al.[252] found high values of cadmium and some histological changes in the testicles of men autopsied after having suffered severe occupational cadmium fume poisoning. The testicle changes were of a rather unspecific nature, and the authors ascribed them to the terminal illness itself. It is difficult to exclude the possibility that these histological changes were associated with cadmium exposure. Favino et al.[75] investigated the fertility of ten cadmium workers and also analyzed androgens in urine. In this investigation one case of impotency with abnormally low testosterone levels in urine was found. Further studies similar to those made by Favino et al.[75] are necessary before any conclusions can be drawn concerning the possible effect of cadmium on the endocrine function of the testicles of humans exposed to cadmium.

VII. EMBRYOTOXIC AND TERATOGENIC EFFECTS AND DOSE-RESPONSE RELATIONSHIPS

A. Animals

Most experiments on the fetal effects of cadmium have been performed on animals given relatively large doses of cadmium parenterally in a single or small number of doses. Such exposure conditions are unlikely to occur for pregnant women, but are of course of toxicological interest. A few studies have also been carried out on animals exposed to cadmium for longer periods, through inhalation or via food. The latter exposure condition is more likely to be encountered by humans. These two aspects of the teratogenic embryotoxic effects of cadmium will be discussed in the following two sections. The teratogenic effects of cadmium on several species have been reviewed by Ferm and Layton.[78]

The passage of cadmium through the placenta is dependent on the stage of gestation[65,171] (Chapter 6, Section IV.D). In early pregnancy of mice and hamsters, cadmium does reach the yolk sac and primitive gut of the embryo which are connected via the vitellin duct.[65] Once the vitellin duct is closed, very little cadmium reaches the fetus.[2] Sonawane et al.[254] found that less than 0.02% of the total dose of cadmium injected to the dam reached the fetus.

1. After Parenteral Exposure
a. Effects on the Fetus

The effect of cadmium on the fetus, as well as its metabolism and placental transfer, are dependent on the dose given and the gestational age at which cadmium is administered.[65,171] Teratogenic effects of cadmium and death of the embryo occur as a consequence of cadmium given in early pregnancy, whereas fetal death is the dominating effect when cadmium is administered shortly before delivery. Early embryogenetic deaths can be seen at resorption sites in the uterus. Thus, e.g., Krause-Fabricius and

Hilscher[149] observed malformation of fetuses in pregnant rats given a single dose of 2.5 mg cadmium chloride (1.5 mg Cd) per kilogram body weight when the cadmium was administered prior to the 15th day of gestation and fetal deaths when given later than day 12. Between days 12 and 15, both malformations and fetal deaths occurred. The number of resorption sites was not reported as no examination of the uterus was carried out.

The first study on teratogenic effects of cadmium was performed by Ferm and Carpenter[77] on golden hamsters. In this study 2 and 4 mg cadmium sulfate (0.9 and 1.8 mg Cd) per kilogram body weight were given intravenously on the 8th day of gestation. This brought about severe facial abnormalities in the embryos. In the same study the authors showed that a simultaneous injection of zinc could protect against the teratogenic effects of cadmium. Selenium given to hamsters close to or at the same time as cadmium also protects against negative teratogenic effects.[117] Later on Ferm and co-workers described additional types of congenital malformations in hamsters given 2 mg cadmium sulfate per kilogram body weight on days 8 and 9 of gestation. The malformations exhibited were limb defects, changes in the ossification centers of the skull, vertebral column, ribs, hyoid bone, and sternum.[76,94] According to Tassinari and Long,[264] the predominant teratogenic effect of cadmium given to pregnant hamsters, on day 8 is seen in the anterior neural segment which forms the primary palate, nasal septum, and snout. Different strains of hamsters have been shown to exhibit differences in types and frequency of various malformations.[95]

The teratogenic effect of cadmium has also been confirmed in rats[47] and mice.[120] Chernoff[47] studied the teratogenic effects of s.c. injection of cadmium in rats. Daily injections of 4 to 12 mg cadmium chloride (2.6 to 7.7 mg Cd) per kilogram body weight were given on 4 consecutive days beginning on day 13 and extending to the 16th day of gestation. This treatment resulted in a dose-dependent rise in the fetal death rate, decrease in fetal rate, and increase in the rate of anomalies, which included micrognathia, cleft palate, clubfoot, and small lungs. The rate of anomalies rose from 4% at a dose of 4 mg $CdCl_2$ (2.6 mg Cd) per kilogram body weight to 70% at 12 mg $CdCl_2$ (7.7 mg Cd) per kilogram body weight. Toxic effects from cadmium on developing rat lung were also reported by Daston[63] who gave large doses of cadmium (8 mg $CdCl_2$ (5 mg Cd) per kilogram body weight) subcutaneously to pregnant rats on days 12 to 15. This treatment resulted in fetal pulmonary hypoplasia, decreased amount of glycogen present in the fetal lung, and a decreased synthesis of pulmonary surfactant. Barr[23] gave different strains of Wistar rats i.p. injections of cadmium chloride, 1.8 mg Cd per kilogram, and noticed marked differences in regard to response in the form of fetal mortality and type of malformations among animals from the different strains. Samarawickrama and Webb[237] gave pregnant rats a single i.v. injection of 1.25 mg Cd per kilogram body weight between days 9 and 15 of gestation. In agreement with other investigators, a high incidence of malformations was observed. As many as 80% of the fetuses had hydrocephalus. Other defects observed were anophthalmia, microophthalmia, gastroschiasis, and umbilical hernia. It is mentioned in the report that 1.1 mg Cd per kilogram body weight produced no malformations, while 1.35 mg Cd per kilogram body weight killed all embryos.

As it is known that maternal zinc deficiency in rats may produce a high incidence of congenital malformations[118] and that cadmium influences the metabolism of zinc, it is likely that a zinc deficiency will increase fetal susceptibility to cadmium. Parzyck et al.[198] studied the effect of 1.5 mg Cd per kilogram body weight given intraperitoneally to rats on days 8 and 12 of gestation. Both zinc deficiency and cadmium exposure brought about malformations. The incidence was higher, 48% of implantation sites among zinc-deficient rats given cadmium on the 8th day of gestation, when compared to rats with a zinc deficiency only, incidence 22%. Fetal death as indicated by resorbed

Table 2

EFFECT OF CADMIUM ON
EMBRYONIC UPTAKE OF
^{65}Zn ON DAY 12 OF
GESTATION IN THE RAT[237]

Cd^{2+} dose (mg/kg)	Embryonic ^{65}Zn up-take[a]	
	mg/kg	% of control
0	4.9	100.0
0.25	3.9	80.0
0.50	2.6	53.1
0.75	2.2	44.9
1.00	1.7	34.7
1.25	1.3	26.5

[a] Uptake measured at 15 min after the i.v. administration of ^{65}ZnCl$_2$ (3 μCi) to the maternal animal.

implantation sites was not seen at all in zinc-deficient animals, and rarely (13%) in cadmium-exposed rats on a zinc-sufficient diet, but was very prevalent (75%) among rats with both zinc deficiency and exposure to cadmium on day 12 of gestation.

Further experimental data on rats, provided by Samarawickrama and Webb[237] and Webb and Samarawickrama,[284] indicate that maternal cadmium exposure gives rise to a fetal zinc deficiency and that this is one cause of the teratogenic effects observed. By giving i.v. cadmium injections to pregnant rats at doses ranging from 0.25 to 1.25 mg Cd per kilogram body weight on day 12 of gestation, the authors were able to show a dose-related decrease in fetal uptake of a dose of ^{65}Zn given 4 hr later (Table 2). Furthermore, it was shown that maternal cadmium exposure, 1.25 mg Cd per kilogram body weight, resulted in decreased activity of a fetal zinc-dependent enzyme, thymidine kinase, which is responsible for the incorporation of thymidine in DNA. Zinc given together with cadmium at an atomic ratio of 2:1 prevented the inhibition of thymidine incorporation into fetal DNA.[284] Additional evidence that cadmium-induced fetotoxicity is related to a cadmium-induced fetal zinc deficiency was reported by Daston.[62] The author was able to show that coadministration of zinc [12 mg ZnCl$_2$ (5 mg Zn) per kilogram body weight] almost completely eliminated the severe lesions in the fetal lung when the pregnant rat was given cadmium [8 mg CdCl$_2$ (5 mg Cd) per kilogram body weight] on gestation days 12 to 15.

A dose-response relationship for cadmium given to mice subcutaneously on the 7th day of gestation has also been established.[120] The results are summarized in Table 3. An increased incidence of malformations and fetal deaths was seen at doses exceeding 0.3 mg CdCl$_2$ (0.2 mg Cd) per kilogram body weight. Detailed information about the type of cadmium-induced limb defects in developing mice is available from Layton and Layton[151] and Messerele and Webster.[170]

Of particular interest is a study on mice conducted by Layton and Ferm.[152] The authors show that pretreatment with an injection of about 0.7 mg Cd (as sulfate) per kilogram body weight before the pregnancy or with mercury during the pregnancy significantly reduced the number of fetal malformations when cadmium [1.3 mg Cd (as sulfate) per kilogram body weight] was injected on day 9 of pregnancy. It was suggested that the preventive effect of pretreatment with either cadmium or mercury was related to the induction of metallothionein synthesis in the dam.

Table 3

INCIDENCE OF
MALFORMATIONS
AMONG MOUSE FETUSES
DUE TO CdCl$_2$
INJECTION[120]

Dose CdCl$_2$ injected (mg/kg)	Malformed (%)	Dead (%)
5	59.0	17.5
2.5	19.5	12.0
0.6	0.1	11.9
0.3	0	9.9
0	0	9.7

b. Effects on the Placenta

At later stages of pregnancy, large doses of cadmium destroy the placenta and thereby cause fetal death. This was shown primarily by Parižek[193,194] who gave pregnant rats s.c. injections of 2.5 to 4.5 mg Cd per kilogram body weight on the 17th to the 20th day of gestation. In 1968, Parizek et al.[196] demonstrated that simultaneous administration of sodium selenide protected the fetus from destructive effects of cadmium in spite of the fact that placental uptake of cadmium increased. The mechanism underlying the fetal death that occurs after s.c. injections of cadmium in pregnant rats has been further elucidated by Levin and Miller.[158,159] By giving fetal and maternal injections of cadmium on day 18 of gestation, it could be shown that the fetal burden of cadmium per se did not give rise to fetal deaths. Instead, fetal death was the result of placental damage brought about by cadmium. After a single maternal s.c. injection of 4.5 mg Cd per kilogram body weight, the placental blood flow decreased by 73% within 18 to 24 hr after injection when compared to controls. The placenta showed marked morphological changes when examined by light and electron microscopy. The most pronounced changes were seen in the trophoblast cell layer II.[66]

Copius Peereboom-Stegeman et al.[57] has drawn attention to the fact that repeated injections of cadmium chloride to pregnant rats produce morphological changes in the placenta which resemble those seen in placentas obtained from smoking mothers.[276] Pregnant rats which had been given 0.2 mg CdCl$_2$ per kilogram body weight daily during their first 19 days of pregnancy had a decreased volume density of maternal and fetal vessels compared to controls and an increased thickness of the trophoblast barrier between the maternal and fetal circulation.[57]

2. After Inhalation or Peroral Exposure

Schroeder and Mitchener[244] exposed three descendant generations of mice to 10 mg/ℓ cadmium in deionized drinking water. Control animals were given the same diet and water without the addition of cadmium. Toxic effects induced by cadmium were apparent as early as in the first generation. The incidence of infantile deaths and runts was significantly higher in exposed rats compared to controls. At weaning time, pairs were randomly selected from the cadmium-exposed animals and from the controls to produce a second F$_2$ generation. Again the incidence of infantile deaths and runts was higher in exposed rats compared to controls. A congenital abnormality, sharp angulation of the distal third of the tail, occurred in four litters of the cadmium-exposed groups, which comprised 16% of the F$_1$ and F$_2$ generations. Three of five pairs of the cadmium F$_2$ generation failed to breed and the experiment was, therefore, discontinued.

Webster[285] gave pregnant mice 10, 20, and 40 mg Cd per liter in drinking water throughout pregnancy. Mean fetal weight as well as mean litter size were less in exposed groups compared to controls. Parenteral administration of iron completely normalized both litter size and mean fetal weight. There was no mention of teratogenic effects. Machemer and Lorke[162] exposed pregnant rats perorally to different doses of cadmium chloride in food or by oral gavage from 6 to 15 days of pregnancy. Daily doses ranged from about 1.2 to 12.5 mg Cd per kilogram body weight in food and from 1.8 to 61.3 Cd mg/kg body weight by gavage. Maternal toxicity was seen only in the pregnant rats exposed to 100 mg Cd per kilogram in food, which corresponds to approximately 12.5 mg/kg body weight, and in those exposed to 6.1 mg/kg or more by oral gavage. Effects on the fetus in the form of increased incidence of reduced fetal and placental weight, and increase in stunted forms and malformations, were only seen in the two groups of pregnant rats given the highest doses by oral gavage — 18 and 61.3 mg Cd per kilogram body weight. Similar results were obtained by Barański et al.[21] They gave pregnant rats cadmium chloride daily from day 7 to day 16 by gavage at doses ranging from 2 to 40 mg Cd per kilogram body weight. Doses exceeding 8 mg Cd per kilogram body weight significantly reduced the fetal weight. Only the highest dose (40 Cd mg/kg body weight) produces a significant decrease in the number of live fetuses per litter.

Possible mutagenic and teratogenic effects of cadmium, in rats, were also examined by Wills et al.[289] The authors exposed four descendant generations of rats to very low cadmium concentrations in the diet; 0.07, 0.10, and 0.125 mg Cd per kilogram, and saw no effects on growth, reproduction, or frequency of malformations. However, exposure levels were so low, in fact close to the natural background levels, that toxic effects were hardly to be expected. Zenick et al.[293] in experiments on orally exposed male rats, which were discussed earlier (Section VI.B), did not find any teratogenic effects in litters when the cadmium-exposed rats were mated with nonexposed females.

Two experiments have been conducted on pregnant rats exposed to cadmium via inhalation. Cvetkova[60] exposed female rats to cadmium sulfate, 3 mg/m^3, during pregnancy. On day 22 of gestation, examination of half of the exposed rats was performed. Fetal weight was found to be lower among the exposed animals, compared to controls, but there was no evidence of mortality. The other half of the rats were allowed to complete gestation. The weight of the newborn of exposed rats was lower than that of controls. The offspring of cadmium-exposed rats displayed an increased prenatal mortality. In a similar experiment Prigge[223] exposed pregnant and nonpregnant rats to cadmium chloride aerosols at three dose levels: 0.2, 0.4, and 0.6 mg Cd per cubic meter, continuously for 21 days. Weight gain was reduced in all three groups of exposed pregnant rats, but in only one group of nonpregnant animals, that were exposed to the highest dose (0.6 mg Cd per cubic meter). Fetal weights were significantly reduced in animals exposed to the highest level. Fetal alkaline phosphatase activity was also elevated in the most highly exposed group.

B. Humans

From the U.S.S.R.[60] it has been reported that pregnant women exposed to high concentrations of cadmium give birth to children with lower birth weights compared to normals. An examination was made of 106 women, aged 18 to 48 years employed in cadmium factories; 61 worked in an alkaline accumulator factory, 21 in a zinc smelter, and 24 in a chemical factory. The control group consisted of 20 women. Levels of cadmium in air varied from 0.1 to 25, 0.02 to 25, and 0.16 to 35 mg/m^3 in the different factories, respectively. In the alkaline accumulator factory and the zinc smelter groups, the weights of newborn children, 27 boys and girls in each group, were significantly lower than the weights of children born to members of the control group. In 4 of 27 children born to women in the zinc smelter group, signs of rachitis and delayed development of teeth were recorded. The author concluded that pregnant women should not work in cadmium-contaminated environments. This report is very interesting but lacks data on, e.g., other effects of cadmium which would be expected to occur in view of the rather excessive exposure. It is, therefore, as yet impossible to establish any dose-response relationships for the teratogenic and embryotoxic effects of cadmium in humans.

C. Summary and Conclusions

It is clear from the experimental studies on animals that injections of cadmium during pregnancy, at doses in the order of mg/kg body weight, give rise to fetal death as well as severe malformations when administered in early pregnancy. Species and strain differences probably exist in regard to the types of malformations induced. It is reasonable to assume that the mechanism underlying the embryotoxic and teratogenic effects of cadmium involves interactions with zinc, especially in view of the fact that only a very limited amount of cadmium passes through the placenta into the fetus after maternal exposure subsequent to the closure of the vitellin duct. Maternal zinc deficiency increases fetal susceptibility to maternal cadmium injections. Furthermore, maternal cadmium injections inhibit the transportation of zinc from the mother to the fetus as well as inhibiting the zinc-dependent enzymes present in the fetus.

Large doses of cadmium given to animals at later stages of pregnancy severely damage the placenta and this may cause fetal death. Recent data suggest that repeated injections of lower doses of cadmium may increase the barrier between the maternal and fetal circulation.

High peroral exposure to cadmium in pregnant animals may also result in reduced fetal weight and, as was observed in certain experiments, malformations also.

There is a paucity of data with respect to teratogenic effects of cadmium in humans. However, it is possible that excessive cadmium exposure, such as that which may occur in certain industries, might influence zinc metabolism and thereby also constitute a hazard for normal growth and development in the fetus.

VIII. OTHER EFFECTS

A. Animals

1. Decreased Weight Gain

Decreased normal weight gains in experimental animals, as a nonspecific effect of long-term peroral exposure to cadmium, have been noticed by many researchers in the field: in rats by Decker et al.,[64] Perry et al.,[211] Prigge,[224] and Mahaffey et al.,[163] in lamb by Doyle et al.,[67] and in swine by Cousins et al.[58] Exposure levels utilized in order to produce significant effects on growth in experimental animals have usually exceeded 10 mg Cd per liter in water or 10 mg/kg in food. Nutritional factors other than cadmium are of great importance in this respect. For example, the negative effects on growth induced by cadmium are markedly aggravated by calcium deficiency. Figure 4 shows a typical example of the growth curves of rats given 50 mg Cd per liter in drinking water and fed a calcium-adequate or calcium-deficient diet.[122] Likewise, additions of other trace elements, such as zinc, copper, and iron, will reduce the toxic effects of cadmium.[263]

2. Nervous System

Large doses of cadmium produce toxic effects in the peripheral nervous system. Gabbiani[89] observed hemorrhages in sensory ganglias after s.c. injections of cadmium chloride, from 2.5 to 28 mg/kg body weight, to rats. Similar lesions were also produced

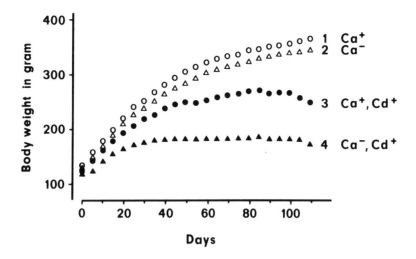

FIGURE 4. Growth of rats given different synthetic diets and water containing cadmium. Group 1: calcium-adequate diet and water. Group 2: calcium-deficient diet and water. Group 3: calcium-adequate diet and water containing 50 mg/ℓ Cd. Group 4: calcium-deficient diet and water containing 50 mg/ℓ Cd. (From Itokawa, Y., Abe, T., Tabei, R., and Tanaka, S., *Arch. Environ. Health*, 28, 149—154, 1974. With permission.)

in mice, hamsters, and guinea pigs.[12,90] The type of effects seen in the peripheral nervous system after cadmium injections are influenced by the age of the animals. Severe changes are usually seen in adult rats.[13] Pretreatment with 5 s.c. injections of cadmium chloride in relatively small doses, 2 mg CdCl₂ (1.2 mg Cd) per kilogram body weight, prevented the ganglionic lesions produced by a single i.v. dose of 8 mg CdCl₂ (5 mg Cd) per kilogram body weight.[91] Pretreatment with zinc acetate was also observed to have preventive properties.[90] The mechanism underlying cadmium-induced injury to sensory ganglias is probably endothelial vascular lesions[93] similar to those seen in testis. Therefore, they are not likely to occur as a result of long-term exposure.

So far only one study has investigated the long-term effects of cadmium on the peripheral nervous system. Sato et al.[238] gave rats cadmium in drinking water for from 18 to 31 months, increasing doses stepwise from 10 to 40 mg/ℓ. These authors noticed peripheral neuropathy with degeneration and accumulation of glycogen particles in the axoplasm. They also described how the 11 exposed animals showed slight weakness in the hindquarters and muscular atrophy. These observations of chronic toxic effects of cadmium on the peripheral nervous system are highly interesting but need to be confirmed since no similar effects have been reported by other researchers, in spite of the large number of chronic toxicity studies that have been carried out. Possible interference due to cadmium or other trace elements in the diet needs to be elucidated. The rats examined were rather old (31 months) and age-related morphological changes of the peripheral nervous system may have been involved.

The central nervous system of newborn mice, rats, and rabbits is more sensitive to cadmium than that of adult animals. Gabbiani et al.[92] produced hemorrhagic encephalopathy in newborn rats and rabbits by s.c. injections of 10 to 20 mg Cd per kilogram body weight. No such changes were seen in mature animals. Likewise, Webster and Valois,[286] after giving mice s.c. injections of cadmium chloride at doses ranging from 2 to 8 mg (1.2 to 5 mg Cd) per kilogram body weight, reported the emergence of petechial hemorrhages, edema, and cellular pyknosis in the immature brain. In rats older than 22 days, no such effects occurred after cadmium injections. Wong and

Klaassen[292] noticed that surviving newborn mice became hyperactive later in life after parenteral administration of cadmium in doses of 4 mg Cd per kilogram body weight. Peroral cadmium exposure of young rats has also been reported to induced behavioral changes. Smith et al.[253] report that young rats given 0.25 mg $CdCl_2$ (0.15 mg Cd) per kilogram body weight by gavage in fact did better than controls on subsequent performance tests.

Effects of cadmium on the peripheral and central nervous system in animals have been reviewed recently by Arvidson.[14]

3. Immunological System

Long-term peroral exposure of mice to cadmium via drinking water gave rise to a decrease in the number of antibody-forming cells in the spleen, as well as a decrease in the antibody production response when animals were challenged with an antigen.[136,138,139] Repeated i.p. injections of cadmium in mice have also been shown to inhibit the cellular-mediated immune system.[38] Cook et al.[53] showed that i.v. cadmium administration of 6 mg cadmium (as cadmium acetate) per kilogram body weight to rats markedly affected the susceptibility to injections of *Escherichia coli*. The authors suggested that impaired immune response in cadmium-exposed rats could be one of the mechanisms involved. Suzuki et al.[261] examined the thymus and spleen in 8-week-old mice given 1.8 mg Cd per kilogram body weight intraperitoneally. The weight of the thymus decreased significantly shortly after exposure, whereas the spleen almost doubled in weight. The authors interpreted this as an acute effect of cadmium on the immunological system. The spleen may also increase considerably in weight as a result of hemolysis. Hemolysis might occur when large doses of cadmium are given parenterally (Section IV.A.).

Similar results in regard to the effects of injections of cadmium on the weight of the thymus and spleen were reported by Kojima and Tamura[134] and Kawamura et al.[131] in mice. Kojima and Tamura[134] were also able to demonstrate that a single injection of 0.75 to 6 mg Cd per kilogram body weight resulted in suppression of the induction of delayed hypersensitive response, as well as suppressed memory T- and B-cell activities, when cadmium was administered 2 days before investigation of the rats' immune response. Kawamura et al.[131] noticed that primary antibody response was suppressed in mice given cadmium chloride (0.3 to 2.4 mg Cd per kilogram body weight) 2 days after immunization with sheep blood cells.

Experimental data on the immunological effects of cadmium have been reviewed by Koller.[137]

4. Endocrine Organs

Secondary to the dramatic effects from cadmium injections on testis, described in Section VI, cadmium also affects the serum levels of testosterone and gonadotropins. A single s.c. injection of cadmium chloride, 10 mg (6 mg Cd) per kilogram body weight, given to male rats, causes a prompt decline in the pituitary follicle-stimulating hormone (FSH) level and an increase in luteinizing hormone (LH).[130] Zylber-Haran et al.[295] noted a change in plasma concentration of testosterone, FSH, and LH in male rats 12 to 280 days after the s.c. administration of 2.2 mg Cd (as $CdCl_2$) per kilogram body weight. This dose gave rise to severe testicular damage. Testosterone levels were markedly lowered in exposed rats compared to controls, whereas the plasma concentrations of FSH and LH were excessively elevated. At the end of the observation period, plasma levels of testosterone and LH tended to normalize, but were still significantly different from control animals.

Zielinska-Psuja et al.[294] reported that long-term peroral exposure to cadmium may influence the concentration of LH in blood. There was, however, no consistent change.

Sometimes exposed rats had higher levels than controls and vice versa. These changes may also be secondary to testicular or hepatic damage. Is is not known whether cadmium directly affects the pituitary as well. Nevertheless, it is interesting to note that Berlin and Ullberg[30] observed an uptake in the pituitary gland after a single dose of radioactive cadmium.

There are also indications that heavy exposure to cadmium might influence the hormones involved in carbohydrate metabolism. For example, Merali and Singhal[167] have reported a decreased release of insulin in vitro from islets of Langerhans obtained from rats which were exposed to daily peroral levels of cadmium at a dose of 1 mg Cd per kilogram body weight for a period of 45 days. Nishiyama and Takata[177] found that daily injections of 2 mg $CdCl_2$ (1.3 mg Cd) per kilogram body weight in rats resulted in a lowering and more rapid clearance of serum corticosterone.

Influence on the thyroid has been demonstrated by Anbar and Inbar[5] who gave mice (weight not given) 0.12 mg Cd intraperitoneally. They found a decrease in the uptake of iodine, but as the dose was large and other metals, both essential and nonessential, had the same effect, no reliable conclusions can be drawn from the experiment. A decrease in the uptake of iodine was also found by Balkrishna,[20] who gave rats single intramuscular injections of cadmium chloride, 10 mg (6 mg Cd) per kilogram.

B. Humans

Over the years, many different symptoms have been reported in human beings exposed to cadmium. These include loss of appetite, loss of weight, fatigue, increase in the erythrocyte sedimentation rate (ESR), etc. These unspecific symptoms can be related to the systemic effects. Valetas[273] reported in more detail on the patients from a French accumulator factory first described by Nicaud et al.[175] In addition to the bone effects and the pain (see Chapter 10), Valetas also mentioned that several workers experienced paresthesia and involuntary muscular contractions. This could be one effect resulting from the abnormal changes in serum electrolytes, such as calcium or potassium, which may constitute a secondary effect caused by severe kidney damage.

More specific effects from cadmium have been the yellow discoloration of the proximal part of the front teeth[10,24,87,226,273] and anosmia.[87] Naturally, discoloration of the teeth is influenced by dental hygiene as well as by the degree of cadmium exposure.

Anosmia was found by Friberg[87] in about one third of a group of workers with a mean exposure time to cadmium oxide dust of 20 years. In a Japanese smelter, Iguchi (1968), cited by Sakurai,[236] found yellow discoloration of teeth in 6 of 16 examined workers exposed for more than 8 years. Air levels of cadmium were mentioned to range from 0.008 to 9.3 mg/m³. Baader[18] in Germany and recently Apostolov[10] from Bulgaria in agreement with Friberg[87] also noted that anosmia is common in workers exposed to cadmium oxide dust for long periods of time. Suzuki et al.[262] and Tsuchiya[269] in Japan found no increase in prevalence of anosmia in workers exposed to cadmium stearate and cadmium oxide fumes.

Probably, cadmium-induced dental discoloration and anosmia are effects mainly seen following very excessive exposure to cadmium.

Symptoms occurring in the nervous system have been reported by Vorobjeva[278] who investigated 160 workers in an accumulator factory in the U.S.S.R. Subjective symptoms consisted of headache, vertigo, sleep disturbances, etc. Physical examination revealed increase in knee-joint reflexes, tremor, dermographia, and sweating. Special attention given to sensory, dermal, optic, and motoric chronaxia revealed that cadmium-exposed workers who reported subjective disturbances also showed changes on these tests. Data on dose-response relationships were not available.

Contact allergy to cadmium is very rare, if it exists at all.[280] Among 1502 eczema patients tested with 2% cadmium chloride solution in a routine patch test, none revealed contact allergy. Cadmium sulfide is sometimes used as a yellow tattoo pigment. The pigment is thereby deposited intradermally. Local phototoxic reactions may take place when the skin is exposed to UV light. The local reaction brought about by cadmium sulfide pigment is probably connected with the marked photoconduction properties of this cadmium compound. Of 24 patients with yellow tattoos examined by Björnberg,[35] 18 experienced swelling when exposed to sunlight.

REFERENCES

1. Adams, R. G., Harrison, J. F., and Scott, P., The development of cadmium-induced proteinuria, impaired renal function, and osteomalacia in alkaline battery workers, *Q. J. Med.,* 38, 425—443, 1969.

2. Ahokas, R. A. and Dilts, P. V., Cadmium uptake by the rat embryo as a function of gestational age, *Am. J. Obstet. Gynecol.,* Sept. 15, 219—222, 1979.

3. Allanson, M. and Deanesly, R., Observations on cadmium damage and repair in rat testes and the effects on the pituitary gonadotrophs, *J. Endocrinol.,* 24, 453—462, 1962.

4. Alsberg, C. L. and Schwartze, E. W., Pharmacological action of cadmium, *J. Pharmacol.,* 13, 504—505, 1919.

5. Anbar, M. and Inbar, M., The effect of certain metallic cations on the iodide uptake in the thyroid gland of mice, *Acta Endocrinol.,* 46, 643—652, 1964.

6. Andersson, K., Elinder, C.-G., Hogstedt, C., Kjellström, T., and Spång, G., Mortality among cadmium workers in a Swedish battery factory, *Toxicol. Environ. Chem.,* 9, 53—62, 1984; *Arbete and Halsa,* Swedish National Board of Occupational Safety and Health, Solna, 1983, 31.

7. Ando, M., Cadmium effect on microsomal drug-metabolizing enzyme activity in rat livers with respect to differences in age and sex, *Environ. Res.,* 27, 446—456, 1982.

8. Andreuzzi, P. and Odescalchi, C. P., Experimental acute intoxication from cadmium chloride in the rabbit. I. Changes in the GOT-activity in the serum, *Boll. Soc. Ital. Biol. Sper.,* 34, 1376—1379, 1958 (in Italian).

9. Aoki, A. and Hoffer, A. P., Reexamination of the lesions in rat testis caused by cadmium, *Biol. Reprod.,* 18, 579—591, 1978.

10. Apostolov, Ch., Effects on the upper respiratory tract after long-term work with cadmium, in *Kadmium-Symposium August 1977 in Jena,* Friedrich-Schiller-Universität, Jena, 1979, 322—325 (in German).

11. Armstrong, B. G. and Kazantzis, G., The mortality of cadmium workers, *Lancet,* June, 1424—1427, 1983.

12. Arvidson, B., Regional differences in severity of cadmium-induced lesions in the peripheral nervous system in mice, *Acta Neuropathol.,* 49, 213—224, 1980.

13. Arvidson, B., Influence of age on the development of cadmium-induced vascular lesions in rat sensory ganglia, *Environ. Res.,* 32, 240—246, 1983.

14. Arvidson, B., Is cadmium toxic to the nervous system?, *T.I.N.S.,* XI—XIV, September 1981.

15. Athanasiu, M. and Langlois, P., Comparison of salts of cadmium and zinc, *Arch. Physiol.* 28, 251—263, 1896 (in French).

16. Axelsson, B. and Piscator, M., Renal damage after prolonged exposure to cadmium. An experimental study, *Arch. Environ. Health,* 12, 360—373, 1966a.

17. Axelsson, B. and Piscator, M., Serum proteins in cadmium-poisoned rabbits with special reference to hemolytic anemia, *Arch. Environ. Health,* 12, 374—381, 1966b.

18. Baader, E. W., Chronic cadmium poisoning, *Dtsch. Med. Wochenschr.,* 76, 484—487, 1951 (in German).

19. Baker, T. D. and Hafner, W. G., Cadmium poisoning from a refrigerator shelf used as an improvised barbecue grill, *Public Health Rep.,* 76, 543—544, 1961.

20. Balkrishna, Changes of thyroid function in response to cadmium administration in rats — studies with ^{131}I, *J. Sci. Ind. Res.,* 21C, 187—189, 1962.

21. Baranski, B., Stetkiewicz, I., Trzcinka-Ochocka, M., Sitarek, K., and Szymczak, W., Teratogenicity, fetal toxicity and tissue concentration of cadmium administered to female rats during organogenesis, *J. Appl. Toxicol.,* 2, 255—259, 1982.

22. Barlow, S. M. and Sullivan, F. M., Cadmium and its compounds, in *Reproductive Hazards of Industrial Chemicals. An Evaluation of Animal and Human Data,* Barlow, S. M., Ed., Academic Press, London, 1982, 137—173.

23. Barr, M., Jr., The teratogenicity of cadmium chloride in two stocks of Wistar rats, *Teratology,* 7, 237-242, 1973.

24. Barthelemy, P. and Moline, R., Chronic intoxication of cadmium hydrate, the early sign: yellow rings on the teeth, *Paris Med.,* 1, 7—8, 1946.

25. Battersby, S., Chandler, J. A., and Morton, M. S., The effect of orally administered cadmium on the ultrastructure of the rat prostate, *Urol. Res.,* 10, 123—130, 1982.

26. Beevers, D. G., Campbell, B. C., Goldberg, A., Moore, M. R., and Hawthorne, V. M., Blood-cadmium in hypertensives and normotensives, *Lancet,* 4, 1222—1224, 1976.

27. Beevers, D. G., Cruickshank, J. K., Yeoman, W. B., Carter, G. F., Goldberg, A., and Moore, M. R., Blood-lead and cadmium in human hypertension, *J. Environ. Pathol. Toxicol.,* 4-2:3, 251—260, 1980.

28. Berlin, M. and Friberg, L., Bone-marrow activity and erythrocyte destruction in chronic cadmium poisoning, *Arch. Environ. Health,* 1, 478—486, 1960.

29. Berlin, M. and Piscator, M., Blood volume in normal and cadmium-poisoned rabbits, *Arch. Environ. Health,* 2, 576—583, 1961.

30. Berlin, M. and Ullberg, S., The fate of ^{109}Cd in the mouse. An autoradiographic study after a single intravenous injection of ^{109}Cd Cl$_2$, *Arch. Environ. Health,* 7, 686—693, 1963.

31. Berlin, M., Fredricsson, B., and Linge, G., Bone-marrow changes in chronic cadmium poisoning in rabbits, *Arch. Environ. Health,* 3, 176—184, 1961.

32. Bernard, A., Buchet, J. P., Roels, H., Masson, P., and Lauwerys, R., Renal excretion of proteins and enzymes in workers exposed to cadmium, *Eur. J. Clin. Invest.,* 9, 11—22, 1979.

33. Beton, D. C., Andrews, G. S., Davies, H. J., Howells, L., and Smith, G. F., Acute cadmium fume poisoning, five cases with one death from renal necrosis, *Br. J. Ind. Med.,* 23, 292—301, 1966.

34. Bierenbaum, M. L., Dunn, J., Fleischman, A. I., and Arnold, J., Possible toxic water factor in coronary heart-disease, *Lancet,* May 3, 1008—1010, 1975.

35. Björnberg, A., Reactions to light in yellow tattoos from cadmium sulfide, *Arch. Dermatol.,* 88, 267—271, 1963.

36. Bonnell, J. A., Emphysema and proteinuria in men casting copper-cadmium alloys, *Br. J. Ind. Med.,* 12, 181—197, 1955.

37. Bonnell, J. A., Kazantzis, G., and King, E., A follow-up study of men exposed to cadmium oxide fume, *Br. J. Ind. Med.,* 16, 135—145, 1959.

38. Bozelka, B. E. and Burkholder, P. M., Inhibition of mixed leukocyte culture responses in cadmium-treated mice, *Environ. Res.,* 27, 421—432, 1982.

39. Calabrese, E. J., Moore, G. S., Tuthill, R. W., and Sieger, Th. L., Eds., *Drinking Water and Cardiovascular Disease,* Pathotox Publishers, Park Forest South, Ill., 1980.

40. Cameron, E. and Foster, C. L., Observations on the histological effects of sub-lethal doses of cadmium chloride in the rabbit, *J. Anat.,* 97, 269—280, 1963.

41. Carroll, R. E., The relationship of cadmium in the air to cardiovascular disease death rates, *J.A.M.A.,* 198, 267—269, 1966.

42. Castenfors, J. and Piscator, M., cited in *Cadmium in the Environment,* 2nd ed., Friberg, L., Piscator, M., Nordberg, G. F., and Kjellström, T., Eds., CRC Press, Boca Raton, Fla., 1974, 118.

43. Commission of the European Communities, *Hardness of Drinking Water and Public Health,* Amavis, R., Hunter, W. J., and Smeets, J. G. P. M., Eds., Pergamon Press, New York, 1976.

44. Commission of the European Communities, *Criteria (Dose/Effect Relationships) for Cadmium,* Pergamon Press, New York, 1978.

45. Chang, L. W., Reuhl, K. R., and Wade, P. R., Pathological effects of cadmium poisoning, in *Cadmium in the Environment, Part 2,* Nriagu, J. O., Ed., John Wiley & Sons, New York, 1982, 783—839.

46. Chapatwala, K. D., Hobson, M., Desaiah, D., and Rajanna, B., Effect of cadmium on hepatic and renal gluconeogenic enzymes in female rats, *Toxicol. Lett.,* 12, 27—34, 1982.

47. Chernoff, N., Teratogenic effects of cadmium in rats, *Teratology,* 8, 29—32, 1973.

48. Chiappino, G. and Baroni, M., Morphological signs of hyperactivity of the renin-aldosterone system in cadmium-induced experimental hypertension, *Med. Lav.,* 60, 297—305, 1969 (in Italian).

49. Chiquoine, A. D., Observations on the early events of cadmium necrosis of the testis, *Anat. Rec.,* 149, 23—36, 1964.

50. Clarkson, T. W., Nordberg, G. F., and Sager, P. R., *Reproductive and Developmental Toxicity of Metals,* Plenum Press, New York, 1983.

51. Clegg, E. J. and Carr, I., Changes in the blood vessels of the rat testis and epididymis produced by cadmium chloride, *J. Pathol. Bacteriol.,* 94, 317—322, 1967.

52. Cole, G. M. and Baer, L. S., "Food poisoning" from cadmium, *U.S. Nav. Med. Bull.*, 43, 398—399, 1944.

53. Cook, J. A., Hoffmann, E. O., and Di Luzio, N. R., Influence of lead and cadmium on the susceptibility of rats to bacterial challenge, *Proc. Soc. Exp. Biol. Med.*, 150, 741—747, 1975.

54. Copius Peereboom-Stegeman, J. H. J. and Jongstra-Spaapen, E. J., The effect of a single sublethal administration of cadmium chloride on the microcirculation in the uterus of the rat, *Toxicology*, 13, 199—213, 1979.

55. Copius Peereboom-Stegeman, J. H. J., Melet, J., Copius Peereboom, J. W., and Hooghwinkel, G. J. M., Influence of chronic Cd intoxication on the alkaline phosphatase activity of liver and kidney; biochemical, histochemical and histological investigations, *Toxicology*, 14, 67—80, 1979.

56. Copius Peereboom-Stegeman, J. H. H., Jongstra-Spaapen, E. J., and Oosting, H., Light and electron microscopic study of small blood vessels in the uterus of the rat after chronic cadmium exposure, in *Industrial Conference on Heavy Metals in the Environment*, Vol. 2, CEP Consultants Ltd., Heidelberg, 1983a.

57. Copius Peereboom-Stegeman, J. H. J., Velde, van der, W. J., and Dessing, J. W. M., Influence of cadmium on placental structure, *Ecotoxicol. Environ. Saf.*, 7, 79—86, 1983b.

58. Cousins, R. J., Barber, A. K., and Trout, J. R., Cadmium toxicity in growing swine, *J. Nutr.*, 103, 964—972, 1973.

59. Cummins, P. E., Dutton, J., Evans, C. J., Morgan, W. D., Sivyer, A., and Elwood, P. E., An in-vivo study of renal cadmium and hypertension, *Eur. J. Clin. Invest.*, 10, 459—461, 1980.

60. Cvetkova, R. P., Materials on the study of the influence of cadmium compounds on the generative function, *Gig. Tr. Prof. Zabol.*, 14, 31—33, 1970 (in Russian with English summary).

61. Dalhamn, T. and Friberg, L., The effect of cadmium on blood pressure and respiration and the use of dimercaprol (BAL) as antidote, *Acta Pharmacol. Toxicol.*, 10, 199—203, 1954.

62. Daston, G. P., Fetal zinc deficiency as a mechanism for cadmium-induced toxicity to the developing rat lung and pulmonary surfactant, *Toxicology*, 24, 55—63, 1982a.

63. Daston, G. P., Toxic effects of cadmium on the developing rat lung. II. Glycogen and phospholipid metabolism, *J. Toxicol. Environ. Health*, 9, 51—61, 1982b.

64. Decker, L. E., Byerrum, R. U., Decker, C. F., Hoppert, C. A., and Langham, R. F., Chronic toxicity studies. I. Cadmium administered in drinking water to rats, *A.M.A. Arch. Ind. Health*, 18, 228—231, 1958.

65. Dencker, L., Danielsson, B., Khayat, A., and Lindgren, A., Disposition of metals in the embryo and fetus, in *Reproductive and Developmental Toxicity of Metals*, Clarkson, T. W., Nordberg, G. F., and Sager, P. R., Eds., Plenum Press, New York, 1983, 607—632.

66. Di Sant'Agnese, P. A., Demesy Jensen, K., Levin, A., and Miller, R. K., Placental toxicity of cadmium in the rat: an ultrastructural study, *Placenta*, 4, 149—164, 1983.

67. Doyle J. J., Pfander, W. H., Crenshaw, D. B., and Snethen, J. M., The induction of chromosomal hypodiploidy in sheep leucocytes by cadmium, *Interface*, 3, 9, 1974.

68. Doyle J. J., Bernhoft, R. A., and Sandstead, H. H., The effects of a low level of dietary cadmium on blood pressure, ^{24}Na, ^{42}K, and water retention in growing rats, *J. Lab. Clin. Med.*, 86, 57—63, 1975.

69. Dudley, R. E. and Klaassen, C. D., Changes in hepatic glutathione concentration modify cadmium-induced hepatotoxicity, *Toxicol. Appl. Pharmacol.*, 72, 530—538, 1984.

70. Dudley, R. E., Svoboda, D. J., and Klaassen, C. D., Acute exposure to cadmium causes severe liver injury in rats, *Toxicol. Appl. Pharmacol.*, 65, 302—313, 1982.

71. Eakin, D. J., Schroeder, L. A., Whanger, P. D., and Weswig, P. H., Cadmium and nickel influence on blood pressure, plasma renin, and tissue mineral concentrations, *Am. J. Physiol.*, 238, 53—61, 1980.

72. Engström, B. and Nordberg, G. F., Effects of detergent formula chelating agents on the metabolism and toxicity of cadmium in mice, *Acta Pharmacol. Toxicol.*, 43, 387—397, 1978.

73. Erickson, J. E. and Pincus, G., Insensitivity of fowl testes to cadmium, *J. Reprod. Fertil.*, 7, 379—382, 1964.

74. Faeder, E. J., Chaney, S. O., King, L. C., Hinners, T. A., Bruce, R., and Fowler, B. A., Biochemical and ultrastructural changes in livers of cadmium-treated rats, *Toxicol. Appl. Pharmacol.*, 39, 473—487, 1977.

75. Favino, A., Candura, F., Chiappino, G., and Cavalleri, A., Study on the androgen function of men exposed to cadmium, *Med. Lav.*, 59, 105—110, 1968.

76. Ferm, V. H., Developmental malformations induced by cadmium, *Biol. Neonate*, 19, 101—432, 1971.

77. Ferm, V. H. and Carpenter, S. J., The relationship of cadmium and zinc in experimental mammalian teratogenesis, *Lab. Invest.*, 18(4), 429—432, 1968.

78. Ferm, V. H. and Layton, W. M., Jr., Teratogenic and mutagenic effects of cadmium, in *Cadmium in the Environment, Part 2*, Nriagu, J. O., Ed., John Wiley & Sons, New York, 1981, 743—756.

79. Fielder, R. J. and Dale, E. A., Toxicity review 7, *Cadmium and Its Compounds,* Her Majesty's Stationery Office, London, 1983.

80. Fingerle, H., Fischer, G., and Classen, H. G., Failure to produce hypertension in rats by chronic exposure to cadmium, *Food Chem. Toxicol.,* 20, 301—306, 1982.

81. Finlayson, J. S., Asofsky, R., Potter, M., and Runner, C. C., Major urinary protein complex of normal mice: origin, *Science,* 149, 981—982, 1965.

82. Folsom, A. R. and Prineas, R. J., Drinking water composition and blood pressure: a review of the epidemiology, *Am. J. Epidemiol.,* 115, 818—832, 1982.

83. Fowler, B. A., Jones, H. S., Brown, H. W., and Haseman, J. K., The morphologic effects of chronic cadmium administration on the renal vasculature of rats given low and normal calcium diets, *Toxicol. Appl. Pharmacol.,* 34, 233—252, 1975.

84. Fox, M. R. S. and Fry, B. E., Jr., Cadmium toxicity decreased by dietary ascorbic acid supplements, *Science,* 169, 989—991, 1970.

85. Fox, M. R. S., Fry, B. E., Jr., Harland, B. F., Schertal, M. E., and Weeks, C. E., Effects of ascorbic acid on cadmium toxicity in the young coturnix, *J. Nutr.,* 101, 1295—1305, 1971.

86. Francavilla, S., Moscardelli, S., Francavilla, F., Casasanta, N., Properzi, G., Martini, M., and Santiemma, V., Acute cadmium intoxication: influence of cyproterone acetate on the testis and epididymis of the rat, *Arch. Androl.,* 6, 1—11, 1981.

87. Friberg, L., Health hazards in the manufacture of alkaline accumulators with special reference to chronic cadmium poisoning. Doctoral thesis, *Acta Med. Scand.,* 138(Suppl. 240), 1—124, 1950.

88. Friberg, L., Iron and liver administration in chronic cadmium poisoning and studies on the distribution and excretion of cadmium. Experimental investigations in rabbits, *Acta Pharmacol.,* 11, 168—178, 1955.

89. Gabbiani, G., Action of cadmium chloride on sensory ganglia, *Experientia,* 22, 261, 1966.

90. Gabbiani, G., Gregory, A., and Baic, D., Cadmium-induced selective lesions of sensory ganglia, *J. Neuropathol. Exp. Neurol.,* 3, 498—506, 1967a.

91. Gabbiani, G., Baic, D., and Deźiel, C., Studies on tolerance and ionic antagonism for cadmium or mercury, *Can. J. Physiol. Pharmacol.,* 45, 443—450, 1967b.

92. Gabbiani, G., Baic, D., and Deźiel, C., Toxicity of cadmium for the central nervous system, *Exp. Neurol.,* 18, 154—160, 1967c.

93. Gabbiani, G., Badonnel, M.-C., Mathewson, S. M., and Ryan, G. B., Acute cadmium intoxication. Early selective lesions of endothelial clefts, *Lab. Invest.,* 30, 686—695, 1974.

94. Gale, T. F. and Ferm, V. H., Skeletal malformations resulting from cadmium treatment in the hamster, *Biol. Neonate,* 23, 149—160, 1973.

95. Gale, T. F. and Layton, W. M., The susceptibility of inbred strains of hamsters to cadmium-induced embryotoxicity, *Teratology,* 21, 181—186, 1980.

96. Girod, C., Concerning the influence of cadmium chloride on the testicle; studies on the *Singe Macacus irus F. Cuv., C. R. Seances Soc. Biol.,* 158, 297—299, 1964a (in French).

97. Girod, C., Studies on the antehypophysical gonadotropic cells, in *Singe Macacus irus F. Cuv.,* after administration of cadmium chloride, *C. R. Seances Soc. Biol. Fil.,* 158, 948—949, 1964b (in French).

98. Glauser, S. C., Bellow, C. T., and Glauser, E. M., Blood-cadmium levels in normotensive and untreated hypertensive humans, *Lancet,* April 3, 717—718, 1976.

99. Gunn, S. A. and Gould, T. C., Cadmium and other mineral elements, in *The Testis,* Vol. 3, Johnson, A. D., Gomes, W. R., and VanDemark, N. L., Eds., Academic Press, New York, 1970, 377—481.

100. Gunn, S. A., Gould, T. C., and Anderson, W. A. D., Zinc protection against cadmium injury to rat testis, *AMA Arch. Pathol.,* 71, 274—281, 1961.

101. Gunn, S. A., Gould, T. C., and Anderson, W. A. D., Cadmium-induced interstitial cell tumors in rats and mice and their prevention by zinc, *J. Natl. Cancer Inst.,* 31, 745—753, 1963a.

102. Gunn, S. A., Gould, T. C., and Anderson, W. A. D., The selective injurious response of testicular and epididymal blood vessels to cadmium and its prevention by zinc, *Am. J. Pathol.,* 42, 685—702, 1963b.

103. Gunn, S. A., Gould, T. C., and Anderson, W. A. D., Strain differences in susceptibility of mice and rats to cadmium-induced testicular damage, *J. Reprod. Fertil.,* 10, 273—275, 1965.

104. Gunn, S. A., Gould, T. C., and Anderson, W. A. D., Protective effect of thiol compounds against cadmium-induced vascular damage to testis, *Proc. Soc. Exp. Biol. Med.,* 122, 1036—1039, 1966.

105. Gunn, S. A., Gould, T. C., and Anderson, W. A. D., Selectivity of organ response to cadmium injury and various protective measures, *J. Pathol. Bacteriol.,* 96, 89—96, 1968a.

106. Gunn, S. A., Gould, T. C., and Anderson, W. A. D., Mechanisms of zinc, cysteine and selenium protection against cadmium-induced vascular injury to mouse testis, *J. Reprod. Fertil.,* 15, 65—70, 1968b.

107. Gunn, S. A., Gould, T. C., and Anderson, W. A. D., Maintenance of the structure and function of the cauda epididymis and contained spermatozoa by testosterone following cadmium-induced testicular necrosis in the rat, *J. Reprod. Fertil.,* 21, 443—448, 1970.

108. Hadley, W. M., Miya, T. S., and Bousquet, F., Cadmium inhibition of hepatic drug metabolism in the rat, *Toxicol. Appl. Pharmacol.,* 28, 284—291, 1974.

109. Hammer, D. I., Finklea, J. F., Creason, J. P., Sandifer, S. H., Keil, J. E., Priester, L. E., and Stara, J. F., Cadmium exposure and human health effects, in *Trace Substances in Environmental Health — V,* Hemphill, D. D., Ed., University of Missouri, Columbia, 1972, 269—283.

110. Hickey, R. J., Schoff, E. P., and Clelland, R. C., Relationship between air pollution and certain chronic disease death rates, *Arch. Environ. Health,* 15, 728—738, 1967.

111. Hodgen, G. D., Gomes, W. R., and VanDemark, N. L., Carbonic anhydrase isoenzymes in rat erythrocytes, kidney and testis, *Fed Proc.,* 28, 773, 1969.

112. Hodgen, G. D., Gomes, W. R., and VanDemark, N. L., *In vitro* and *in vivo* effects of cadmium chloride on isoenzymes of carbonic anhydrase in rat testes and erythrocytes, *Biol. Reprod.,* 2, 197—201, 1970.

113. Hoffmann, E. O., Cook, J. A., Di Luzio, N. R., and Coover, J. A., The effects of acute cadmium administration in the liver and kidney of the rat, *Lab. Invest.,* 32, 655—661, 1975.

114. Holden, H., Cadmium toxicology (letter), *Lancet,* 2, 57, 1969.

115. Holden, H., A mortality study of workers exposed to cadmium fumes, in *Cadmium 79, Proc. 2nd Int. Cadmium Conf. Cannes,* Cadmium Association, London, 1980a, 211—215.

116. Holden, H., Further mortality studies on workers exposed to cadmium fume, in *Occupational Exposure to Cadmium,* Cadmium Association, London, 1980b, 23—24.

117. Holmberg, R. E. and Ferm, V. H., Interrelationships of selenium, cadmium, and arsenic in mammalian teratogenesis, *Arch. Environ. Health,* 18, 873—877, 1969.

118. Hurley, L. S., Gowan, J., and Swenerton, H., Teratogenic effects of short-term and transitory zinc deficiency in rats, *Teratology,* 4, 199—204, 1971.

119. Inskip, H., Beral, V., and McDonall, M., Mortality of Shipham residents: 40-year follow-up, *Lancet,* April 17, 896—899, 1982.

120. Ishizu, S., Minami, M., Suzuki, A., Yamada, M., Sato, M., and Yamamura, K., An experimental study on teratogenic effect of cadmium, *Ind. Health,* 11, 127—139, 1973.

121. Ito, T. and Sawauchi, K., Inhibitory effects on cadmium-induced testicular damage by pretreatment with smaller cadmium dose, *Okajimas Folia Anat. Jpn.,* 42, 107—117, 1966.

122. Itokawa, Y., Abe, T., Tabei, R., and Tanaka, S., Renal and skeletal lesions in experimental cadmium poisoning, *Arch. Environ. Health,* 28, 149—154, 1974.

123. Johnson, A. D., Gomes, W. R., and VanDemark, N. L., Early actions of cadmium in the rat and domestic fowl testis. I. Testis and body temperature changes caused by cadmium and zinc, *J. Reprod. Fertil.,* 21, 383—393, 1970.

124. Johnson, M. H., The effect of cadmium chloride on the blood-testis barrier of the guinea-pig, *J. Reprod. Fertil.,* 19, 551—553, 1969.

125. Kanisawa, M. and Schroeder, H. A., Renal arteriolar changes in hypertensive rats given cadmium in drinking water, *Exp. Mol. Pathol.,* 10, 81—98, 1969.

126. Kar, A. B. and Das, R. P., Testicular changes in rats after treatment with cadmium chloride, *Acta Biol. Med. Ger.,* 5, 153—172, 1960.

127. Kar, A. B. and Das, R. P., Effect of cadmium chloride on fertility of rats, *Indian J. Vet. Sci.* 32, 210—215, 1962.

128. Kar, A. B. and Das, R. P., The nature of protective action of selenium on cadmium-induced degeneration of the rat testis, *Proc. Natl. Inst. Sci. India Part B Biol. Sci.,* 29, 297—305, 1963.

129. Kar, A. B., Das, R. P., and Karkun, J. N., Ovarian changes in prepubertal rats after treatment with cadmium chloride, *Acta Biol. Med. Ger.,* 3, 372—379, 1959.

130. Kar, A. B., Dasgupta, R. R., and Das, R. P., Effect of cadmium chloride on gonadotrophin content of the pituitary of male and female rats, *J. Sci. Ind. Res. (India) Sect. C Biol. Sci.,* 19, 225—230, 1960.

131. Kawamura, R., Shimizu, F., Fujimaki, H., and Kubota, K., Effects of single exposure to cadmium on the primary humoral antibody response, *Arch. Toxicol.,* 54, 289—296, 1983.

132. Kazantzis, G., Flynn, F. V., Spowage, J. S., and Trott, D. G., Renal tubular malfunction and pulmonary emphysema in cadmium pigment workers, *Q. J. Med.,* 32, 165—192, 1963.

133. Khera, A. K., Wibberley, D. G., Edwards, K. W., and Waldron, H. A., Cadmium and lead levels in blood and urine in a series of cardiovascular and normotensive patients, *Int. J. Environ. Stud.,* 14, 309—312, 1980.

134. Kojima, A. and Tamura, S.-I., Acute effects of cadmium on delayed-type hypersensitivity in mice, *Jpn. J. Med. Sci. Biol.,* 34, 281—291, 1981.

135. Kolakowski, J., Baranski, B., and Opalska, B., Effect of long-term inhalation exposure to cadmium oxide fumes on cardiac muscle ultrastructure in rats, *Toxicol. Lett.,* 19, 273—278, 1983.

136. Koller, L. D., Immunosuppression produced by lead, cadmium and mercury, *Am. J. Vet. Res.,* 34, 1457—1458, 1973.

137. Koller, L. T., Immunological effects of cadmium, in *Cadmium in the Environment, Part 2,* Nriagu, J. O., Ed., John Wiley & Sons, New York, 1981, 719—728.

138. Koller, L. D., Exon, J. H., and Roan, J. G., Antibody suppression by cadmium, *Arch. Environ. Health,* 30, 598—601, 1975.

139. Koller, L. D., Exon, J. H., and Roan, J. G., Humoral antibody response in mice after single dose exposure to lead or cadmium, *Proc. Soc. Exp. Biol. Med.,* 151, 339—342, 1976.

140. Kopp, S. J., Baker, J. C., D'Agrosa, L. S., and Hawley, P. L., Simultaneous recording of His bundle electrogram, electrocardiogram, and systolic tension from intact modified Langendorff rat heart preparations. I. Effects of perfusion time, cadmium and lead, *Toxicol. Appl. Pharmacol.,* 46, 475—487, 1978.

141. Kopp, S. J., Glonek, T., Erlanger, M., Perry, E. F., Perry, H. M., Jr., and Bárány, M., Cadmium and lead effects on myocardial function and metabolism, *J. Environ. Pathol. Toxicol.,* 4—2, 3, 205—227, 1980a.

142. Kopp, S. J., Perry, H. M., Jr., Glonek, T., Erlanger, M., Perry, E. F., Bárány, M., and D'Agrosa, L. S., Cardiac physiologic-metabolic changes after chronic low-level heavy metal feeding, *Am. J. Physiol.,* 239, H22—H30, 1980b.

143. Kopp, S. J., Glonek, T., Perry, H. M., Jr., Erlanger, M., and Perry, E. F., Cardiovascular actions of cadmium at environmental exposure levels, *Science,* 217, 837—839, 1982.

144. Kopp, S. J., Perry, H. M., Jr., Perry, E. F., and Erlanger, M., Cardiac physiologic and tissue metabolic changes following chronic low-level cadmium and cadmium plus lead ingestion in the rat, *Toxicol. Appl. Pharmacol.,* 69, 149—160, 1983.

145. Kormano, M., Microvascular supply of the regenerated rat testis following cadmium injury, *Virchows Arch.,* 349, 229—235, 1970.

146. Kotsonis, F. N. and Klaassen, C. D., Toxicity and distribution of cadmium administered to rats at sublethal doses, *Toxicol. Appl. Pharmacol.,* 41, 667—680, 1977.

147. Kotsonis, F. N. and Klaassen, C. D., The relationship of metallothionein to the toxicity of cadmium after prolonged oral administration to rats, *Toxicol. Appl. Pharmacol.,* 46, 39—54, 1978.

148. Krasovskii, G. N., Varshavskaya, S. P., and Borisov, A. I., Toxic and gonadotropic effects of cadmium and boron relative to standards for these substances in drinking water, *Environ. Health Perspect.,* 13, 69—75, 1976.

149. Krause-Fabricius, G. and Hilscher, W., Effects of cadmium on the embryogeny, in *Kadmium-Symposium August 1977 in Jena,* Friedrich-Schiller-Universität, Jena, 1979, 166—169, (in German).

150. Laskey, J. W., Rehnberg, G. L., Laws, S. C., and Hein, J. F., Reproductive effects of low acute doses of cadmium chloride in adult male rats, *Toxicol. Appl. Pharmacol.,* 73, 250—255, 1984.

151. Layton, W. M., Jr. and Layton, M. W., Cadmium-induced limb defects in mice: strain associated differences in sensitivity, *Teratology,* 19, 229—236, 1979.

152. Layton, W. M., Jr. and Ferm, V. H., Protection against cadmium-induced limb malformations by pretreatment with cadmium or mercury. *Teratology,* 21, 357—360, 1980.

153. Lemen, R. A., Lee, J. S., Wagoner, J. K., and Blejer, H. P., Cancer mortality among cadmium production workers, *Ann. N.Y. Acad. Sci.,* 271, 273—279, 1976.

154. Lener, J., Determination of Trace Amounts of Cadmium in Biological Materials from the Viewpoint of Their Biological Consequences and Their Significance in Hygiene, Doctoral Thesis, Charles University, Prague, 1968 (in Czech).

155. Lener, J. and Bibr, B., Cadmium content in some foodstuffs in respect of its biological effects, *Vitalst. Zivilisationskr.,* 15, 139—141, 1970.

156. Lener, J. and Bibr, B., A contribution to the study of cadmium effects in the sphere of experimental hypertension, *Cesk. Hyg.,* 18, 282—286, 1973 (in Czech with English and Russian summaries).

157. Lener, J. and Musil, J., Cadmium influence on the excretion of sodium of kidneys, *Experientia,* 26, 902, 1970.

158. Levin, A. A. and Miller, R. K., Fetal toxicity of cadmium in the rat: maternal vs. fetal injections, *Teratology,* 22, 1—5, 1980.

159. Levin, A. A. and Miller, R. K., Fetal toxicity of cadmium in the rat: decreased utero-placental blood flow, *Toxicol. Appl. Pharmacol.,* 58, 297—306, 1981.

160. Lufkin, N. H. and Hodges, F. T., Cadmium poisoning, *U.S. Nav. Med. Bull.,* 43, 1273—1276, 1944.

161. Lui, E. M. K., Slaughter, S. R., Philpot, R. M., and Lucier, G. W., Endocrine regulation of cadmium-sensitive cytochrome P-450 in rat liver, *Mol. Pharmacol.,* 22, 795—802, 1982.

162. Machemer, L. and Lorke, D., Embryotoxic effect of cadmium on rats upon oral administration, *Toxicol. Appl. Pharmacol.,* 58, 438—443, 1981.

163. Mahaffey, K. R., Capar, S. G., Gladen, B. C., and Fowler, B. A., Concurrent exposure to lead, cadmium, and arsenic. Effects on toxicity and tissue metal concentrations in the rat, *J. Lab. Clin. Med.,* 98, 463—481, 1981.

164. McKenzie, J. M. and Kay, D. L., Urinary excretion of cadmium, zinc and copper in normotensive and hypertensive women, *N.Z. Med. J.,* July, 68—70, 1973.

165. Meek, E. S., Cellular changes induced by cadmium in mouse testis and liver, *Br. J. Exp. Pathol.*, 40, 503—506, 1959.

166. Merali, Z. and Singhal, R. L., Influence of chronic exposure to cadmium on hepatic and renal cyclic amp-protein kinase system, *Toxicology,* 4, 207—214, 1975.

167. Merali, Z. and Singhal, R. L., Biochemistry of cadmium in mammalian systems: pancreotoxic and hepatotoxic manifestations, in *Cadmium in the Environment, Part 2,* Nriagu, J. O., Ed., John Wiley & Sons, New York, 1981, 617—641.

168. Merali, Z., Kacew, S., and Singhal, R. L., Response of hepatic carbohydrate and cyclic AMP metabolism to cadmium treatment in rats, *Can. J. Physiol. Pharmacol.*, 53, 174—184, 1974.

169. Mertz, D. P., Koschnick, R., and Wilk, G., The renal excretion of cadmium in normotensive and hypertensive humans, *Z. Klin. Chem. Biochem.*, 10, 21—24, 1972 (in German).

170. Messerele, K. and Webster, W. S., The classification and development of cadmium-induced limb defects in mice, *Teratology,* 25, 61—70, 1982.

171. Miller, R. K. and Shaikh, Z. A., Prenatal metabolism: metals and metallothionein, in *Reproductive and Developmental Toxicity of Metals,* Clarkson, T. W., Nordberg, G. F., and Sager, P. R., Eds., Plenum Press, New York, 1983, 151—204.

172. Mohammed-Brahim, B., Buchet, J. P., Bernard, A., and Lauwerys, R., *In vitro* effects of lead, mercury and cadmium on the enzymatic activity of red blood pyrimidine 51-nucleotidase, *Toxicol. Lett.,* 20, 195—199, 1984.

173. Morselt, A. F. W., Peereboom-Stegeman, C., Puvion, E., and Maarschalkerweerd, V. J., Investigation for the mechanism of cadmium toxicity at cellular level. II. An electron microscopical study, *Arch. Toxicol.,* 52, 99—108, 1983.

174. Murata, I., Hirono, T., Saeki, Y., and Nakagawa, S., Cadmium enteropathy, renal osteomalacia ("Itai-itai" disease in Japan), *Bull. Soc. Int. Chir.,* 1, 34, 1970.

175. Nicaud, P., Lafitte, A., and Gros, A., Symptoms of chronic cadmium intoxication, *Arch. Mal. Prof. Med. Trav. Secur. Soc.,* 4, 192—202, 1942 (in French).

176. Niemi, M. and Kormano, M., An angiographic study of cadmium-induced vascular lesions in the testis and epididymis of the rat, *Acta Pathol. Microbiol. Scand.,* 63, 513—521, 1965.

177. Nishiyama, S. and Takata, T., Effects of cadmium on the level of serum corticosterone and adrenocortical function in male and female rats, *Res. Commun. Chem. Pathol. Pharmacol.,* 37, 65—79, 1982.

178. Nogawa, K., Itai-Itai disease and follow-up studies, in *Cadmium in the Environment, Part 2,* Nriagu, J. O., Ed., John Wiley & Sons, New York, 1981, 1—38.

179. Nogawa, K. and Kawano, S., A survey of the blood pressure of women suspected of Itai-itai disease, *J. Juzen Med. Soc.,* 77, 357, 1969 (in Japanese).

180. Nogawa, K., Kawano, S., and Nishi, M., Mortality study of inhabitants in a cadmium-polluted area with special reference to low molecular weight proteinuria, in *Heavy Metals in the Environment,* CEP Consultants Ltd., Edinburgh, 1981, 538—540.

181. Nordberg, G. F., Effects of acute and chronic cadmium exposure on the testicles of mice. With special reference to protective effects of metallothionein, *Environ. Physiol. Biochem.,* 1, 171—187, 1971.

182. Nordberg, G. F., Effects of long-term cadmium exposure on the seminal vesicles of mice, *J. Reprod. Fertil.,* 45, 165—167, 1975.

183. Nordberg, G. F. and Piscator, M., Influence of long-term cadmium exposure on urinary excretion of protein and cadmium in mice, *Environ. Physiol. Biochem.,* 2, 37—49, 1972.

184. Nordberg, G. F., Piscator, M., and Nordberg, M., On the distribution of cadmium in blood, *Acta Pharmacol. Toxicol.,* 30, 289—295, 1971.

185. Nordberg, G. F., Slorach, S., and Stenström, T., Cadmium poisoning caused by a cooled soft-drink machine, *Läkartidningen,* 70, 601—604, 1973 (in Swedish).

186. Ohanian, E. V. and Iwai, J., Effects of cadmium ingestion in rats with opposite genetic predisposition to hypertension, *Environ. Health Perspect.,* 28, 261—266, 1979.

187. Ohanian, E. V. and Iwai, J., Etiological role of cadmium in hypertension in an animal model, *J. Environ. Pathol. Toxicol.,* 3, 229—241, 1980.

188. Omaye, S. T. and Tappel, A. L., Effect of cadmium chloride on the rat testicular soluble selenoenzyme, glutathione peroxidase, *Res. Commun. Chem. Pathol. Pharmacol.,* 12, 695—711, 1975.

189. Østergaard, K., Renal cadmium concentration in relation to smoking habits and blood pressure, *Acta Med. Scand.,* 203, 379—383, 1978.

190. Parizek, J., Effect of cadmium salts on testicular tissue, *Nature (London),* 177, 1036—1037, 1956.

191. Parizek, J., The destructive effect of cadmium ion on testicular tissue and its prevention by zinc, *J. Endocrinol.,* 15, 56—63, 1957.

192. Parizek, J., Sterilization of the male by cadmium salts, *J. Reprod. Fertil.,* 1, 294—309, 1960.

193. Parizek, J., Vascular changes at sites of oestrogen biosynthesis produced by parenteral injection of cadmium slats, the destruction of placenta by cadmium salts, *J. Reprod. Fertil.,* 7, 263—265, 1964.

194. Parížek, J., The peculiar toxicity of cadmium during pregnancy — an experimental "toxaemia of pregnancy" induced by cadmium salts, *J. Reprod. Fertil.*, 9, 111—112, 1965.

195. Parížek, J. and Zahor, Z., Effect of cadmium salt on testicular tissue, *Nature (London)*, 177, 1036, 1956.

196. Parížek, J., Oštádalová, I., Benes, I., and Babicky, A., Pregnancy and trace elements: the protective effect of compounds of an essential trace element — selenium — against the peculiar toxic effects of cadmium during pregnancy, *J. Reprod. Fertil.*, 16, 507—509, 1968a.

197. Parížek, J., Oštádalová, I., Benes, I., and Babicky, A., The effect of a subcutaneous injection of cadmium salts on the ovaries of adult rats in persistent oestrus, *J. Reprod. Fertil.*, 17, 559—562, 1968b.

198. Parzyck, D. C., Shaw, S. M., Kessler, M. V., Vetter, R. J., van Sickle, D. C., and Mayes, R. A., Fetal effects of cadmium in pregnant rats on normal and zinc-deficient diets, *Bull. Environ. Contam. Toxicol.*, 19, 206—214, 1978.

199. Pate, F. M., Johnson, A. D., and Miller, W. J., Testicular changes in calves following injection with cadmium chloride, *J. Anim. Sci.*, 31, 559—564, 1970.

200. Pence, D. H., Miya, T. S., and Schnell, R. C., Cadmium alteration of hexabarbital action: sex-related differences in the rat, *Toxicol. Appl. Pharmacol.*, 39, 89—96, 1977.

201. Perry, H. M., Jr. and Erlanger, M. W., Hypertension and tissue metal levels after intraperitoneal cadmium, mercury and zinc, *Am. J. Physiol.*, 220, 808—811, 1971.

202. Perry, H. M., Jr. and Erlanger, M. W., Elevated circulating renin activity in rats following doses of cadmium known to induce hypertension, *J. Lab. Clin. Med.*, 82, 399—405, 1973.

203. Perry, H. M., Jr. and Erlanger, M. W., Metal-induced hypertension following chronic feeding of low doses of cadmium and mercury, *J. Lab. Clin. Med.*, 83, 541—547, 1974.

204. Perry, H. M., Jr. and Erlanger, M. W., Effect of diet on increases in systolic pressure induced in rats by chronic cadmium feeding, *J. Nutr.*, 112, 1983—1989, 1982.

205. Perry, H. M., Jr. and Kopp, S. J., Does cadmium contribute to human hypertension?, *Sci. Total Environ.*, 26, 223—232, 1983.

206. Perry, H. M., Jr. and Yunice, A., Acute pressor effects of intra-arterial cadmium and mercuric ions in anesthetized rats, *Proc. Soc. Exp. Biol. Med.*, 120, 805—808, 1965.

207. Perry, H. M., Jr., Erlanger, M., Yunice, A., Schoepfle, E., and Perry, E. F., Hypertension and tissue metal levels following intravenous cadmium, mercury and zinc, *Am. J. Physiol.*, 219, 755—761, 1970.

208. Perry, H. M., Jr., Perry E. F., and Purifoy, J. F., Antinatriuretic effect of intramuscular cadmium in rats, *Proc. Soc. Exp. Biol. Med.*, 136, 1240—1244, 1971.

209. Perry, H. M., Jr., Perry, E. F., and Erlanger, H. M., Reversal of cadmium-induced hypertension by selenium or hard water, in *Trace Substances in Environmental Health — VIII*, Hemphill, D. D., Ed., University of Missouri, Columbia, 1974, 51—57.

210. Perry, H. M., Jr., Erlanger, M. W., and Perry, E. F., Limiting conditions for the induction of hypertension in rats by cadmium, in *Trace Substances in Environmental Health — X*, Hemphill, D. D., Ed., University of Missouri, Columbia, 1976, 459—467.

211. Perry, H. M., Jr., Erlanger, M., and Perry, E. F., Elevated systolic pressure following chronic low-level cadmium feeding, *Am. J. Physiol.*, 232, 114—121, 1977a.

212. Perry, H. M., Jr., Erlanger, M., and Perry, E. F., Hypertension following chronic, very low dose cadmium feeding, *Proc. Soc. Exp. Biol. Med.*, 156, 173—176, 1977b.

213. Perry, H. M., Jr., Erlanger, M., and Perry, E. F., Increase in the systolic pressure of rats chronically fed cadmium, *Environ. Health Perspect.*, 28, 251—260, 1979.

214. Perry, H. M., Jr., Perry, E. F., and Erlanger, M. W., Possible influence of heavy metals in cardiovascular disease: introduction and overview, *J. Environ. Pathol. Toxicol.*, 4-2:3, 195—203, 1980.

215. Petering, H. G., Choudhury, H., and Stemmer, K. L., Some effects of oral ingestion of cadmium on zinc, copper, and iron metabolism, *Environ. Health Perspect.*, 28, 97—106, 1979.

216. Piscator, M., in *Cadmium in the Environment*, Friberg, L., Piscator, M., and Nordberg, G., Eds., Chemical Rubber Co., Cleveland, 1971.

217. Piscator M. and Axelsson, B., Serum proteins and kidney function after exposure to cadmium, *Arch. Environ. Health*, 21, 604—608, 1970.

218. Piscator, M. and Larsson, S. E., Retention and toxicity of cadmium in calcium-deficient rats, in *Proc. 17th Int. Congr. Occup. Health*, available through the Secretariat, Ave Rogue Saenz, Pena, 110-2 piso, Oficio 8, Buenos Aires, 1972.

219. Piscator, M., Adamsson, E., Elinder, C.-G., Petersson, B., and Steninger, P., Studies on cadmium-exposed women, in *Abstr. 18th Int. Congr. Occup. Health*, Brighton, England, 1975, 281—282.

220. Pond, W. G. and Walker, E. F., Jr., Cadmium-induced anemia in growing rats: prevention by oral or parenteral iron, *Nutr. Rep. Int.*, 5, 365—370, 1972.

221. Porter, M. C., Miya, T. S., and Bousquet, W. F., Cadmium: inability to induce hypertension in the rat. Short communication, *Toxicol. Appl. Pharmacol.*, 27, 692—695, 1974.

222. Porter, M. C., Miya, T. S., and Bousquet, W. F., Cadmium and vascular reactivity in the rat, *Toxicol. Appl. Pharmacol.*, 34, 143—150, 1975.

223. Prigge, E., Inhalative cadmium effects of pregnant and fetal rats, *Toxicology*, 10, 297—309, 1978a.

224. Prigge, E., Early signs of oral and inhalative cadmium uptake in rats, *Arch. Toxicol.*, 40, 231—247, 1978b.

225. Prigge, E., Baumert, H. P., and Muhle, H., Effects of dietary and inhalative cadmium on hemoglobin and hematocrit in rats, *Bull. Environ. Contam. Toxicol.*, 17, 585—590, 1977.

226. Princi, F., A study of industrial exposures to cadmium, *J. Ind. Hyg. Toxicol.*, 29, 315—324, 1947.

227. Prodan, L., Cadmium poisoning. II. Experimental cadmium poisoning, *J. Ind. Hyg. Toxicol.*, 14, 174—196, 1932.

228. Réme and Peres, In regard to a collective intoxication caused by cadmium, *Arch. Mal. Prof. Med. Trav. Secur. Soc.*, 20, 783—785, 1959 (in French).

229. Revis, N. W. and Zinsmeister, A. R., The relationship of blood cadmium level to hypertension and plasma norepinephrine level: a Romanian study, *Proc. Soc. Exp. Biol. Med.*, 167, 254—260, 1981.

230. Richardson, M. E. and Fox, M. R. S., Dietary cadmium and enteropathy in the Japanese quail, *Lab. Invest.*, 31, 722—731, 1974.

231. Roe, F. J. C., Dukes, C. F., Cameron, K. M., Pugh, R. C. B., and Mitchley, B. C. V., Cadmium neoplasia: testicular atrophy and Leydig cell hyperplasia and neoplasia in rats and mice following the subcutaneous injection of cadmium salts, *Br. J. Cancer*, 18, 674—681, 1964.

232. Roy, A. K. and Neuhaus, O. W., Proof of the hepatic synthesis of a sex-dependent protein in the rat, *Biochim. Biophys. Acta*, 127, 82—87, 1966.

233. Roy, A. K., Neuhaus, O. W., and Harmison, C. R., Preparation and characterization of a sex-dependent rat urinary protein, *Biochim. Biophys. Acta*, 127, 72—81, 1966.

234. Saksena, S. K. and Salmonsen, R., Effects of cadmium chloride on ovulation and on induction of sterility in the female golden hamster, *Biol. Reprod.*, 29, 249—256, 1983.

235. Saksena, S. K., Dahlgren, L., Lau, I. F., and Chang, M. C., Reproductive and endocrinological features of male rats after treatment with cadmium chloride, *Biol. Reprod.*, 16(5), 609—613, 1977.

236. Sakurai, H., Epidemiological studies, in *Cadmium Studies in Japan — A Review*, Tsuchiya, K., Ed., Elsevier, Amsterdam, New York, 1978, 133—267.

237. Samarawickrama, G. P. and Webb, M., Acute effects of cadmium on the pregnant rat and embryo-fetal development, *Environ. Health Perspect.*, 28, 245—249, 1979.

238. Sato, K., Iwamasa, T., Tsuru, T., and Takeuchi, T., An ultrastructural study of chronic cadmium chloride-induced neuropathy, *Acta Neuropathol. (Berl.)*, 41, 185—190, 1978.

239. Schmid, B. P., Hall, J. L., Goulding, E., Fabro, S., and Dixon, R., *In vitro* exposure of male and female mice gametes to cadmium chloride during the fertilization process, and its effects on pregnancy outcome, *Toxicol. Appl. Pharmacol.*, 69, 326—332, 1983.

240. Schroeder, H. A., Cadmium hypertension in rats, *Am. J. Physiol.*, 207, 62—66, 1964.

241. Schroeder, H. A., Cadmium as a factor in hypertension, *J. Chronic Dis.*, 18, 647—656, 1965.

242. Schroeder, H. A., Cadmium, chromium, and cardiovascular disease, *Circulation*, 35, 570—582, 1967.

243. Schroeder, H. A. and Buckman, J., Cadmium hypertension, its reversal by a zinc chelate, *Arch. Environ. Health*, 14, 693—697, 1967.

244. Schroeder, H. A. and Mitchener, M., Toxic effects of trace elements on the reproduction of mice and rats, *Arch. Environ. Health*, 23, 102—106, 1971.

245. Schroeder, H. A. and Vinton, W. H., Jr., Hypertension induced in rats by small doses of cadmium, *Am. J. Physiol.*, 202, 515—518, 1962.

246. Schroeder, H. A., Kroll, S. S., Little, J. W., Livingston, P. O., and Myers, M. A. G., Hypertension in rats from injection of cadmium, *Arch. Environ. Health*, 13, 788—789, 1966.

247. Schroeder, H. A., Nason, A. P., and Balassa, J. J., Trace metals in rat tissues as influenced by calcium in water, *J. Nutr.*, 93, 331—336, 1967.

248. Schroeder, H. A., Nason, A. P., and Mitchener, M., Action of a chelate of zinc on trace metals in hypertensive rats, *Am. J. Physiol.*, 214, 796—800, 1968.

249. Schwartze, E. W. and Alsberg, C. L., Studies on the pharmacology of cadmium and zinc with particular reference to emesis, *J. Pharmacol. Exp. Ther.*, 21, 1—22, 1923.

250. Setchell, B. P. and Waites, G. M. H., Changes in the permeability of the testicular capillaries and of the "blood-testis" barrier after injection of cadmium chloride in the rat, *J. Endocrinol.*, 47, 81—86, 1970.

251. Shigematsu, I., Kitamura, S., Takeuchi, J., Minowa, M., Nagai, M., Usui, T., and Fukushima, M., A retrospective mortality study on cadmium-exposed populations in Japan, in *Cadmium 81, Proc. 3rd Int. Cadmium Conf. Miami,* Cadmium Association, London, 1982, 115—118.

252. Smith, J. P., Smith, J. C., and McCall, A. J., Chronic poisoning from cadmium fume, *J. Pathol. Bacteriol.*, 80, 287—296, 1960.

253. Smith, M. J., Pihl, R. O., and Garber, B., Postnatal cadmium exposure and long-term behavioral changes in the rat, *Neurobehav. Toxicol. Teratol.*, 4, 283—287, 1982.

254. Sonawane, B. R., Nordberg, M., Nordberg, G. F., and Lucier, G. W., Placental transfer of cadmium in rats: influence of dose and gestational age, *Environ. Health Perspect.*, 12, 97—102, 1975.

255. Sorahan, T., A mortality study of nickel-cadmium battery workers, in *Cadmium 81, Proc. 3rd Int. Cadmium Conf. Miami,* Cadmium Association, London, 1982, 138—141.

256. Sporn, A., Dinu, I., and Stoenescu, L., Influence of cadmium administration on carbohydrate and cellular energetic metabolism in the rat liver, *Rev. Roum. Biochim.*, 7, 299—305, 1970.

257. Staessen, J., Bulpitt, C. J., Roels, H., Bernard, A., Fagard, R., Joossens, J. V., Lauwerys, R., Lijnen, P., and Amery, A., Urinary cadmium and lead and their relationship to blood pressure in a population with low average exposure, *Br. J. Ind. Med.*, 41, 241—248, 1984.

258. Stowe, H. D., Wilson, M., and Goyer, R. A., Clinical and morphological effects of oral cadmium toxicity in rabbits, *Arch. Pathol.*, 94, 389—405, 1972.

259. Sugawara, Ch. and Sugawara, N., Cadmium toxicity for rat intestine, especially on the absorption of calcium and phosphorus, *Jpn. J. Hyg.*, 6, 511—516, 1974.

260. Sutherland, D. J. B., Tsang, B. K., and Singhal, R. L., Testicular and prostatic cyclic AMP metabolism following chronic cadmium treatment and subsequent withdrawal, *Environ. Physiol. Biochem.*, 4, 205—213, 1974.

261. Suzuki, K. T., Yamada, Y. K., and Shimizu, F., Essential metals and metallothionein in cadmium-induced thymic atrophy and splenomegaly, *Biochem. Pharmacol.*, 11, 1217—1222, 1981.

262. Suzuki, S., Suzuki, T., and Ashizawa, M., Proteinuria due to inhalation of cadmium stearate dust, *Ind. Health*, 3, 73—85, 1965.

263. Task Group on Metal Interaction, Factors influencing metabolism and toxicity of metals: a consensus report, *Environ. Health Perspect.*, 25, 3—41, 1978.

264. Tassinari, S. M. and Long, S. Y., Normal and abnormal mid-facial development in the cadmium-treated hamster, *Teratology*, 25, 101—113, 1982.

265. Teare, F. W., Jasansky, P., Renaud, L., and Read, P. R., Acute effect of cadmium on hepatic drug-metabolizing enzymes in the rat, *Toxicol. Appl. Pharmacol.*, 41, 57—65, 1977.

266. Thind, G. S. and Fischer, G. M., Cadmium and zinc distribution in cardiovascular and other tissues of normal and cadmium-treated dogs, *Exp. Mol. Pathol.*, 22, 326—334, 1975.

267. Thind, G. S., Biery, D. N., and Bovee, K. C., Production of arterial hypertension by cadmium in the dog, *J. Lab. Clin. Med.*, 81, 549—556, 1973.

268. Tsang, B. K., Sutherland, D. J. B., Merali, Z., and Singhal, R. L., The influence of chronic cadmium on the cyclic AMP-adenyl cyclase protein kinase system of rat testes and prostate glands, *Fed. Proc.*, 33, 227, 1974.

269. Tsuchiya, K., Proteinuria of workers exposed to cadmium fume. The relationship to concentration in the working environment, *Arch. Environ. Health*, 14, 875—880, 1967.

270. Tsuchiya, K., Environmental pollution by cadmium and its health effect in Japan, in *Working Papers of Japanese Participants for a Planning Conference — United States-Japan,* Direct Cooperative Program, East-West Center, Honolulu, 1971, 65.

271. Tulley, R. T. and Lehmann, H. P., Method for the simultaneous determination of cadmium and zinc in whole blood by atomic absorption spectrophotometry and measurement in normotensive and hypertensive humans, *Clin. Chim. Acta*, 122, 189—202, 1982.

272. U.S. Public Health Service, Public Health Report, Cadmium poisoning, *Div. Ind. Hyg. Natl. Inst. Health*, 57, 601—612, 1942.

273. Valetas, P., Cadmiosis or Cadmium Intoxication, Doctoral thesis, School of Medicine, University of Paris, 1946 (in French).

274. Vallee, B. L. and Ulmer, D. D., Biochemical effects of mercury, cadmium, and lead, *Ann. Rev. Biochem.*, 41, 91—128, 1972.

275. Vander, A. J., Cadmium enhancement of proximal tubular sodium reabsorption, *Am. J. Physiol.*, 203, 1005—1007, 1962.

276. Velde, van der, W. J., Copius Peereboom-Stegeman, J. H. J., Treffers, P. E., and James, J., Structural changes in the placenta of smoking mothers: a quantitative study, *Placenta*, 4, 231—240, 1983.

277. Voors, A. W., Shuman, M. S., and Gallagher, Ph. N., Zinc, and cadmium autopsy levels for cardiovascular disease in geographical context, in *Trace Substances in Environmental Health — VI*, Hemphill, D. D., Ed., University of Missouri, Columbia, 1973, 215—222.

278. Vorobjeva, R. S., Investigation of the nervous system function in workers exposed to cadmium oxide, *Nevropatol. Psikkiatr.*, 57, 385—388, 1957b (in Russian).

279. Vorobjeva, R. S. and Eremeeva, E. P., Cardiovascular function in workers exposed to cadmium, *Gig. Sanit.*, 10, 22—25, 1980 (in Russian with English summary).

280. Wahlberg, J. E., Routine patch testing with cadmium chloride, *Contact Dermititis*, 3, 293—296, 1977.

281. Waites, G. M. H. and Setchell, B. P., Changes in blood flow and vascular permeability of the testis, epididymis and accessory reproductive organs of the rat after the administration of cadmium chloride, *J. Endocrinol.,* 34, 329—342, 1966.

282. Ward, R. J., Fisher, M., and Tellez-Yudilevich, M., Significance of blood cadmium concentrations in patients with renal disorders or essential hypertension and the normal population, *Ann. Clin. Biochem.,* 15, 197—200, 1978.

283. Watanabe, T., Shimada, T. and Endo, A., Mutagenic effects of cadmium on the oocyte chromosomes of mice, *Nippon Eisegaku Zasshi,* 32(3), 472—481, 1977.

284. Webb, M. and Samarawickrama, G. P., Placental transport and embryonic utilization of essential metabolites in the rat at the teratogenic dose of cadmium, *J. Appl. Toxicol.,* 1, 5, 270—277, 1981.

285. Webster, W. S., Cadmium-induced fetal growth retardation in the mouse, *Arch. Environ. Health,* Jan./Febr., 36—42, 1978.

286. Webster, W. S. and Valois, A. A., The toxic effects of cadmium on the neonatal mouse CNS, *J. Neuropathol. Exp. Neurol.,* 40, 247—257, 1981.

287. Whanger, P. D., Cadmium effects in rats on tissue iron, selenium, and blood pressure; blood and hair cadmium in some Oregon residents, *Environ. Health Perspect.,* 28, 115—121, 1979.

288. White, I. G., The toxicity of heavy metals to mammalian spermatozoa, *Aust. J. Exp. Biol. Med. Sci.,* 33, 359—366, 1955.

289. Wills, H. J., Groblewski, G. E., and Coulston, F., Chronic and multigeneration toxicities of small concentrations of cadmium in the diet of rats, *Exotoxicol. Environ. Saf.,* 5, 452—464, 1981.

290. Wilson, R. H., DeEds., F., and Cox, A. J., Effects of continued cadmium feeding, *J. Pharmacol. Exp. Ther.,* 71, 222—235, 1941.

291. Wisniewska-Knypl, J. M., Jablonska, J., and Myslak, Z., Binding of cadmium on metallothionein in man: an analysis of a fatal poisoning by cadmium iodide, *Arch. Toxicol.,* 28, 46—55, 1971.

292. Wong, K.-L. and Klaassen, C. D., Neurotoxic effects of cadmium in young rats, *Toxicol. Appl. Pharmacol.,* 63, 330—337, 1982.

293. Zenick, H., Hastings, L., and Goldsmith, M., Chronic cadmium exposure: relation to male reproductive toxicity and subsequent fetal outcome, *J. Toxicol. Environ. Health,* 9, 377—387, 1982.

294. Zielinska-Psuja, B., Lukaszyk, A., and Senczuk, W., The anti-reproductive effects of long-term oral administration of cadmium on the adult male rat, *Int. J. Androl.,* 2, 150—161, 1979.

295. Zylber-Haran, E. A., Gershman, H., Rostenmann, E., and Spitz, I. M., Gonadotrophin, testosterone and prolactin interrelationships in cadmium-treated rats, *J. Endocrinol.,* 92, 123—130, 1982.

Chapter 12

CARCINOGENIC AND GENETIC EFFECTS

Carl-Gustaf Elinder and Tord Kjellström

TABLE OF CONTENTS

I. CARCINOGENIC EFFECTS

A. Animals
1. After Parenteral Administration

Parenteral administration of cadmium (subcutaneous, intramuscular, or subperiosteal) may give rise to sarcomas at the injection site and interstitial cell tumors of the testis. The first evidence of an association between cadmium exposure and carcinogenesis in animals was reported by Haddow et al.[28] The authors were carrying out a series of tests on iron-containing compounds and iron complexes to determine carcinogenicity in rats. A preparation of rat ferritin induced not only a high incidence of sarcomata at the site of repeated s.c. injections, but also testicular atrophy, Leydig's cell hyperplasia, and benign Leydig's cell tumors. The ferritin used in these experiments was prepared from the precipitation of rat liver protein by a cadmium salt and contained cadmium as an essential part of its crystalline structure. Since it was known that cadmium injections in rats cause testicular atrophy (see Chapter 11, Section VI.A), Haddow and co-workers suspected that the cadmium compound may have caused sarcomata and Leydig's cell tumors.

Heath et al.[30] injected 20 rats intramuscularly with 14 to 28 mg cadmium powder suspended in fowl serum. Of the 20 rats, 15 developed rhabdomyosarcomata and fibrosarcomata. This finding supported the earlier observation by Haddow and co-workers. Kazantzis[39] injected ten rats subcutaneously with 25 mg cadmium sulfide in physiological saline. Six of the ten rats developed sarcomata within a year.

To confirm their original observation, Haddow et al.[29] conducted an experiment in which rats were injected subcutaneously at the same site with 20 + 20 +(8 × 2 mg) cadmium-precipitated rat ferritin (total cadmium dose 0.25 mg) or with 10 × 0.5 mg cadmium sulfate (total cadmium dose 2.0 mg). The rat ferritin treatment gave rise to 7 injection-site sarcomata in 20 exposed rats following an average latent interval of 22 months. The cadmium sulfate treatment caused 14 injection-site sarcomata in 20 rats following an average latent interval of 13 months. No tumors of this kind were observed in 16 control rats.

A similar experiment on mice showed no evidence of injection-site tumors.[29] Rats and mice were also studied for testicular effects by Roe et al.[66] The latter group of authors noted testicular atrophy, Leydig's cell hyperplasia, and neoplasia in cadmium-exposed animals. The studies by Heath et al.[30] and Kazantzis[39] did not describe the changes that may have occurred in the testicles, but the occurrence of testicular interstitial tumors after single cadmium chloride injections has also been demonstrated by Gunn et al.[25] Roe et al.,[66] as well as by Lucis et al.[52]

Further studies on cadmium-induced testicular tumors[25] showed that they could actually be prevented by the simultaneous administration of zinc. Twenty-five rats received a single s.c. injection of cadmium chloride (3.4 mg Cd per kilogram). Seventeen rats received the same cadmium dose and three additional injections of zinc acetate, 65.4 mg Zn per kilogram, corresponding to a molar excess of zinc of 100 times the given cadmium dose. Twenty rats comprised controls. All the rats were sacrificed after 11 months. Interstitial cell tumors composed of Leydig's cells were found in 17 of the 25 cadmium-treated rats, 2 of the 17 cadmium-zinc-treated rats, and 1 of the control rats.[25] The same treatment was given to three groups of mice. Of the 26 cadmium-treated mice, 20 developed interstitial cell tumors, whereas none of the 25 cadmium-zinc-treated or 25 control mice developed this type of tumor.

A further study by Gunn et al.[26] on 22 cadmium-treated rats, 17 cadmium-zinc-treated rats, and 18 control rats (same doses as in earlier report) resulted in 88% interstitial cell tumors in the cadmium-treated rats, 18% in the cadmium-zinc-treated rats, and none in the controls. The incidence s.c. injection-site sarcomata was 41, 12, and

0%, for the three groups, respectively.[26] No injection-site tumors developed when zinc only was injected.

One study was designed to evaluate whether the occurrence of injection-site tumors is dependent on the type of tissue being exposed.[27] Rats were given injections of cadmium chloride at four different sites: tissues derived from ectoderm, endoderm, epithelial mesoderm, or mesenchymal mesoderm. Each rat was injected once at all four sites. For 11 rats the total dose was 1.3 mg and for 30 rats the total dose was 0.67 mg. The other groups of rats were injected with zinc or cobalt. No tumors occurred in these latter groups, but in the cadmium groups there were seven tumors all of which developed in mesenchymal mesodermal structures (subcutaneous, intramuscular, or subperiosteal injections). The incidence of tumors was 36% in the high dose group and 10% in the low dose group.

In an extension of the study described above,[27] 49 rats were given either single s.c. or single i.m. injections of 1.8 mg cadmium chloride. The incidence of tumors was 44% in the s.c. group and 12% in the i.m. group. It was pointed out that all the injection-site tumors[27] were fibrosarcomas and that the number of fibroblast cells exposed to the injected cadmium may be a determining factor in the development of a tumor.

One study observed metastases from the primary tumor.[40] Ten rats were given a single s.c. injection of 25 mg cadmium sulfide. Within 1 year, six of these rats had developed injection-site tumors. The tumor cells were seen to infiltrate the connective tissue and muscle of the chest and abdominal wall. Metastatic deposits with cells similar in appearance to the cells in the primary tumors were seen in the regional lymph nodes or in the lungs.[40] Tumors were also seen after i.m. injection of cadmium sulfide. Furthermore, Kazantzis and Hanbury[40] obtained similar tumors in eight out of ten rats injected subcutaneously with 25 mg cadmium oxide.

It has been shown that inert films and foils can give rise to local sarcomas originating from the fibrous tissue capsule around the implant.[60] It may be postulated that injection of insoluble metal salts such as cadmium sulfide may cause tumors in a similar way, but this would not explain the lack of tumor formation as a result of injection of other metal salts.[24]

Levy et al.[49] set out to confirm or dispute, by way of animal experiments, the reported finding of an increased mortality in prostatic cancer among cadmium workers. Three groups of 25 rats were exposed to weekly s.c. injections of $CdSo_4$ (0.022, 0.044, and 0.087 mg Cd per rat and week; equivalent to about 0.01, 0.02, and 0.04 mg Cd per kilogram body weight and day). The tumor incidence in these groups was compared to the incidence in a control group of 75 rats (injected with water only) for up to 2 years. At this time the cadmium concentration in whole kidney was about 140, 270, and 480 mg Cd per kilogram wet weight, in the three exposed groups, respectively. Due to drying of specimens, the true wet weight values would have been somewhat lower. In any case, the total dose would have been high enough to cause some damage in the kidney cortex. Sarcomatas at the injection site were found in four of the high exposure rats, one each of the medium and low cadmium exposure rats, and in none of the control rats. No prostatic cancers or hyperplasias were reported and there were only a few other scattered causes (one in each group) up to age 2 years. The estimated life span of a rat is 4 years,[1] and it may be that the follow-up period was not long enough for prostatic cancer to develop. The prostatic cancer rate in humans at half of their life span is very low, more than a thousand times lower than that after age 70.[53]

Poirier et al.[63] examined the influence of calcium and magnesium on the carcinogenicity of cadmium chloride in rats. One single s.c. injection of 2.4 or 5 mg cadmium (as $CdCl_2$) per kilogram body weight was given and the animals were followed up for about 2 years. Magnesium and calcium were administered to different subgroups of rats via the diet or via repeated s.c. injections. In rats given cadmium chloride alone, injection-site tumors (mostly fibrosarcomas) were seen in 33 and 34% of the rats in the high and low dose group, respectively. In addition, testicular tumors (mostly interstitial cell tumors) developed in 86 and 85% of the rats, respectively. Rats given 0.9% NaCl developed no injection-site tumors, but on the other hand had a rather high cumulative incidence of testicular tumors (30%). Simultaneous injection of magnesium acetate given at the same site as cadmium completely prevented the development of injection-site tumors for both $CdCl_2$ doses, but had no effect on the final yields of testicular tumors. Another interesting observation in this study was that the cadmium-exposed groups also had an increased cumulative incidence of pancreatic islet cell tumors (8.5 vs. 2.2%). There was no mention of prostatic cancer in any of the animals.

2. After Peroral Exposure

In contrast to the convincing data on animals given injections of cadmium, there is no evidence from animal investigations to show that long-term peroral exposure to cadmium will result in an increased incidence of malignant tumors. Schroeder et al.[70] exposed mice to cadmium acetate, 5 mg/ℓ in drinking water over whole lifetime (normally about three years)[1] and found no excess increase in the number of tumors in 48 male and 39 female rats compared to similar size groups of control rats, after approximately 2.5 years. However, the tumor mortality rate in the control group (32%) was high and the cadmium-exposed mice had a significantly shorter life span than controls in spite of the total dose being very low (average kidney cadmium level was only 3 mg/kg wet weight). The validity of these data for the interpretation of human carcinogenicity is therefore uncertain.

Levy et al.[50] gave 3 groups of 50 mice weekly doses of cadmium sulfate via a stomach tube (0.44, 0.88, and 1.75 mg Cd per kilogram body weight and week) for 18 months. There was no increase of mortality in the exposed mice as compared to 150 control mice and neither was there any increase of tumor incidence, but 18 to 29% of the mice had lung adenomas and 23 to 36% had liver adenomas. Pathological changes in the prostate were looked for, but only occasional foci of slight epithelial hyperplasia were found. None of the changes reported were associated with the cadmium dose.

As pointed out earlier for rats exposed to s.c. injections,[49] the follow-up period in this experiment by Levy et al.[50] may also have been too short (half a life span of mice) to record any increased occurrence of prostatic cancer.

Similar negative data were also obtained using rats followed up for half a life span or less. Schroeder et al.[71] found the same incidence of tumors in 50 male and 46 female rats given cadmium acetate (5 mg Cd per liter in drinking water for periods exceeding 2 years) as that found in control groups. In a later experiment, with the same experimental design, Kanisawa and Schroeder[37] reported the occurrence of 7 malignant tumors in 47 cadmium-exposed male rats in comparison to 2 malignant tumors in 34 controls. In spite of this observation, the authors concluded that ingestion of cadmium cannot be considered carcinogenic at the dose levels administered. Due to the small number of rats examined, the statistical significance of the study is poor and the conclusion reached is questionable. In order to obtain a significant difference between exposed and control rats with 80% probability the relative risk would have to be 4.8.

Loeser[51] exposed male and female rats, in a 2-year study, to four different levels of cadmium in the diet; 1, 3, 10, and 50 mg/kg. An additional group of rats was given a cadmium-free diet. Number of rats in the cadmium-exposed groups ranged from 46 to 50 and the control groups comprised 94 and 98 females and males, respectively. There was no indication of an increased incidence of either benign or malignant neoplasias among the exposed animals compared to controls.

A 2-year experiment on rats[48] also failed to find any evidence of an increased tumor incidence. Three groups of 30 rats exposed to cadmium sulfate via gastric tube (0.09, 0.18, and 0.35 mg Cd per kilogram body weight and week) were compared with 90 control rats. The average cadmium concentration in whole kidney reached only 5.2 mg Cd per kilogram wet weight. It was not explained why such a low dose was used. Particular emphasis was laid on examinations of the prostate, but no tumors were found. "Spontaneous" cancers in other tissues were found in 30% of the "high cadmium exposure" group, and in 10, 13 and 11% in the medium exposure, low exposure, and control groups, respectively.

The peroral exposure experiments are difficult to interpret in a conclusive way as the cadmium doses used have been too low to reach doses common in human beings with excessive cadmium exposure and the follow-up periods may have been too short to include the typical latency periods for cancer. Nomiyama[58] suggested that when cadmium is given via the oral route it will be distributed in the body bound to metallothionein and this complex may not have the same carcinogenic potential as nonmetallothionein-bound cadmium ions, which are present shortly after an injection.

3. Effects Due to Inhalation or Intratracheal Instillations

As a part of a program to study the deposition, retention, and absorption of inhaled cadmium compounds,[59] some long-term experiments were carried out. In a pilot study, ten rats had been exposed to $CdCl_2$ aerosols at a Cd concentration of 20 $\mu g/m^3$ for 18 months.[31] In the lungs of all animals, adenomatous hyperplasia was found, and five animals had developed lung tumors, including four adenomas and one adenocarcinoma. Based on this unexpected finding, a more detailed study was designed by the same group. This highly interesting study confirmed that inhalation of cadmium aerosols may cause lung cancer in Wistar rats.[81] Three groups of 40 rats each were continuously exposed to $CdCl_2$ aerosols for 18 months. The concentrations of cadmium in air were 12.5, 25, and 50 $\mu g/m^3$, respectively. An additional group of 41 rats served as controls. The experiment was continued for 13 months after the end of exposure, giving a total observation period of 31 months. At that point a careful histological examination of each rat's lung was carried out. A dose-related increase in the incidence of lung cancer was seen in the exposed groups, 15, 53, and 71% for the three exposure groups, respectively. No cases of lung cancer occurred in the control group (Table 1). There were also a number of cancers in other tissues, but they were not associated with cadmium exposure.

The importance of these data can scarcely be overestimated. Only a few animal studies have been performed which attempt to evaluate the potential carcinogenicity of respirable metal-containing particles, and the study by Takenaka et al.[81] is the first and only published report of long-term exposure of animals to a respirable cadmium-containing aerosol and with a lifetime follow-up. The dose levels used in this experiment are comparably low, in fact at or below the hygienic standard for cadmium (50 $\mu g/m^3$) still in use in most countries. The average cadmium concentrations in the rats' lungs were 5.6, 4.7, and 10.4 mg Cd per kilogram wet weight in the three groups, respectively. In "normal" human lungs at middle age the average is about 0.5 mg/kg (Volume I, Chapter 5) and in workers with cadmium poisoning it may be as high as 12 to 25 mg per kilogram Cd per kilogram.[77] In view of these data, cadmium in the form of an inhaled cadmium chloride aerosol must definitely be regarded as a highly carcinogenic agent in animals. It is noteworthy that the West German Hygienic Standards Committee in 1983 on the basis of the study by Takenaka et al.[81] and the other available evidence declared cadmium compounds in industry to be carcinogenic, which means that the exposure levels have to be minimized.[18] In 1984, Sanders and Mahaffey[69] reported no evidence of lung cancer in rats given intratracheal instillations of cadmium oxide. One hundred and ninety rats (Fischer-344) were divided into four groups: con-

Table 1
LUNG CHANGES IN WISTAR RATS AFTER EXPOSURE TO CdCl$_2$ AEROSOLS[81]

Exposure groups	Initial no. of rats	No. of rats examined histologically	Adenomatous hyperplasia	Adenoma	No. of rats with lung changes Carcinomas				
					Total (%)	Adeno	Epidermoid	Mucoepidermoid	Combined epidermoid and adeno
Control	41	38[a]	1	1	0	0	0	0	0
12.5 μg Cd/m^3	40	39[b]	6	1	6(15.4%)	4	2	0	0
25 μg Cd/m^3	40	38[c]	5	0	20(52.6%)	15	4	0	1
50 μg Cd/m^3	40	35[d]	3	1	25(71.4%)	14	7	3	1

[a] Two rats died during the first 18 months; another rat was not examined because of autolysis.

[b] One rat was not examined because of autolysis.

[c] Two rats were not examined because of autolysis.

[d] Three rats died during the first 18 months; two other rats were not examined because of autolysis.

trols given instillations of NaCl; one group instillations of 25 μg cadmium oxide; a second group two instillations of 25 μg cadmium oxide; and a third group three instillations of 25 μg cadmium oxide. The rats were followed for their whole life span, that is to say, up to 900 days. Two lung cancers (5%) were seen in the group that was given two instillations of 25 μg cadmium oxide, but none in the two other exposed groups of rats, and not in the control group. Overall the cumulative cancer incidence was very similar in the control group as well as in the exposed groups. There was, however, a slight increase in fibroadenomas of the mammary gland: 7% in the control group and 16, 12, and 23% in the three exposed groups, respectively. In addition, the prevalence of rats without tumors was somewhat higher in the control group (16%) compared to the exposed groups (5 to 7%).

The data obtained by Sanders and Mahaffey[69] thus, appear to be contradictory to the results reported by Takenaka et al.[81] There are, however, several differences in the experimental protocol of these studies which may explain this phenomenon. Takenaka et al.[81] exposed the rats to cadmium chloride aerosols, whereas Sanders and Mahaffey used cadmium oxide. Another, and probably a more crucial difference between the studies, is the way of exposure and exposure intensity. Takenaka et al.[81] exposed rats to comparably low levels of cadmium chloride in the atmosphere more or less continuously for 18 months, whereas Sanders and Mahaffey installed cadmium oxide at one, two, and three occasions at dose levels close to (75%) the LD_{50}. This dose is known to cause pulmonary edema.[69]

A carefully conducted study on cancer incidence in rats continuously exposed to low levels of cadmium oxide would certainly be of great interest.

B. Humans
1. Studies of Occupationally Exposed Workers

Several epidemiological studies have emerged which deal with cancer incidence among workers occupationally exposed to cadmium.[3,5,5a,19a,33,34,42,43,47,78,79,79a,82] Several of the studies provide evidence of a slightly increased mortality in cancer of the prostate and lung. The available data are, however, difficult to interpret, not only because the mortality experience is different in one cadmium plant compared to another, but also since different reports from the same industry sometimes provide more or less conflicting data. An attempt to summarize the data with regard to prostatic and lung cancer among cadmium workers has been made in Tables 2 and 3.

The work by Kipling and Waterhouse[42] was initiated on the basis of the observations reported by Potts.[64] In a survey of 70 men exposed to cadmium oxide dust in an alkaline battery plant for at least 10 years, Potts recorded 8 fatalities, 3 of them due to prostatic cancer. Kipling and Waterhouse[42] carried out a more comprehensive study of 248 workers at the same plant. These workers had been exposed to cadmium oxide dust for more than 1 year. The groups formerly studied by Potts were also included in the cohort. The observed mortality due to different causes of death in the cadmium-exposed cohort was compared to the expected number as estimated from a regional cancer register (Table 4). There were four deaths in prostatic cancer compared to an expected number of 0.58, which was a highly significant increase.

Humperdinck[35] reported on causes of death for 17 cadmium workers in a German battery plant who died between 1949 and 1966. There were five fatalities in neoplastic diseases, two cases of lung cancer, and one each of liver, stomach (cardia), and prostatic cancer. The study was not conducted as an epidemiological study and no definite conclusions can be drawn from the report. Holden[32] in a short communication mentioned that one case of lung cancer and another of prostatic cancer had occurred among 42 men exposed for 2 to 4 years to cadmium fumes possibly in the form of cadmium oxide. Again, no epidemiological analysis was carried out.

Table 2

SUMMARY OF MORTALITY AND INCIDENCE DATA FOR PROSTATE CANCER REPORTED IN STUDIES OF CADMIUM WORKERS[19a]

Study population	Type of industry	First study			Second study			Third study			Fourth study		
		SMR	Obs./exp.	Ref.	SMR	Obs./exp.	Ref.	SMR	Obs./exp.	Ref.	SMR	Obs./exp.	Ref.
1	Ni-Cd battery plant in the U.K.	667**	4/0.6	42	117	7/6.0	78	235	4/1.7	4[a]	400**	8/2.0	79a[b]
2	Cd-smelter in the U.S.	444*	4/0.9	47[c]	136	3/2.2	82						
3	Ni-Cd battery plant in Sweden	167	2/1.2	43[d]	167	4/2.4	3[e]	148	4/2.7	19a[f]			
4	Cd-Cu alloy plant in Sweden	148	4/2.7	43									
5	Cd-Cu alloy plant in the U.K.	66	1/1.5	33[a]	267*	8/3.0	34[g]	191	9/4.7	4[a]			
6	17 different plants using Cd in the U.K.	0	0/2.9	4,5[a]	0	0/1.4	5a[h]						

Note: *Significant at $p < 0.05$ and **significant at $p < 0.01$.

[a] "High or medium exposure".

[b] Incidence, "high exposure".

[c] "20 years of latency".

[d] Incidence.

[e] "High exposure, 10 years latency".

[f] "≥ 5 years of high exposure", 10 years latency".

[g] "Vicinity workers".

[h] "High or medium exposure" but restricted to workers with at least ten years of exposure.

Table 3

SUMMARY OF MORTALITY DATA FOR LUNG CANCER REPORTED IN STUDIES OF CADMIUM WORKERS

Study population	Type of industry	First study			Second study			Third study			Fourth study		
		SMR	Obs./exp.	Ref.	SMR	Obs./exp.	Ref.	SMR	Obs./exp.	Ref.	SMR	Obs./exp.	Ref.
1	Ni-Cd battery plant in the U.K.	114	5/4.4	42	121	77/63.4	78	131	26/19.9	4[a]	127*	89/70.2	79a
2	Cd-smelter in the U.S.	235*	12/5.1	47	165*	20/12.2	82						
3	Ni-Cd battery plant in Sweden	143	2/1.4	43[b]	158	6/3.8	3[c]	175	7.40	19a[a]			
5	Cd-Cu alloy plant in the U.K.	81	10/12.4	33[a]	138	36/26.1	34[e]	101	47/46.4	4[a]			
6	17 different plants using Cd in the U.K.	112	32/28.6	4,5[a]	131	18/13.7	5a[f]						

Note: *Significant at $p < 0.05$.

[a] "High or medium exposure".
[b] Incidence.
[c] "High exposure, 10 years latency".
[d] "≥ 5 years of high exposure, 10 years latency".
[e] "Vicinity workers".
[f] "High medium exposure" but restricted to workers with at least 10 years of exposure.

Table 4

COMPARISON OF EXPECTED AND
OBSERVED NUMBERS OF CANCER
CASES AMONG 248 WORKERS
EXPOSED TO CADMIUM OXIDE[42]

Site of cancer	No. of cases		SMR
	Observed	Expected	
All sites	12	13.1	92
Bronchus	5	4.4	114
Bladder	1	0.5	200
Prostate	4	0.6	667[a]
Testis	0	0.1	0

[a]   Significant at $p < 0.01$.

Table 5

OBSERVED AND EXPECTED DEATHS ACCORDING
TO CAUSE AMONG WHITE MALES WHO HAD BEEN
EMPLOYED 2 YEARS OR MORE IN A CADMIUM-
SMELTING FACTORY[47]

Cause of death	Observed	Expected	SMR
Malignant neoplasm	27	17.5	154[a]
Digestive system	6	5.8	104
Respiratory system	12	5.1	235[a]
Prostatic cancer	4	1.1	347
Other unspecified	5	6.6	136
Heart disease	24	43.5	55[b]
Nonmalignant respiratory disease	8	5.0	159
Influenza and pneumonia	3	2.4	122
Other respiratory diseases	5	2.6	193
Other known causes	30	33.2	90
Unknown causes	3	0	
Total	92	99.3	

[a]   Significant at $p < 0.05$.
[b]   Significant at $p < 0.01$.

Lemen et al.[47] studied mortality in a cohort of 292 white male workers exposed to cadmium oxide fumes and dust in a smelter for at least 2 years between 1940 and 1969. At the time of measurement the cadmium concentration in air in the smelter varied considerably, with individual values usually varying between 75 and 1105 $\mu g/m^3$. No overall average was estimated. In 1947 the same factory had exposure levels in the range 40 to 31,300 $\mu g/m^3$. The workers were also exposed to low concentrations of arsenic, 0.3 to 1.4 $\mu g/m^3$, which may be of importance since arsenic is known to cause lung cancer in humans.[85]

In 1974, when the follow-up was terminated, 180 members of the cohort were alive, 92 had died, and another 20 could not be traced.[47] For the purpose of calculation, this latter group was considered to be alive. Observed mortality, due to different causes of death, was compared to expected mortality as estimated from the whole U.S. white male population (Table 5). Significant excess mortality was found for cancer (all sites) and respiratory tract cancer. There were four deaths due to prostatic cancer compared to an expected number of 1.15, which is close too, but does not quite attain statistical significance. When a latency period of 20 years from onset of exposure was taken into account, statistical significance, $p < 0.05$, was attained; 4 cases of prostatic cancer compared to an expected number of 0.88.

On the basis of data available in 1976, the International Agency for Cancer Research (IARC)[36] concluded that occupational exposure to cadmium in some form, possibly the oxide, increases the risk of prostatic cancer and that one study[47] suggested an increased risk of respiratory tract cancer.

Since 1976, several additional epidemiological studies have been published: Kjellström et al.,[43] Holden,[33,34] Sorahan,[78] Armstrong and Kazantzis,[4,5,5a] Andersson et al.[3] and Sorahan and Waterhouse,[79,79a,82] Thun et al.,[82] and Elinder et al.[19a]

Kjellström et al.[43] analyzed the mortality by life-table technique in two cohorts of cadmium-exposed workers in Sweden, one from a cadmium-nickel battery factory and the other from a cadmium-copper alloy plant. The cohort in the battery plant comprised 269 workers who had been exposed to cadmium oxide dust for more than 5 years. Exposure periods ranged from 5 to 51 years, i.e., employment from as far back as 1920. Prior to 1947, the cadmium exposure levels were on average above 1 mg/m³. In the 1950s the levels were in the order of 0.3 mg/m³ and between 1962 and 1974 about 50 μg/m³. The workers had also been exposed to concentrations of nickel hydroxide dust approximately five times higher than these levels.

There was a total of 43 deaths between 1949 and 1975.[43] Eight of the deaths were caused by cancer, a figure which was not in excess of the expected number based on the total Swedish male population. Cancer morbidity in the battery factory was also investigated. Between 1959 and 1975, 15 cancers had been diagnosed compared to an expected incidence of 16.4, based on data from the Swedish Cancer Register. With regard to specific causes, a significant increase could be shown for cancer of the nasopharynx only — two cases were found — whereas only 0.2 were expected. This was primarily thought to be the result of nickel hydroxide exposure. There was also a tendency for an increased incidence of prostatic, lung, and colorectal cancer, observed cases being 2, 2, and 5, respectively, compared to expected figures of 1.2, 1.4, and 2.2.

Andersson et al.[3] reported on a longer follow-up of the mortality in the same battery factory studied by Kjellström et al.[43] Among 531 men who had been exposed to cadmium and nickel in the factory for more than 1 year, there was a lower than expected total mortality: 105 fatalities before the age of 80 compared to 122 as estimated from National Swedish Statistics. Two hundred and seventy-three workers had been exposed to cadmium in workroom air exceeding 0.3 mg/m³ for at least 5 years. Assuming a 10-year latency period, these men had nonsignificant excess mortalities in cancer of all sites (26 observed, 21.6 expected), in lung (6 observed, 3.8 expected), intestines (5 observed, 2.6 expected), prostate (4 observed, 2.4 expected), and bladder (2 observed, 0.7 expected), in spite of a hypomortality for all causes of death (SMR = 93).

In 1985 Elinder et al[19a] updated the mortality in the same Swedish battery factory up to 1983. Another two cases of lung cancer had occurred and there was a total of seven lung cancers in workers who had experienced at least 5 years of high exposure compared to an expected number of 4.0 (SMR = 175).

The other cohort examined by Kjellström et al.[43] consisted of 94 workers who had been exposed to cadmium oxide dust and fumes at concentrations in the order of 100 to 400 μg/m³ during the 1960s. There were four deaths due to prostatic cancer in the exposed group compared to the expected number of 2.7 based on data from the Swedish Cancer Register at the National Board of Health and Welfare. An internal reference group of 328 workers in the same plant, not exposed to cadmium, was also investigated. Among these men, a total of 4 deaths in prostatic cancer had occurred, which was lower than the expected number of 6.4.

Holden[33] has reported on the mortality among 347 workers in 2 factories (A and B) manufacturing cadmium-copper alloys in the U.K. The criteria for selection of the cohort were not mentioned in the report. Many of the workers were employed before 1953, and the duration of exposure for several of the workers exceeded 15 years. Before 1953 cadmium levels in air sometimes exceeded 1 mg/m^3. During the 1950s and 1960s cadmium levels in air were in the order of 50 to 150 μg/m^3. Earlier medical surveys of cadmium-exposed workers in these two plants were conducted by Bonnell,[9] Kazantzis,[38] and Bonnell et al.[10] (see Chapter 8). Mortality was followed up until 1978 and compared with the expected number of deaths as calculated from normal death rates in England and Wales. There were 176 workers still alive, 154 had died, 4 had emigrated, and 13 could not be traced. There was a nonsignificant excess of malignant neoplasms, 34 observed compared to an expected number of 30.0. There was no significant increase in mortality in either cancer of the respiratory tract or cancer of the urogenital system. There was only one case of prostatic cancer. However, there were 4 cases of cancer in the reticuloendothelial system compared to an expected number of 1.2. The diagnoses were two cases of lymphatic leukemia, one case of Hodgkin's disease, and one of reticulosarcoma.

With regard to cancer of the respiratory tract, there were 8 cases in factory B compared to an expected number of 4.5, whereas in factory A there were only 2 cases compared to an expected number of 7.8. In another report Holden[34] extended the mortality study and also included "vicinity workers", i.e., men who worked in the same building as the smelter workers, but who were not directly involved in the manufacturing of copper-cadmium alloy. Among 624 "vicinity workers" who had been employed for at least 12 months, there was a significant excess of prostatic cancer, 8 cases compared to an expected number of 3.0.

Armstrong and Kazantzis,[4] in their extensive study on mortality of cadmium-exposed workers in the U.K., also presented detailed data from the cadmium-copper alloy plant formerly studied by Holden.[33,34] Among smelter and vicinity workers exposed to cadmium for more than 1 year before 1970, there were altogether 47 deaths in lung cancer compared to an expected number of 46.4, and an excess of prostatic cancer, 9 deaths compared to an expected figure of 4.7.

Sorahan[78] reported on mortality among nickel-cadmium battery workers in the U.K. This study is a more extensive follow-up of the same group of workers reported on by Kipling and Waterhouse.[42] The cohort was comprised of 3024 workers who had been employed at the plant for more than 1 month between 1923 and 1975. There were 2560 males and 464 females and 1284 of them were employed before 1947. The cadmium concentration in workroom air was first estimated in 1949 and was found to range from 0.6 to 2.8 mg/m^3. After installation of an exhaust ventilation system in 1950, cadmium in air was reduced to below 0.5 mg/m^3. Subsequently, levels of cadmium in air were reduced stepwise and since 1975 the plant has operated at cadmium levels in air below 50 μg/m^3. It is, however, evident from the report that only a minor part of the workers examined had actually been engaged in jobs considered to entail a "high" or "slight" exposure to cadmium. Most of them were considered to have had "minimal" exposure only.

Data on the cause of death for the 659 workers who died between 1946 and 1980 were recorded from the National Health Recorder. The follow-up could not be completed because 66 had emigrated and 82 could not be traced. The remaining 2219 were alive on June 30, 1980.

Among the male workers, 152 persons had died of cancer, compared to an expected figure of 151.4. Expected number of cancers were estimated from the National Rates of England and Wales. Altogether there were 77 fatalities in cancer of the lung and bronchus (ICD Code No. 162-163), compared to an estimated number of 63.4. The observed number of deaths in prostatic cancer was 7 compared to an expected number of 6.0. Mortality among workers with "high" exposure to cadmium was also reported. This group was, however, defined in a curious way; "those with a period of high exposure employment, alive 15 years after the first employment and having left the company by that date". According to our interpretation, this definition implies that workers who had been exposed for more than 15 years would not be included, even though this group may be the most prone to suffer from long-term health consequences as a result of cadmium exposure. Preferably, the group defined by Sorahan[78] should be termed a "leaving" cohort. The group will not include the most heavily exposed workers. Excess cancer mortality among heavily exposed workers may well have existed without being discovered due to the unusual methodology.

In another report, which included mortality data from the same plant, Armstrong and Kazantzis[4] reported increased mortality in both cancer of the respiratory tract and cancer of the prostate. Among workers considered to have been exposed, on any occasion, to either "medium" or "high" concentrations of cadmium during their work in the battery plant, there were 26 deaths in lung cancer compared to an expected number of 19.9. There were 4 deaths in prostatic cancer in the same group compared to an expected figure of 1.7. However, neither of these mortality figures attained statistical significance.

In 1983 Sorahan and Waterhouse[79] published another report on the causes of death of workers in the same nickel and cadmium battery factory. The period of observation was extended to January 31, 1981; therefore, the number of expected and observed cases was somewhat higher. For the whole cohort there was now a significant excess in lung cancer, 89 observed vs. 70.2 expected. Among men employed during the preamalgamation period 1923 to 1946, and thus heavily exposed to cadmium, there were 52 cases of lung cancer and 42.4 expected. Since the cohort also included workers with "minimal exposure", a further analysis was used taking exposure periods and levels of exposure of different workers into consideration. Using such a procedure it could be shown that the ten workers who had died from prostatic cancer had a significantly higher degree of exposure when compared to other workers in the same factory.

In 1985 Sorahan and Waterhouse[79a] reported on the incidence of prostatic cancer in a subgroup of 458 workers employed for at least 1 year in a job involving high exposure to cadmium oxide dust. There were 8 observed cancers compared to 2.0 expected (SMR = 400, $p < 0.01$). These eight cases included the 4 cases previously reported by Kipling and Waterhouse.[42]

Armstrong and Kazantzis[5] have investigated the mortality rate for cadmium-exposed workers in 17 other plants in the U.K. The cohort was comprised of workers who were employed before 1970 and were born before 1940. Altogether 6995 subjects were included in the study reported in 1983. Most of the workers (4453) were involved in primary cadmium production. The remaining 2542 were engaged in the production of cadmium alloys (1559), pigment and oxides (531), and stabilizers (452). Most of the members of the cohort (5623) had been exposed to cadmium to a limited extent only. In a later publication[41] it is reported that 49 smelter workers previously rated as being "medium" exposed to cadmium for more than 10 years had median urinary and blood cadmium concentrations of 3.4 μg Cd per gram creatinine and 3.5 μg Cd per liter, respectively. These levels are low in comparison to what is frequently found in cadmium-exposed workers (Chapter 6) and no evidence of renal dysfunction was found in this group.[41]

Only 199 men were considered to have ever been subject to "high" exposure levels of cadmium.[5] Among these men, 13 had died of cancer, which should be compared to the expected number of 10.4. There were 5 cases of lung cancer, compared to an expected number of 4.4. Other types of cancer numbered two or less. When including

Table 6

OBSERVED AND EXPECTED NUMBER OF DEATHS WITH
CERTAIN DIAGNOSIS. TOTAL OF ALL STUDIES ON
CADMIUM-EXPOSED POPULATIONS IN THE U.K.
ESTIMATE BASED ON WORKERS WHO HAD BEEN
EXPOSED TO EITHER "MEDIUM" OR "HIGH" CADMIUM
LEVELS DURING WORK (AGES 15 TO 84)[4]

Cause of death	ICD no.	Observed	Expected	SMR
All neoplasms	140—239	237	230.7	103
Malignant neoplasms				
Trachea, bronchus, and lung	162	105	93.8	112
Prostate	185	13	9.4	138
Hypertensive disease	400—404	18	17.7	102
Cerebrovascular disease	430—438	74	85.1	87
Disease of the respiratory system	490—492	112[a]	68.7	163[b]
Nephritis and nephrosis	580—584	11	7.3	150

[a]   Includes 11 bronchitis deaths which were primarily misleadingly coded as "acute cadmium poisoning".[4,33]

[b] $p < 0.01$.

both "ever-high" and "ever-medium" in the cohort, there was a slight increase in lung cancer, a total of 32 cases compared to 28.6 expected, but a deficit in prostatic cancer, no cases vs. expected 2.9. When only workers who had at least 10 years of exposure were considered there were 18 cases of lung cancer compared to 13.7 expected (SMR = 131) and no prostatic cancer compared to 1.4 expected.

Table 6 presents results from the pooled population of cadmium-exposed workers in the U.K.,[4] which have been reported separately by Holden,[33,34] Sorahan,[78] and Armstrong and Kazantzis.[5] Observed and expected number of deaths in workers considered to have ever experienced "medium" or "high" exposure to cadmium are included. With regard to cancer of the lung and prostate there are tendencies towards an excess in both, although neither attained statistical significance.

An update and extension of the mortality among U.S. cadmium production workers have also recently been reported.[82] Cause-specific mortality rates for 602 white males with at least 6 months of cadmium exposure between 1940 and 1969 were determined and compared to the estimated figures from the U.S. white population. The previously reported increased incidence of lung cancer was confirmed: in the new follow-up there were 20 observed cases vs. 12.2 expected ($p < 0.05$). The earlier observation of prostatic cancer was, however, weakened because one previous case was excluded from the cohort because he had less than 6 months of cadmium exposure. In the new follow-up there were 3 cases of prostatic cancer vs. an expected number of 2.2 ($p > 0.05$).

Of particular interest in the new study is a dose-response analysis which reveals that cadmium production workers with long-term exposure to high levels of cadmium ran a considerably higher risk of developing lung cancer compared to workers with less exposure. Among workers with a cumulative exposure estimate of less than 584 (mg-days/m³) there were 2 deaths in lung cancer compared to an expected number of 3.8 (SMR = 53). Seven of the workers having a cumulative exposure index between 584 and 2920 mg-days/m³ died of lung cancer compared to 4.6 expected (SMR = 152). Among workers who had a cumulative exposure estimate exceeding 2921 mg-days/m³ there were 7 fatalities in lung cancer compared to an expected number of 2.5 (SMR = 280). The cumulative exposure indexes of less than 584 mg-days/m³ and more than 2921 mg-days/m³ are equivalent to 40 years of exposure to less than 40 μg Cd per cubic meter and more than 200 μg Cd per cubic meter, respectively.

2. Studies of General Population Groups

Several investigators have utilized a case reference approach aimed at investigating the possible association between cadmium and cancer. Morgan[54-56] studied renal and hepatic cadmium levels among patients who had died from different diseases and noticed that patients who died of diseases of the lung and emphysema had significantly higher cadmium concentrations in these organs compared to persons who died of other diseases.

In 36 controls, i.e., patients with causes of deaths other than cancer, e.g., chronic liver, kidney, and lung disease, the average cadmium concentration in liver was 170 mg Cd per kilogram ash weight corresponding to about 2.2 mg/kg wet weight. The figures for 23 lung cancer patients were 240 mg Cd per kilogram ash weight which corresponds to about 3.2 mg/kg wet weight. The corresponding figures for cadmium concentration in kidneys obtained from noncancer patients were 2512 mg Cd per kilogram and from lung cancer patients 3478 mg Cd per kilogram ash weight, which corresponds to about 27.6 and 38.2 mg/kg on a wet weight basis. Smoking is, however, a major confounding factor in this context and the differences seen in the concentration of cadmium in tissue can probably be explained by the different smoking habits among the groups.

Kolonel[44] carried out a case control study and claimed that a relationship existed between cadmium exposure and renal cancer. The degree of cadmium exposure, based on occupation, diet, and smoking habits, was calculated in retrospect for patients with renal cancer and compared to similar estimates for patients with colon cancer and for noncancer patients. The calculations of dietary cadmium intake incorporated serious errors, however, and the estimates of occupational exposure were very uncertain. The daily intake of cadmium via food was estimated from erroneous data on the cadmium concentration in different footstuffs (particularly milk; see Chapter 3) and, therefore, appeared excessively high. Thus, the daily intake of cadmium was estimated to be 135 μg for the renal cancer group and 128 and 129 μg for the colon cancer group and the noncancer group, respectively. Compared to more recent data on daily intake of cadmium (about 20 μg/day, see Chapter 3), these estimates are unrealistic. The evaluation of occupational exposure is also very doubtful as it was based on crude classification of jobs as being "high risk" for cadmium exposure and not on the actual data obtained from measurements: jobs classified as being "high risk" were, e.g., electroplating, alloy smelting, and welding. Needless to say this type of work does not necessarily involve cadmium exposure. The only data that appear to be reasonable in the report are the findings of a slightly higher prevalence of smokers (62%) in the renal cancer group compared to the colon cancer group (52%) and the noncancer group (58%). This does not mean that cadmium in cigarettes caused renal cancer.

In 1977 Kolonel and Winkelstein[45] reported a similar case control study on the relationship between cadmium and prostatic cancer. Occupational exposure to cadmium was more common among the cases than in the controls, but the differences were not significant. As in the previous study,[44] exposure to cadmium was classified in a very uncertain way and because of this the data given in this short report are of no value.

Bako et al.[6] found a positive correlation between the age-adjusted incidence of prostatic cancer in different geographical areas of Canada and certain environmental measurements of cadmium, i.e., in water and soil. Taking into consideration the various possible confounding factors such as socioeconomic variations, race, etc. that may be involved, the authors were reluctant to suggest any causal relationship between environmental cadmium exposure and increased incidence of prostatic cancer.

Studies on causes of death in cadmium-exposed areas of Japan have been made by Shigematsu et al.[72] and Kjellström and Matsubara (unpublished data). Shigematsu et al.[72] examined the age-adjusted mortality for several different causes of death in four cadmium-polluted areas. Average cadmium content in rice in the examined areas

Table 7

POPULATION SIZE AND PERIOD OF
INVESTIGATION IN THE STUDY BY
SHIGEMATSU ET AL.[72] ON CAUSES OF DEATH
IN CERTAIN AREAS OF JAPAN

| Prefecture | Sex | Average population | | Period of investigation |
		Cd polluted	Nonpolluted	
Akita	M	29,027	24,653	1958—1977[a]
	F	31,362	26,030	(20 years)
Miyagi	M	30,595	19,650	1971—1976[a]
	F	33,251	20,973	(6 years)
Toyama	M	16,191	21,183	1948—1977[b]
	F	17,513	22,349	(30 years)
Nagasaki	M	2,407	17,522	1952—1977[a]
	F	2,393	18,120	(26 years)

[a]   Source of mortality data — Vital Statistics.
[b]   Source of mortality data — death certificates.

ranged from 0.2 to 0.7 mg/kg. Nonpolluted areas in the same prefecture as the polluted areas, and in which average cadmium concentrations in rice ranged from 0.02 to 0.1 mg/kg, served as controls. Table 7 presents the number of people included in the cohorts from the four polluted and the four control areas. The polluted areas examined are located in the prefectures of Akita, Miyagi, Toyama, and Nagasaki. The overall cancer mortality rates and the death rates in stomach and liver cancer were similar in the polluted and the control areas of the same prefecture. Death due to prostatic cancer was, however, more common in two of the polluted areas (Akita and Nagasaki) (Figure 1). Furthermore, hyperplasia of the prostata as a cause of death was significantly elevated in one area (Akita). If proper autopsies were not carried out, it is possible that hyperplasia of the prostate was confused with prostatic cancer. It should be noted that autopsies are rare in Japan, particularly in rural areas.

Kjellström and Matsubara (unpublished data) have carried out a similar study to that by Shigematsu et al.,[72] including cohorts from a total of nine different polluted areas: in Akita, Myagi, Toyama, Fukushima, Ishikawa, Hyoga, Oita, and two from Gumma prefectures. Each polluted area was matched with one or several adjacent control areas for which there was no known cadmium pollution, but which had similar geographic and meteorological conditions. The total population in the combined polluted areas, aged 35 to 84, was about 100,000. The same figure applies for the combined control areas. The purpose of combining all the exposed areas was to obtain a sample large enough to allow conclusions to be drawn. Results from one area only will almost never be significant because of the small number of cases from each specific cause of death.

The age-adjusted deaths from different causes among inhabitants of the combined cadmium-polluted areas were compared to the similar age-adjusted death rates among inhabitants in the combined nonpolluted areas (Kjellström and Matsubara, unpublished data). The overall cancer mortality was not significantly different for cadmium- and nonpolluted areas, age-standardized mortality rate ratios (AMRR) being 95.7 and 108 for males and females, respectively. An AMRR of 100 indicates that the age-standardized mortality rates are the same in both exposed and control areas. Leukemia (AMRR = 160), cancer of the bladder (AMRR = 144), cancer of the kidney (AMRR = 136), and cancer of the prostate (AMRR = 134) were, however, reported to be more

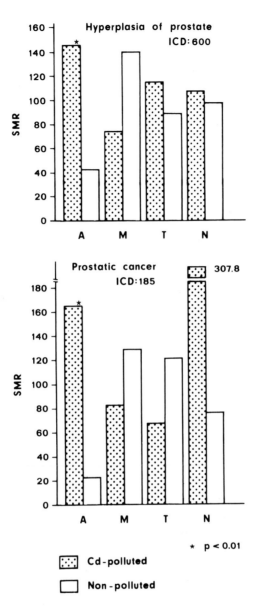

FIGURE 1. Age-adjusted mortality rates (prostatic cancer and hyperplasia of prostate) in polluted and nonpolluted areas. A, M, T, and N denote different prefectures: Akita, Miyagi, Toyama, and Nagasaki, respectively. (From Shigematsu, I., Kitamura, S., Takeuchi, J., Minowa, M., Nagai, M., Usui, T., and Fukushima, M., *Cadmium 81, Proc. 3rd Int. Cadmium Conf. Miami,* Cadmium Association, London, 1982, 115—118. With permission.)

common among male inhabitants of the polluted areas compared to male inhabitants of the combined nonpolluted areas. Certain other cancers, such as cancer of the liver (AMRR = 44) and cancer of the pancreas (AMRR = 70), were less common among males in the polluted areas. Among females in the combined polluted areas there was excess mortality in cancer of the kidney (AMRR = 233), cancer of the lung (AMRR = 117), and cancer of the breast (AMRR = 114). In contrast to what was found in men, the female mortaltiy in leukemia (AMRR = 98) and cancer of the bladder (AMRR =

78) was lower among inhabitants of polluted areas compared to inhabitants of control areas.

There are some problems when interpreting the two studies conducted by Shigematsu et al.[72] and Kjellström and Matsubara (unpublished data), one of which is the classification of exposure. The exposed cohorts comprised people living in "polluted areas". For practical purposes these areas have been chosen from administrative areas of the prefecture, villages, or towns, where increased levels of cadmium in rice are know to occur. Obviously this does not mean that all inhabitants of the "polluted areas" have eaten polluted rice and thereby were actually exposed to more than normal amounts of cadmium. For example, Shigematsu et al.[72] in their study classified 60,000 inhabitants of Akita prefecture, and another 64,000 inhabitants of Miyagi prefectures as being cadmium exposed, whereas in the monograph *Cadmium Studies in Japan* edited by Tsuchiya[83] only about 800 inhabitants in each of these areas were suspected of being exposed to cadmium to such an extent that health effects might occur (see further Chapter 3, Table 6).

With regard to Toyama and Nagasaki, the exposed groups were comprised of about 34,000 and 4,700, compared to the 7,600 and 2,400 for earlier estimates (Chapter 3, Table 6). Kjellström and Matsubara (unpublished data) state that about 100,000 subjects included in the cohort were people living in polluted areas. Based on data from Tsuchiya[83] (Chapter 3, Table 6), the number of persons in these areas within the risk zone for negative health effects, due to cadmium pollution, can be estimated to be around 36,000. When only a relatively small proportion of the exposed group has in fact been exposed to any significant extent, there is a risk that any possible effect on the causes of death will be extenuated and thereby difficult, if not impossible, to prove statistically.

Other problems in connection with macroepidemiological studies of these types are the difficulties involved in selecting an appropriate control area. Large differences in causes of death are known to exist in different districts in Japan and differences in, e.g., the frequency in which autopsies are carried out between exposed and control areas, may easily invalidate comparison. The macroepidemiological studies from cadmium-exposed areas in Japan can, therefore, not be regarded as conclusive, even though it is noteworthy that increased mortality in prostatic cancer was found for certain areas in both of the studies.

Another Japanese study on the causes of death among persons with cadmium-induced tubular damage[57] is of interest as it is based on persons with known excessive long-term exposure to cadmium (see further Chapter 11, Section V.B.2.b). Among 81 male patients, 3 deaths due to cancer were reported compared to an expected figure of 2.9, and among 124 female patients, 4 cancers had occurred compared to an expected number of 2.5.

C. Discussion and Conclusion

Studies on animals have provided conclusive evidence that injection of cadmium can cause sarcoma at the site of injection as well as interstitial tumors of the testis. A recent study has shown that inhalation of cadmium chloride aerosol has a strong carcinogenic effect in rats. No significant increase in the occurrence of tumors has been found after long-term peroral exposure of animals, but none of the studies published is free of methodological deficiencies.

In some of the epidemiological studies on cadmium-exposed workers in smelters and in nickel-cadmium battery factories, a significant excess has been found for cancer of the lung and prostate. As the excess of these cancers is only 30 to 70%, studies of individual factories often fail to reach statistical significance, and negative results have also been reported. If data from the most recent published report from each of the six examined populations are added together, the conjoined results become statistically significant for both prostatic and lung cancer. The total observed number of prostatic cancers in the most recent studies are 28 compared to 15.7 expected, SMR = 178, $p <$ 0.05 (Table 2). For lung cancer (Table 3) the total number of cases in the most recent reports is 181 and the expected figure 146.5, SMR = 124, $p < 0.05$.

The epidemiology of prostatic cancer and its causal mechanism are quite complex and national and regional rates are frequently very variable. Several constitutional factors such as race as well as "life style" factors, e.g., marital status, dietary habits, and alcohol consumption, are factors which have been associated with the disease. There is, however, no apparent reason why cadmium-exposed workers should have a higher prevalence of confounding factors and therefore be biased towards a greater prevalence of prostatic cancer compared to other occupational groups. At present the most reasonable interpretation is that exposure to cadmium may contribute to the development of prostatic cancer among exposed workers. This was also the conclusion reached at a meeting on the Carcinogenicity of Metals organized by the Scientific Committee on the Toxicology of Metals of the Permanent Commission and International Association of Occupational Health.[8]

In view of experimental data, the possibility that cadmium may cause lung cancer is probably of even greater interest. A significant excess mortality in lung cancer has been found in two populations (Table 3). Smoking habits and simultaneous exposure to other known or suspected carcinogenic agents such as asbestos and nickel (especially in the nickel-cadmium battery plants) have to be considered and therefore the human data at hand do not provide conclusive evidence. Nevertheless, there are good reasons for suspecting that high inhalative cadmium exposure may induce lung cancer in humans.

In this context it should also be mentioned that smokers inhale substantial amounts of cadmium from cigarettes (see Chapters 5 and 8) and it can be speculated that cadmium in cigarettes may play a role in the development of lung cancer in smokers. There are also many other, more potent, carcinogenic substances in cigarette smoke. The carcinogenic effects from these other substances are likely to overshadow the possible carcinogenic role of cadmium.

In order to draw more definite conclusions about the possible role of cadmium in carcinogenesis, additional epidemiological studies on cancer mortality and morbidity among cadmium-exposed workers would be of great value. Additional experimental studies, especially on the carcinogenic effects of cadmium due to inhalation, are urgently needed.

II. MUTAGENIC EFFECTS

Mutagenic action of cadmium has been investigated in various ways, e.g., by examining the effects of cadmium on the synthesis of RNA and DNA, mutagenic effects of cadmium in microorganisms, plants, insects, cells of mammals, and humans in vivo and in vitro. A comprehensive review covering the mutagenic effects of cadmium has recently been provided by Degraeve.[13] The present text has been restricted to a discussion of the mutagenic effects of cadmium on mammalian cells in vivo, and to mutagenic effects of cadmium seen in experimentally exposed animals and in environmentally or occupationally exposed humans. Teratogenic effects and effects from cadmium on the oöcytes are discussed in Section VII of Chapter 11.

A. Mammalian Cells In Vitro

Mutagenic effects of cadmium, as revealed by chromosome changes, have mainly been seen in cells exposed to cadmium sulfate, cadmium sulfide, and to cadmium chloride. Shiraishi et al.[76] examined chromosomal aberrations in human leukocytes, cultured for 4 to 8 hr in a medium containing 62 μg/ℓ of cadmium sulfide. A marked increase in frequency of chromatid breaks, translocation and, dicentric chromosomes was observed. Röhr and Bauchinger[67] studied fibroblast cell cultures from Chinese hamsters which had been exposed to cadmium sulfate at concentrations ranging from 1 μg to 10 mg Cd per liter. At concentrations exceeding 100 μg/ℓ, the mitotic index was significantly reduced and at concentrations exceeding approximately 0.5 mg Cd per liter, chromosome damage was seen.

Paton and Allison[62] investigated the effects of cadmium chloride on human lymphocytes and fibroblasts cultured for periods ranging from 24 to 72 hr. Cadmium concentrations ranged from 0.3 to 5.6 μg Cd per liter. No gross chromosomal abnormalities were observed. Similar negative results were obtained by Deknudt and Deminatti,[15] who performed studies on human lymphocytes cultured in a medium containing considerably higher cadmium chloride concentrations, 0.6 and 6 mg Cd per liter.

On the other hand, Andersen et al.[2] found that the average chromosome length in human lymphocytes was reduced compared to controls when the cells were cultured for 4 hr in a medium containing 10^{-5} *M* of cadmium chloride (1.1 mg Cd per liter). After a longer period of cultivation (48 hr), the average length of the cells was normalized. The authors provide evidence that this finding could be related to the formation of metallothionein.[2] When metallothionein was present, the toxicity of cadmium was less.

Deaven and Campbell[12] investigated the influence of different types of media on the threshold concentration of cadmium chloride in connection with the induction of chromosomal aberrations in cultured Chinese hamster cells. With 15% of newborn calf serum in the medium, chromosomal aberrations in the cells occurred when the cadmium concentration in the medium exceeded 0.1 mg Cd per liter. Supplementing the medium with 20% fetal calf serum, in place of the newborn calf serum, increased the threshold concentration of cadmium required in order to induce chromosomal damage to 0.2 mg/ℓ. The same authors[12] also showed that the degree of growth rate inhibition induced by cadmium chloride was influenced by the type of medium used. It is clear from this study that standardized experimental protocols should be used in order to facilitate a comparison of the results of the effects of cadmium on cultured cells, i.e., the results which can be derived from various reports.

B. Animals

The mutagenic potential of cadmium in vivo has not been sufficiently assessed. Deknudt[14] and Deknudt and Gerber[16] exposed mice to cadmium in the diet for 1 month, 60 mg/kg food, and observed no abnormalities in cells obtained from the bone marrow. Degraeve[13] refers to Doyle et al.[19] who exposed lambs to the same concentration of cadmium in diet as mice, i.e., 60 mg/kg, for 198 days. In contrast to the findings for mice, Doyle et al.[19] noticed an increased frequency of peripheral lymphocytes with hypodiploidy in lambs.

Intraperitoneal injection of cadmium chloride to male mice, at doses ranging from 0.5 to 7 mg/kg body weight, did not significantly decrease the number of living embryos in female mice mated with treated mice,[20,23,65] nor did it increase the number of chromosome aberrations in spermatocytes.[23] However, in mice, Felten[21] observed chromatid breaks and deletions in cells taken from the bone marrow as well as gaps and breaks in spermatocytes 6 hr after exposure to parenteral administration of cadmium chloride at doses ranging from 0.6 to 2.8 mg Cd per kilogram body weight.

The effects of cadmium on female reproduction and on female germ cells have also been tested. Suter[80] gave female mice a single i.p. injection of cadmium chloride, 1.3 mg Cd per kilogram body weight as cadmium chloride prior to mating. No detectable increase in dead implants was recorded when exposed animals were compared to a control group given distilled water only. Shimada et al.[73] and Watanabe et al.[85] gave nonpregnant female mice and hamsters s.c. injections of cadmium chloride shortly before ovulation at doses ranging from 0.6 to 3.6 mg Cd per kilogram. Cytogenetic preparation of unfertilized oöcytes was performed 12 hr after the injections. At doses exceeding 1.3 mg Cd per kilogram, numerical chromosomal anomalies such as hypodiploidy, hyperdiploidy, and diploidy were seen. In an extension of these studies, Watanabe and Endo[84] examined blastocysts from cadmium chloride-treated mice (1.0 and 2.0 mg Cd per kilogram body weight). They observed a significant increase in the number of triploidy in blastocysts from treated mice compared to controls given saline. As discussed in Chapter 11, Section VI.C, injections of large doses of cadmium (3 to 6 mg Cd per kilogram body weight) in female hamsters produce periods of infertility. When the hamsters became pregnant, the litter size and the morphological appearance of the pups were normal.[68]

C. Humans

Conflicting results have been reported for humans exposed to cadmium. Shiraishi and Yoshida[75] examined lymphocyte chromosomes in blood samples obtained from seven female patients suffering from Itai-itai disease (Chapter 10) and from a control group of seven similar-aged females. A dramatic increase in chromosomal aberrations was seen among the Itai-itai patients, especially with regard to the frequency of cells with chromatid breaks and translocation. Further examination of 12 Itai-itai patients and 9 controls gave similar results.[74]

These results were not confirmed in a study by Bui et al.,[11] who examined cultured lymphocytes from four Itai-itai patients and from four controls. These authors also examined lymphocytes obtained from five cadmium-exposed workers in a battery factory and from five controls and found no evidence of an increase in frequency of structural chromosomal aberrations among cadmium-exposed subjects compared to controls. The groups studied were small and it is, therefore, difficult to obtain statistical significance even though true differences may exist between the groups.

Bauchinger et al.[7] analyzed chromosomes in peripheral lymphocytes from 24 workers exposed to cadmium and lead in a cadmium-zinc smelter and compared the number of cells with chromosomal aberrations to results obtained from 15 controls. Chromatid changes were more prevalent among the workers, especially with regard to breaks and gaps. The exposure to cadmium, however, does not appear to have been very extensive since blood cadmium levels were in the order of 4 μg Cd per liter, with a scatter indicative of low and moderate exposure.

In a similar study, Deknudt and Léonard[17] found an increased frequency of complex chromosomal anomalies in 23 workers exposed to cadmium and lead, and in 12 workers exposed mainly to zinc and to lower levels of cadmium and lead when compared to 12 control persons working in the administrative department. Data on cadmium in blood are presented in the report but appear to be unrealistically high. Since there was a combination of exposure to cadmium and lead, it is not clear whether the chromosome aberrations were causally related to cadmium.

O'Riordan et al.[61] performed chromosome analyses on lymphocytes from two groups of workers in a cadmium pigment-producing plant. The two groups consisted of 40 men who had worked with cadmium for exposure periods ranging from 6 weeks to 34 years. The second group of workers were 13 similarly aged men working at the same plant, but not considered to be exposed to cadmium. The frequency of chromosomal aberrations in the high and low exposure group of workers was also compared to results obtained earlier from the general population. There were no statistically significant differences in frequency of aberrations between the groups. However, there were more cells with gaps (4.2%) in the exposed group of workers compared to the low exposure group (3.1%) and in comparison to what was found in the general population (2.2%). O'Riordan et al.[61] concluded that the study produced no evidence of an association between cadmium exposure and chromosome aberrations.

D. Discussion and Conclusions

Data from experimental systems (mammalian cells in vitro and animals) show that cadmium, in certain forms, has mutagenic properties. Doses used in order to elicit these effects have generally been very high.

With regard to human exposure, data are conflicting. There are some indications for increased prevalence of chromosome aberrations in peripheral lymphocytes, but at the same time negative findings have also been published.

Smokers have recently been shown to have more chromosome aberrations in comparison to nonsmokers[22,46] and, therefore, smoking is an important confounding factor. Unfortunately, smoking habits were not considered in any of the human studies mentioned above and, therefore, no definite conclusions can be drawn. Differences in the frequency of chromosomal aberrations in various occupational groups may easily occur or disappear if smoking habits are dissimilar.

REFERENCES

1. Altman, P. L. and Dittmer, B. S., Eds., *Biology Data Book,* 2nd ed., Vol 1, Federation of American Society for Experimental Biology, Bethesda, Md., 1972, 229—233.
2. Andersen, O., Rønne, M., and Nordberg, G. F., Effects of inorganic metal salts on chromosome length in human lymphocytes, *Hereditas,* 98, 65—70, 1983.
3. Andersson, K., Elinder, C.-G., Hogstedt, C., Kjellström, T., and Spång, G., Mortality among cadmium workers in a Swedish battery factory, *Toxicol. Environ. Chem.,* 9, 53—62, 1984; *Arbete and Hälsa,* Swedish National Board of Occupational Safety and Health, Solna, 1983, 31.
4. Armstrong, B. G. and Kazantzis, G., A mortality study of cadmium workers in England, A report to the International Lead Zinc Research Organization, TUC Centenary Institute of Occupational Health, London School of Hygiene and Tropical Medicine, London, 1982.
5. Armstrong, B. G. and Kazantzis, G., The mortality of cadmium workers, *Lancet,* June, 1425—1427, 1983.
5a. Kazantzis, G. and Armstrong, B. G., A mortality study of cadmium workers from seventeen plants in England, in Edited Proc. Fourth Int. Cadmium Conf., Munich, March 2—4, 1983, 39—142.
6. Bako, G., Smith, E. S. O., Hanson, J., and Dewar, R., The geographical distribution of high cadmium concentrations in the environment and prostate cancer in Alberta, *Can. J. Public Health,* 73, 92—94, 1982.
7. Bauchinger, M., Schmid, E., Einbrodt, H. J., and Dresp, J., Chromosome aberrations in lymphocytes after occupational exposure to lead and cadmium, *Mutat. Res.,* 40, 57—62, 1976.
8. Belman, S. and Nordberg, G. F., Eds., Proceedings of a workshop conference on the role of metals in carcinogenesis, Atlanta, Georgia, March 24—28, 1980, *Environ. Health Perspect.,* 40, 1—42, 1981.
9. Bonnell, J. A., Emphysema and proteinuria in men casting copper-cadmium alloys, *Br. J. Ind. Med.,* 12, 181—197, 1955.
10. Bonnell, J. A., Kazantzis, G., and King, E., A follow-up study of men exposed to cadmium oxide fume, *Br. J. Ind. Med.,* 16, 135—145, 1959.
11. Bui, The-Hung, Lindsten, J., and Nordberg, G. F., Chromosome analysis of lymphocytes from cadmium workers and itai-itai patients, *Environ. Res.,* 9, 187—195, 1975.
12. Deaven, L. L. and Campbell, E. W., Factors affecting the induction of chromosomal aberrations by cadmium in Chinese hamster cells, *Cytogenet. Cell Genet.,* 26, 251—260, 1980.
13. Degraeve, N., Carcinogenic, teratogenic and mutagenic effects of cadmium, *Mutat. Res.,* 86, 115—135, 1981.
14. Deknudt, Gh., Mutagenicity of heavy metals,, *Mutat. Res.,* 53, 176, 1978.
15. Deknudt, Gh. and Deminatti, M., Chromosome studies in human lymphocytes after *in vitro* exposure to metal salts, *Toxicology,* 10, 67—75, 1978.

16. Deknudt, Gh. and Gerber, G. B., Chromosomal aberrations in bone-marrow cells of mice given a normal or a calcium-deficient diet supplemented with various heavy metals, *Mutat. Res.*, 68, 163—168, 1979.

17. Deknudt, Gh. and Leonard, A., Cytogenetic investigations on leucocytes of workers from a cadmium plant, *Environ. Physiol. Biochem.*, 5, 319—327, 1975.

18. Deutsche Forschungsgemeinschaft, Maximal workplace concentrations and biological tolerance values, 1983 Mitteilung 19 der Senatskommission zur Prüfung Gesundheitsschädlichen Arbeitsstoffe, DF6 German Research Academy, Verlag Chemie, Berlin, 1983 (in German).

19. Doyle, J. J., Pfander, W. H., Crenshaw, D. B., and Snethen, J. M., The induction of chromosomal hypodiploidy in sheep leucocytes by cadmium, *Interface*, 3, 9, 1974.

19a. Elinder, C-G., Kjellström, T., Hogstedt, C., Andersson, K., and Spång, G., Cancer mortality of cadmium workers, *Br. J. Ind. Med.*, in press, 1985.

20. Epstein, S. S., Arnold, E., Andrea, J., Bass, W., and Bishop, Y., Detection of chemical mutagens by the dominant lethal assay in the mouse, *Toxicol. Appl. Pharmacol.*, 23, 288—325, 1972.

21. Felten, T. L., A preliminary report of cadmium-induced chromosomal changes in somatic and germinal tissues of C57Bl/6J male mice, *Genetics*, 88, 26—27, 1979.

22. Fredga, K., Dävring, L., Sunner, M., Bengtsson, B. O., Elinder, C. -G., Sigtryggsson, P., and Berlin, M., Chromosome changes in workers (smokers and nonsmokers) exposed to automobile fuels and exhaust gases, *Scand. J. Work Environ. Health*, 8, 209—221, 1982.

23. Gilliavod, N. and Leonard, A., Mutagenicity tests with cadmium in the mouse, *Toxicology*, 5, 43—47, 1975.

24. Gilman, J. P., Metal carcinogenesis. II. A study on the carcinogenic activity of cobalt, copper, iron and nickel compounds, *Cancer Res.*, 22, 158—162, 1962.

25. Gunn, S. A., Gould, T. C., and Anderson, W. A. D., Cadmium-induced interstitial cell tumors in rats and mice and their prevention by zinc, *J. Natl. Cancer Inst.*, 31, 745,—753, 1963.

26. Gunn, S. A., Gould, T. C., and Anderson, W. A. D., Effect of zinc on cancerogenesis by cadmium, *Proc. Soc. Exp. Biol. Med.*, 115, 653—657, 1964.

27. Gunn, S. A., Gould, T. C., and Anderson, W. A. C., Specific response of mesenchymal tissue to cancerogenesis by cadmium, *Arch. Pathol.*, 83, 493—499, 1967.

28. Haddow, A., Dukes, C. E., and Mitchley, B. C. V., Carcinogenicity of iron preparations and metal-carbohydrate complexes, *Rep. Br. Emp. Cancer Campgn.*, 39, 74—76, 1961.

29. Haddow, A., Roe, F. J. C., Dukes, C. E., and Mitchley, B. C. V., Cadmium neoplasia: sarcomata at the site of injection of cadmium sulphate in rats and mice, *Br. J. Cancer*, 18, 667—673, 1964.

30. Heath, J. C., Daniel, M. R., Dingle, J. T., and Webb, M., Cadmium as a carcinogen, *Nature (London)*, 193, 592—593, 1962.

31. Heering, H., Oberdörster, G., Hochrainer, D., and Baumert, H. P., Learning behavior and memory of chronically Cd-exposed rats and organ distribution of Cd, Document XII/ENV/64/79, Commission of the European Community, Luxembourg, 1979, 61—62.

32. Holden, H., Cadmium toxicology (letter), *Lancer*, 2, 57, 1969.

33. Holden, H., A mortality study of workers exposed to cadmium fumes, in *Cadmium 79, Proc. 2nd Int. Cadmium Conf. Cannes*, Cadmium Association, London, 1980a, 211—215.

34. Holden, H., Further mortality studies on workers exposed to cadmium fume, in *Occupational Exposure to Cadmium*, Cadmium Association, London, 1980b, 23—24.

35. Humperdinck, K., Cadmium and lung cancer, *Med. Klin.*, 63, 948—951, 1968 (in German).

36. IARC, The Evaluation of Carcinogenic Risk of Chemicals to Man: Cadmium and Nickel, IARC Monographs, Vol. 11, International Agency for Research on Cancer, Lyon, 1976, 39—75.

37. Kanisawa, M. and Schroeder, H. A., Life term studies on the effect of trace elements on spontaneous tumors in mice and rats, *Cancer Res.*, 29, 892—895, 1969.

38. Kazantzis, G., Respiratory function in men casting cadmium alloys. I. Assessment of ventilatory function, *Br. J. Ind. Med.*, 13, 30—36, 1956.

39. Kazantzis, G., Induction of sarcoma in the rat by cadmium sulphide pigment, *Nature (London)*, 198, 1213—1214, 1963.

40. Kazantzis, G. and Hanbury, W. J., The induction of sarcoma in the rat by cadmium sulphide and by cadmium oxide, *Br. J. Cancer*, 20, 190—199, 1966.

41. Kazantzis, G. and Armstrong, B. G., Renal function in relation to low levels of cadmium exposure in a group of smelter workers, *Environ. Health Perspect.*, 54, 193—199, 1984.

42. Kipling, M. D. and Waterhouse, J. A. H., Cadmium and prostatic carcinoma (letter), *Lancet*, 1, 730—731, 1967.

43. Kjellström, T., Friberg, L., and Rahnster, B., Mortality and cancer morbidity among cadmium-exposed workers, *Environ. Health Perspect.*, 28, 199—204, 1979.

44. Kolonel, L. N., Association of cadmium with renal cancer, *Cancer*, 37, 1782—1787, 1976.

45. Kolonel, L. and Winkelstein W., Jr., Cadmium and prostatic carcinoma, *Lancet*, 2, 566—567, 1977.

46. Lambert, B., Lindblad, A., Nordenskjöld, M., and Werelius, B., Increased frequency of sister-chromatid exchanges in cigarette smokers, *Hereditas,* 88, 147—148, 1978.

47. Lemen, R. A., Lee, J. S., Wagoner, J. K., and Blejer, H. P., Cancer mortality among cadmium production workers, *Ann. N.Y. Acad. Sci.,* 271, 273—279, 1976.

48. Levy, L. S. and Clack, J., Further studies on the effect of cadmium on the prostate gland. I. Absence of prostatic changes in rats given oral cadmium sulphate for two years, *Ann. Occup. Hyg.,* 17, 205—211, 1975.

49. Levy, L. S., Roe, F. J. C., Malcolm D., Kazantzis, G., Clack, J., and Platt, H. S., Absence of prostatic changes in rats exposed to cadmium, *Ann. Occup. Hyg.,* 16, 111—118, 1973.

50. Levy, L. S., Clack, J., and Roe, F. J. C., Further studies on the effect of cadmium on the prostate gland. II. Absence of prostatic changes in mice given oral cadmium sulphate for eighteen months, *Ann. Occup. Hyg.,* 17, 213—220, 1975.

51. Loeser, E., A 2-year oral carcinogenicity study with cadmium on rats, *Cancer Lett.,* 9, 191—198, 1980.

52. Lucis, O. J., Lucis, R., and Shaikh, Z. A., Cadmium and zinc in pregnancy and lactation, *Arch. Environ. Health,* 25, 14—22, 1972.

53. Mason, T. J., McKay, F. W., Hoover, R., Blot, W. J., and Fraumeni, J. F., Atlas of Cancer Mortality for U.S. Counties 1950-1969, DHEW Publ. (NIH) 75-780, U.S. Department of Health, Education and Welfare, Washington, D.C., 1975.

54. Morgan, J. M., Tissue cadmium concentrations in man, *Arch. Intern. Med.,* 123, 405—408, 1969.

55. Morgan, J. M., Cadmium and zinc abnormalities in bronchogenic carcinoma, *Cancer,* 25, 1394—1398, 1970.

56. Morgan, J. M., Tissue cadmium and zinc content in emphysema and bronchogenic carcinoma, *J. Chronic Dis.,* 24, 107—110, 1971.

57. Nogawa, K., Kawano, S., and Nishi, N., Mortality study of inhabitants in a cadmium-polluted area with special reference to low molecular weight proteinuria, in *Proc. Heavy Metals in the Environment,* CEP Consultants Ltd., Edinburgh, 1981, 538—540.

58. Nomiyama, K., Carcinogenicity of cadmium, *Jpn. J. Ind. Health,* 24, 13—23, 1982.

59. Oberdoerster, G., Baumert, H. P., Hochrainer, D., and Stoeber, W., The clearance of cadmium aerosols after inhalation exposure, *Am. Ind. Hyg. Assoc. J.,* 40, 443—450, 1979.

60. Oppenheimer, B. S., Oppenheimer, E. T., and Stout, A. P., The latent period in carcinogenesis by plastics in rats and its relation to the presarcomatous stage, *Cancer,* 11, 204—213, 1958.

61. O'Riordan, M. L., Hughes, E. G., and Evans, H. J., Chromosome studies on blood lymphocytes of men occupationally exposed to cadmium, *Mutat. Res.,* 58, 305—311, 1978.

62. Paton, G. R. and Allison, A. C., Chromosome damage in human cell cultures induced by metal salts, *Mutat. Res.,* 16, 332—336, 1972.

63. Poirier, L. A., Kasprzak, K. S., Hoover, K. L., and Wenk, M. L., Effects of calcium and magnesium acetates on the carcinogenicity of cadmium chloride in Wistar rats, *Cancer Res.,* 43, 4575—4581, 1983.

64. Potts, C. L., Cadmium proteinuria — the health of battery workers exposed to cadmium oxide dust, *Ann. Occup. Hyg.,* 8, 55—61, 1965.

65. Ramaya, L. K. and Pomerantzeva, M. D., Investigation of cadmium chloride mutagenic effect on germ cells of male mice, *Genetika,* 13, 59—63, 1977.

66. Roe, F. J. C., Dukes, C. F., Cameron, K. M., Pugh, R. C. B., and Mitchley, B. C. V., Cadmium neoplasia: testicular atrophy and Leydig cell hyperplasia and neoplasia in rats and mice following the subcutaneous injection of cadmium salts, *Br. J. Cancer,* 18, 674—681, 1964.

67. Röhr, G. and Bauchinger, M., Chromosome analyses in cell cultures of the Chinese hamster after application of cadmium sulphate, *Mutat. Res.,* 40, 125—130, 1976.

68. Saksena, S. E. and Salmonsen, R., Effects of cadmium chloride on ovulation and on induction of sterility in the female golden hamster, *Biol. Reprod.,* 29, 249—256, 1983.

69. Sanders, C. L. and Mahaffey, J. A., Carcinogenicity of single and multiple intratracheal instillations of cadmium oxide in the rat, *Environ. Res.,* 33, 227—233, 1984.

70. Schroeder, H. A., Balassa, J. J., and Vinton, W. H., Jr., Chromium, lead, cadmium, nickel and titanium in mice: effect on mortality, tumors and tissue levels, *J. Nutr.,* 83, 239—250, 1964.

71. Schroeder, H. A., Balassa, J. J., and Vinton, W. H., Jr., Chromium, cadmium and lead in rats: effects on life span, tumors and tissue levels, *J. Nutr.,* 86, 51—66, 1965.

72. Shigematsu, I., Kitamura, S., Takeuchi, J., Minowa, M., Nagai, M., Usui, T., and Fukushima, M., A retrospective mortality study on cadmium-exposed populations in Japan, in *Cadmium 81, Proc. 3rd Int. Cadmium Conf. Miami,* Cadmium Association, London, 1982, 115—118.

73. Shimada, T., Watanabe, T., and Endo, A., Potential mutagenicity of cadmium in mammalian oöcytes, *Mutat. Res.,* 389—396, 1976.

74. Shiraishi, Y., Cytogenetic studies in 12 patients with Itai-itai disease, *Humangenetik,* 27, 31—44, 1975.

75. Shiraishi, Y. and Yoshida, T. H., Chromosomal abnormalities in cultured leucocyte cells from Itai-itai disease patients, *Proc. Jpn. Acad.,* 48, 248—251, 1972.

76. Shiraishi, Y. Kurahashi, H., and Yoshida, T. H., Chromosomal aberrations in cultured human leucocytes induced by cadmium sulfide, *Proc. Jpn. Acad.,* 48, 133—137, 1972.

77. Smith, J. P., Smith, J. C., and McCall, A. J., Chronic poisoning from cadmium fume, *J. Pathol. Bacteriol.,* 80, 287—295, 1960.

78. Sorahan, T., A mortality study of nickel-cadmium battery workers, in *Cadmium 81, Proc. 3rd Int. Cadmium Conf. Miami,* Cadmium Association, London, 1982, 138—141.

79. Sorahan, T. and Waterhouse, J. A. H., Mortality study of nickel-cadmium battery workers by the method of regression models in life tables, *Br. J. Ind. Med.,* 40, 293—300, 1983.

79a. Sorahan, T. and Waterhouse, J. A. H., Cancer of prostate among nickel-cadmium battery workers, *Lancet,* February 23, 459, 1985.

80. Suter, K. E., Studies on the dominant-lethal and fertility effects of the heavy metal compounds methyl-mercuric hydroxide, mercuric chloride and cadmium chloride in male and female mice, *Mutat. Res.,* 30, 365—374, 1975.

81. Takenaka, S., Oldiges, H., König, H., Hochrainer, D., and Oberdörster, G., Carcinogenicity of cadmium chloride aerosols in Wistar rats, *J. Natl. Cancer Inst.,* 70, 367—373, 1983.

82. Thun, M. J., Schnorr, T. M., Smith, A. B., Halperin, W. E., and Lemen, R. A., Mortality among a cohort of U.S. cadmium production workers — an update, *J. Natl. Cancer Inst.,* 74, 325—333, 1985.

83. Tsuchiya, K., Ed., *Cadmium Studies in Japan — A Review,* Elsevier, Amsterdam, 1978.

84. Watanabe, T. and Endo, A., Chromosome analysis of preimplantation embryos after cadmium treatment of oöcytes at meiosis. I, *Environ. Mutagen.,* 4, 563—567, 1982.

85. Watanabe, T., Shimada, T., and Endo, A., Mutagenic effects of cadmium on mammalian oocyte chromosomes, *Mutat. Res.,* 67, 349—356, 1979.

86. WHO, Arsenic, Environmental Health Criteria 18, World Health Organization, Geneva, 1981.

Chapter 13

CRITICAL ORGANS, CRITICAL CONCENTRATIONS, AND WHOLE BODY DOSE-RESPONSE RELATIONSHIPS

Tord Kjellström

TABLE OF CONTENTS

I. INTRODUCTION

The terms "critical organ", "critical effect", and "critical concentration" have been used in toxicological risk assessments in order to set priorities for the establishment of hygienic standards or chemical safety standards. Unfortunately, different definitions have been used at different times and by different scientists. This has led to and still causes some confusion. The Scientific Committee on the Toxicology of Metals of the Permanent Commission and International Association on Occupational Health[31] defined the three terms and we will use these definitions in the present evaluations.

The "critical organ" was defined as "that particular organ which first attains the critical concentration of a metal under specified circumstances of exposure and for a given population". The "critical concentration for a cell" was defined as "the concentration at which adverse functional changes, reversible or irreversible, occur in the cell". Those changes would thus be termed the "critical effect". For a whole organ the "critical organ concentration" would be the mean concentration in the organ at the time any of its cells reaches the critical concentration.

These definitions still depend on a value judgment as to what functional change is "adverse", but as long as the choice of definition is clearly stated, misunderstandings should not arise. The definitions are also limited by the measureability of the changes. There may be adverse changes that cannot as yet be measured or these changes may occur at metal concentrations in the critical organ, which are below the detection limit of available analytical methods. Therefore, what is currently the critical organ may be superseded when new data are on hand; this is also true of the critical concentration. For different exposure routes and populations with different "sensitivity" to the metal, these terms may also vary. These concepts have been further discussed by the Task Group on Metal Interaction.[30]

The "dose-effect relationship" describes the increasing severity of effect when the dose increases in an individual. It can also be expressed for a group of exposed people as the increasing severity of "average" effect when the average dose increases. The critical effect would occur at the lowest dose. The dose-effect relationship is of value when selecting the effects to be studied in a screening program for exposed people.

It has been pointed out by the Task Group on Metal Toxicity[31] that the critical organ and the critical concentration are specific for the individual. This means that the dose-effect relationship will also vary between individuals. In general, the critical organ does not vary among individuals, but the critical concentration certainly will vary.

The frequency distribution of individual critical concentrations forms the basis for the organ dose-response relationship. To describe this phenomenon in a quantitative way, longitudinal studies on the metal concentration in the critical organ and the occurrence of adverse functional changes are needed. Some studies of this type have been carried out on cadmium-exposed animals, but for humans only cross-sectional studies are available. These can, however, give an indication of the variation of critical concentrations.

The Task Group on Metal Toxicity[31] also defined "subcritical effects" and "subcritical concentrations", i.e., effects that are measurable but do not impair cellular function. A subcritical effect that occurs at a subcritical concentration may be a valuable "warning sign" that the level of the toxic metal in the critical organ is getting close to the critical concentration.

As both the critical concentration and the subcritical concentration are defined on an individual basis, it is essential to express the variability of these variables when discussing dose-response relationships and safety standards. Friberg and Kjellstrom[6] proposed that the term "population critical concentration" (PCC) should be used

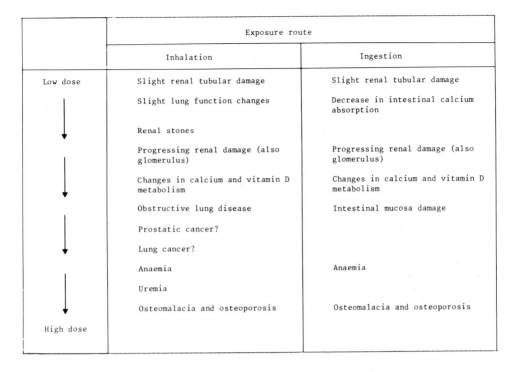

	Exposure route	
	Inhalation	Ingestion
Low dose	Slight renal tubular damage	Slight renal tubular damage
	Slight lung function changes	Decrease in intestinal calcium absorption
	Renal stones	
	Progressing renal damage (also glomerulus)	Progressing renal damage (also glomerulus)
	Changes in calcium and vitamin D metabolism	Changes in calcium and vitamin D metabolism
	Obstructive lung disease	Intestinal mucosa damage
	Prostatic cancer?	
	Lung cancer?	
	Anaemia	Anaemia
	Uremia	
High dose	Osteomalacia and osteoporosis	Osteomalacia and osteoporosis

FIGURE 1.   Dose-effect relationship for chronic effects of cadmium.

when discussing the critical concentration for groups of people. The PCC can then be defined in terms of the proportion of the population that has a critical concentration below a certain level. A PCC of 50 (PCC-50) would be the concentration at which half (50%) of the population has reached individual critical concentrations, and a PCC-10 would be the concentration at which 10% of the population has reached individual critical concentrations. This concept is similar to the LD_{50} or ED_{50} and it would help to clarify the meaning of the terms used to describe the toxic effects of cadmium occurring in populations.

II. CRITICAL ORGANS FOR DIFFERENT TYPES OF EFFECTS AND EXPOSURE ROUTES

As is clear from Chapters 8 (respiratory effects), 9 (renal effects), and 10 (bone effects), the lungs, the kidneys, and the bones are the organs which are particularly affected by cadmium, following long-term exposure. Acute or subacute exposure may also cause damage to the lungs, kidneys, and bones. In addition, high acute cadmium exposures may affect the testicles, the sensory ganglia, the ovaries, the intestinal mucosa, the liver, the fetus, the placenta, etc. (Chapter 11).

The dose-effect relationship for acute effects of cadmium is not well known, but there is evidence (Chapter 8, Section II.B) of the lungs being the critical organ after very high acute human exposure to cadmium fumes. After very high oral exposure to cadmium, the intestinal mucosa may be the critical organ (Chapter 11, Section II.A). In both these situations these local effects develop rapidly. After the cadmium has been absorbed, systemic effects are likely to develop in the other organs listed above.

The dose-effect relationship for chronic effects of cadmium is different for inhalation exposure and ingestion exposure as shown in Figure 1.

It is important to point out that in most long-term animal studies, where renal and other effects have been looked for, the renal effects have preceded or occurred simultaneously with the other effects. Following high inhalation exposure, respiratory effects may occur earlier than the renal effects, but such high levels of exposure in a human would give rise to acute or subacute effects. Similarly, high oral exposure may cause acute or subacute effects. From the point of view of prevention, it is of greater importance to consider the effects which occur after long-term low level exposure.

In the studies of workers with long-term exposure to cadmium in air (Chapter 8, Section III.B), respiratory effects without concurrent renal effects have seldom been reported. For example, Friberg[5] found that of 43 heavily cadmium-exposed workers, 12 had decreased physical working capacity and signs of emphysema on the lung function tests. Of these 12, 11 had proteinuria. There were an additional 25 workers with proteinuria but no respiratory effects.

As has been discussed in Chapter 10 and the Appendix, the bone effects of cadmium on humans hardly ever occur in the absence of renal tubular damage. It is quite possible that in those cases where renal effects were not reported, the methods used to measure them were not sufficiently sensitive.

Chapter 10 (Section V) gives an approximate estimate of the organ dose-response relationship for bone effects of cadmium on humans. Based on autopsy data and the limited amount of intake data available, it was concluded that an intake from food in excess of 1 mg Cd per day may be needed to produce severe bone effects, exemplified by the Itai-itai disease, in a population with diets low in certain minerals and vitamins. An intake of 1 mg Cd per day is much greater than the intake (0.14 mg Cd per day and higher) reported to be associated with renal effects (Chapter 9, Section V.B). The intake level at which early "subclinical" bone effects occur is not known.

It should also be remembered that the intake needed to produce effects is dependent on the absorption rates. The gastrointestinal absorption is "normally" about 3 to 7% (Chapter 6, Section II.B.6), but it may be as high as 20% in a person with iron deficiency (Chapter 6, Section II.B.3). Such high absorption rates will reduce considerably the intake needed to cause effects.

Some studies have shown a significant increase of mortality in prostatic cancer, and lung cancer has been reported among cadmium workers (Chapter 12, Section I.B.1). In groups of workers with a long latency period (15 years or more), mortality in these cancer forms was about twice as high as that expected. None of the studies reviewed provided detailed dose data, but it appears that the workers with the long latency periods had usually been exposed to cadmium in air at levels of about 100 to 1000 $\mu g/m^3$. It may well be that a small increase in the cancer mortality may occur at lower exposure levels. The data available are not sufficient to make possible even an approximate estimate of the dose-response relationship for these types of cancer. A high prevalence of renal effects has been found among workers with much lower air cadmium exposure levels than 100 $\mu g/m^3$ (Chapter 9, Section IV.B.1). Available data indicate that renal effects do develop at lower doses than those needed to produce prostatic cancer or lung cancer. At present, therefore, it seems reasonable to consider renal tubular damage as the critical effect, but it cannot be precluded that when more information is available certain types of cancer may be seen as the critical effects of cadmium.

If an increased cancer risk occurs at the cadmium exposure levels where renal tubular damage does not occur, the risk must be very low.

Thus, the critical organ in most exposed individuals, for both air exposure and food exposure to cadmium, is likely to be the kidney. The following discussion will, therefore, concentrate on the exposure data, renal levels, and renal effects. If these renal effects are prevented, then it is likely that the other effects of cadmium will also be prevented.

III. INTERINDIVIDUAL VARIATION OF CRITICAL CONCENTRATION AND POPULATION CRITICAL CONCENTRATION (PCC)

In the section on organ dose-effect and dose-response relationships for renal effects (Chapter 9, Section VI), the data were reviewed in detail. The problem of interpreting human data, because of the decreased renal concentration of cadmium after the damage has occurred, was discussed (Chapter 9, Section VI). The human in vivo neutron activation data presented by Ellis and co-workers[2] and Roels and co-workers[26] make possible quantitative estimates of the distributions of the critical concentration.

In these two studies it was assumed that the critical effect on the kidney could be detected by measuring urinary β_2-microglobulin. An operational definition of what constituted pathologically increased β_2-microglobulin level was used. This level was 200 $\mu g/g$ creatinine, which is rare in a population with "normal" kidneys (Chapter 9, Section IV.B.5). It should be pointed out that the cadmium-induced increase of low molecular weight proteinuria usually is not reversible (Chapter 9, Section IV.D).

Studies of β_2-microglobulin in control groups have found that the distribution is lognormal and that the geometric mean levels are usually in the range 50 to 100 $\mu g/g$ creatinine or 50 to 100 $\mu g/\ell$ urine.[12,14,18,28,29] The 97.5 percentile for people under age 60 is usually in the range 200 to 600 $\mu g/\ell$ urine (Chapter 9, Section IV.B.5).

In Chapter 9, Table 11, the change in the dose-response relationship based on different effect criteria can be seen. Holden[8] used two cut-off levels of urinary β_2-microglobulin to calculate response rates, 200 and 1000 $\mu g/\ell$ urine. After about 5 years of cadmium exposure, a response rate of 10% was reached if 200 $\mu g/\ell$ was used. In the same group of workers it took about 15 years of exposure to reach a 10% response rate at 1000 $\mu g/\ell$. Thus, it may be that a threefold increase of the dose is needed to reach the 10% response rate for the more severe proteinuria. Obviously, the operational definition of what consititutes an "undesirable" tubular proteinuria is crucial for the estimation of critical concentration. It should also be mentioned that the dose of cadmium may not be accurately determined solely on the basis of air levels of cadmium and exposure duration, as some workers will receive significant cadmium exposure from contamination of cigarettes and/or food in the workplace.[7]

The distribution of "critical concentrations" as reported in the study by Ellis and co-workers[2,3] is depicted in Chapter 9, Figure 30.

Ellis and co-workers[3] did not calculate the concentrations of cadmium in kidney cortex, but used whole kidney cadmium levels. Assuming that the weight of one kidney is 145 g and that the kidney cortex concentration is 1.25 times the whole kidney concentration (Chapter 6, Section IV.B.2), the average critical concentration in this report can be calculated to be 331 mg Cd per kilogram wet weight (38.4 × 125/0.145). This is based on the relationship between liver cadmium concentrations and prevalence of urinary β_2-microglobulin above 200 $\mu g/g$ creatinine (or urinary total protein above 250 mg/g creatinine) and the relationship between kidney cadmium and liver cadmium in 30 controls and 31 active cadmium workers with "normal" urinary proteins. This method was proposed by Friberg and Kjellström.[6] The inclusion of active workers only is necessary because liver values would start to decrease as soon as exposure ceases.[13]

The dose-response relationship can be expressed as a cumulative normal distribution, which in turn can be mathematically approximated by a logistical function:

$$\text{Logit } P = \ln P/(1 - P) = A + B \times X$$

where P is the probability of effect (response rate); X is the cadmium dose; and A and B are constants. Ellis and co-workers[3] estimated A to be −5.53 and B to be 0.144, if X is the cadmium amount in one kidney (mg). With the recalculation procedure indicated above, B would be 0.017 if X was expressed as milligrams per kilogram kidney cortex. Ellis and co-workers did not report whether tests of the distribution shape (normal or log-normal) were carried out. An earlier study by Ellis and co-workers[2] reported a PCC-50 of 275 mg Cd per kilogram (recalculated according to Chapter 6, Section IV.B.2). This was based on the same data but different calculation methods were used. Roels and co-workers[26,27] reported a range of the critical concentration from 180 to 320 mg Cd per kilogram (recalculated). From their data the best estimate of the PCC-50 may be the midpoint, 250 mg Cd per kilogram. The logistic function by Ellis and co-workers[3] will give a PCC-10 at 200 mg Cd per kilogram in renal cortex. Roels and co-workers[27] reported a PCC-10 at 180 mg Cd per kilogram (recalculated). All these data are based on cross-sectional studies. The time duration of renal tubular damage in individuals was not considered. A worker may have a recorded renal cortex cadmium concentration of, e.g., 200 mg/kg, and have renal tubular damage, but the damage may have already developed at a time when the renal cortex concentration was, e.g., 160 mg/kg.[26]

Taking all the data as a whole, the most likely true PCC-50 is in the range 250 to 300 mg Cd per kilogram and the corresponding PCC-10 is in the range 180 to 200 mg Cd per kilogram. This range of concentrations in renal cortex agrees reasonably well with concentrations at which renal effects have been reported in animal experiments (Chapter 9, Section VI.A).

IV. WHOLE BODY DOSE-RESPONSE RELATIONSHIPS

A. Based on Epidemiological Studies

It was shown in Chapter 9, Section IV.B, that the cadmium-induced renal tubular damage can be diagnosed using tests for low molecular weight proteinuria. Sensitive immunoassay methods are available to measure β_2-microglobulin and retinol-binding protein (Chapter 9, Section IV.B.4). These have been used in epidemiological studies of both industrial workers and people exposed in the general population. The operational definitions for a pathological increase of the proteinuria vary, and this has to be considered when different epidemiological studies are compared. It is also of great importance to use a dose estimate which is not influenced by the kidney effects studied. Some reports [9,20,22] compare urinary cadmium concentrations and the prevalence of tubular proteinuria. As it is known that the urinary cadmium levels increase when the body burden increases (Chapter 6, Section VI.C) and the levels also increase as a result of the renal damage (Chapter 9, Section IV.A.5), it is difficult to interpret these kinds of comparisons. To some degree urinary cadmium may be seen as an indicator of the renal effects, and a good correlation with urinary low molecular weight proteinuria should be expected.

For populations of industrial workers with air as the predominant route of exposure, a dose estimate based on cadmium levels in air and duration of exposure is preferable. For general populations where food is the primary route of exposure, the daily intake via food and duration of exposure should be used for dose estimates.

1. In Industry

A number of epidemiological studies carried out in industries are summarized in Chapter 9, Section V.A. From these studies it was concluded that after 10 years of exposure to 30 to 50 μg Cd per cubic meter, there was an apparent increase in the prevalence of tubular proteinuria. Cadmium oxide fumes appear to give higher response rates than cadmium oxide, stearate, or sulfide dust (Chapter 9, Table 11). The dose estimates are, however, rather approximate and it is therefore difficult to draw firm conclusions.

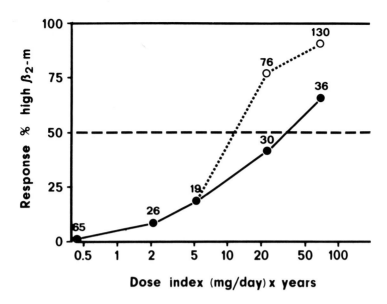

FIGURE 2. Prevalence (background adjusted) of high β_2-microglobulin (<290 $\mu g/\ell$ adjusted to sp. gr. = 1.023) as a function of dose index in a Swedish battery factory. The dose index is the product of exposure duration (years), average exposure level ($\mu g/m^3$) for each year, and a factor (6.2). Numbers of persons in each group indicated. The dotted line indicates maximum possible response if all retired and dead workers with dose index above 12.5 had a high β_2-microglobulin.[12]

The only study in Chapter 9, Table 11, giving detailed dose-effect and dose-response relationships was published by Kjellström and co-workers.[16] The dose-effect relationship is depicted in Chapter 9, Figure 19. Dose was approximated as the time since first employment. The cadmium concentrations in workroom air were known and a more accurate dose estimate was calculated by multiplying air cadmium levels and actual years of cadmium exposure duration.[12] Figure 2 shows the dose-response curve for the workers who were still alive and who were examined and includes the maximum curve that would be obtained if all previous workers who could not be located or had died were assumed to have had the effect and were included in the data.

The average daily inhaled amount was reached multiplying air concentrations by 6.2,[12] which takes into account the volume of air inhaled in a working day and the number of working days in a year. It should be pointed out that at the two lowest dose levels in Figure 2, 95% of the dust in this factory was respirable.[16] The higher dose levels occurred in workers whose exposure started many years ago, and it is possible that a higher proportion of the dust at that time was nonrespirable and that the actual absorbed dose was less than what is seen in Figure 2. A dose index of 2 mg/day × years (Figure 2) is equivalent to about 320 $\mu g/m^3$ × years, or 10 years exposure at 32 μg Cd per cubic meter. At this dose level about 10% of the exposed group would have tubular proteinuria as defined by Kjellström and co-workers[16] (urinary β_2-microglobulin greater than 290 $\mu g/\ell$, adjusted to sp. gr. 1.023). In the study by Lauwerys and co-workers[19] in which the total air cadmium level was 134 $\mu g/m^3$ (88 $\mu g/m^3$ respirable dust) (Chapter 9, Table 11), a 10% response rate was found in workers exposed for 1 to 20 years.

2. General Environment Exposure

Epidemiological studies of renal effects in Japan and Belgium were reviewed in

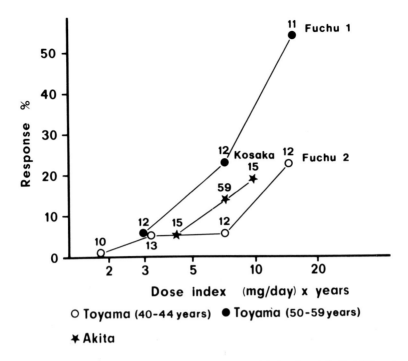

FIGURE 3.   Prevalence (background adjusted) of high β_2-microglobulin (different criteria for the three studies, Chapter 9, Section V.B) as a function of "dose index" in groups of Japanese farmers. The dose index in Fuchu was calculated as twice the product of the years of exposure in the polluted area, the average cadmium concentration in rice consumed by the family, and the average daily intake of rice (350 g). In Kosaka the dose index was calculated as the product of the exposure duration in the polluted area and average daily fecal cadmium amount. Numbers of persons in each group indicated.[12]

Chapter 9, Section V.B. Different criteria for proteinuria have been used in different studies, but it is clear that in Japanese areas with the highest cadmium levels in rice and the longest periods of exposure (Chapter 9, Table 12), renal effects are more common.

Only a few sets of data are suitable for a quantitative dose-response analysis. Three different Japanese populations were studied by Kjellström,[12] Kjellström and co-workers,[15] and Kojima and co-workers.[17] In these studies a dose index was used, based on the product of the estimated average daily cadmium intake via food and the exposure duration in years (Figure 3). To calculate daily cadmium intake in the two former studies (people exposed in the Fuchu area), estimates of average cadmium levels in the rice consumed by each family and assumptions about average daily rice intake (350 g) and cadmium intake from other food were used.[12] In the latter study of the Kosaka area individual data on cadmium in feces were used to estimate daily intake.[12]

Figure 3 shows the dose-response relationship for the three studies. The criteria for pathological proteinuria were based on control groups matched for age and sex. The Fuchu 1 study included women aged 50 to 59; Fuchu 2 included women aged 40 to 44; and the Kosaka study included men and women aged 50 to 69. These age and sex differences may, to some extent, explain the differences in response rates at higher doses.

Hutton[9] combined all the data points shown in Figure 3 and carried out probit regression analysis (Figure 4). From this analysis Hutton calculated the daily intake

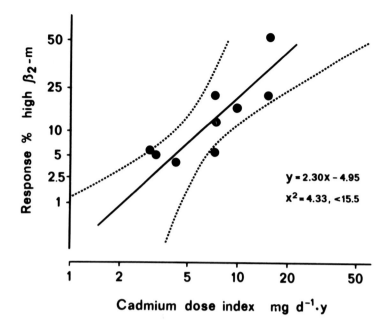

FIGURE 4. Dose-response relationship for cadmium-induced increase in urinary β_2-microglobulin levels based on a probit analysis.[4] Upper and lower 95% fiducial limits indicated. Original data from Kjellström, modified from Hutton.[9]

Table 1

ESTIMATED DAILY CADMIUM INTAKE (μg/day) SUFFICIENT TO CAUSE EXCESSIVE β_2-MICROGLOBULIN EXCRETION AFTER 50 YEARS EXPOSURE IN VARYING PROPORTIONS OF THE POPULATION

	Percentage of the population with elevated (above 500—700 μg/ℓ)				β_2-microglobulin levels		
	0.1%	1%	2.5%	5%	10%	20%	50%
Japanese (53 kg)	19	41	60	82	118	183	426
European (70 kg)	25	55	79	108	156	242	562
	(7—91)	(25—123)	(44—142)	(70—167)	(109—223)	(152—386)	(200—1439)

Note: Estimates for the European population have the lower and upper 95% fiducial limits attached in parentheses. Estimates based on data from Kjellström, 1977.[12]

From Hutton, M., MARC Rep. No. 29, Monitoring Assessment and Research Centre, Chelsea College, London, 1983. With permission.

necessary, over a period of 50 years, to reach certain response rates (Table 1). The data in Figure 4 refer to Japanese people, who on average have a lower body weight than, e.g., European and North Americans. The latter can, therefore, sustain a greater cadmium intake before reaching the same response rate (Table 1).

Hutton[9] also analyzed data from Tsuchiya (Figure 5) and Nogawa and co-workers[21] (Figure 6). The operational definition for tubular proteinuria is similar in Figures 4 and 5 (β_2-microglobulin greater than 97.5 percentile in control group, about 500 μg/g creatinine), but in Figure 6 it is not so strict (RBP greater than 4 mg/ℓ). At a dose of about 10,000 μg/day × years, the average response rate is about 25% in the former

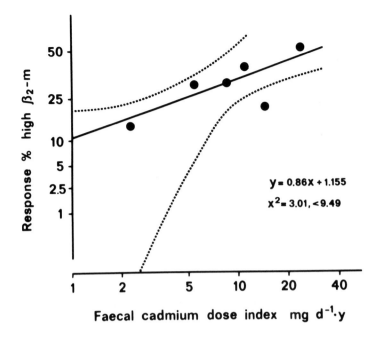

FIGURE 5.   Dose-response relationship for cadmium-induced increase in urinary β_2-microglobulin levels, with upper and lower 95% fiducial limits. Raw data supplied by Tsuchiya. (Modified from Hutton, M., MARC Rep. No. 29, Monitoring Assessment and Research Centre, Chelsea College, London, 1983. With permission.)

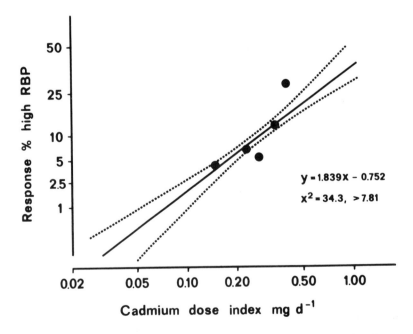

FIGURE 6.   Dose-response relationship for cadmium-induced increase in urinary retinol-binding protein levels, with upper and lower 95% fiducial limits. Data extracted from Nogawa and co-workers.[21] (Modified from Hutton, M., MARC Rep. No. 29, Monitoring Assessment and Research Centre, Chelsea College, London, 1983. With permission.)

Table 2
ESTIMATED DAILY CADMIUM INTAKE (μg/day) SUFFICIENT TO CAUSE
EXCESSIVE RETINOL-BINDING PROTEIN EXCRETION AFTER 50 YEARS
OR MORE EXPOSURE IN VARYING PROPORTIONS OF THE
POPULATION

	Percentage of the population with elevated RBP levels (above 4000 μg/l)						
	0.1%	1%	2.5%	5%	10%	20%	50%
Japanese (53 kg)	28	73	115	171	269	468	1340
European (70 kg)	37	96	152	226	355	618	1767
	(21—166)	(69—134)	(121—191)	(197—260)	(324—390)	(516—740)	(1143—2732)

Note: Estimates for the European population include the lower and upper 95% fiducial limits in parentheses.

Estimates based on data from Nogawa and co-workers.[21]
From Hutton, M., MARC Rep. No. 29, Monitoring Assessment and Research Centre, Chelsea College, London, 1983. With permission.

studies and about 5% in the latter study. The data presented by Tsuchiya gave very wide tolerance limits, but those presented by Nogawa and co-workers can be tabulated (Table 2) in the same way as those of Kjellström and co-workers (Table 1). It should be pointed out that in the original data the lowest response rates were about 5%. Any estimates below this rate are extrapolations.

In using these data to evaluate, e.g., acceptable daily intake (ADI) values, it is necessary to decide on what level tubular proteinuria is "undesirable" in an individual and what prevalence rate of such proteinuria is "acceptable" in a population.

B. Based on the Kinetic Model and Critical Concentration Distributions

As discussed in Chapter 7, a kinetic model in combination with variations in critical concentrations make it possible to calculate the dose-response relationship for metals which accumulate in the body.[12,23-25] The kinetic model for cadmium makes it possible to calculate the cadmium concentrations in kidney cortex for people with a defined daily cadmium intake for each year of life. For people with cadmium intake from food only, we have assumed that the variation of the intake with age follows the average energy intake via food (Chapter 7, Section II.B).

A calculation for nonsmokers with intake of cadmium from food is shown in Table 3. Age 45 was chosen because at that age the concentration of cadmium reaches its maximum level during the lifetime of the exposed person (Chapter 5, Section IV.B). Table 3 shows the mean cadmium concentrations in renal cortex resulting from a certain average food intake of cadmium.

Autopsy studies from Sweden[1] have shown that the renal cortex cadmium concentrations in different age and smoking categories follow log-normal distributions with geometric standard deviations usually at about 2. The same seems to hold true in data from other countries (Chapter 5, Section IV.B), although smokers and nonsmokers were not studied separately. These distributions are the result of variations in actual intake and variations in the metabolism of cadmium in individuals (Chapter 7).

The distribution of individual critical concentrations has not been studied to the same extent and only limited data based on in vivo neutron activation analysis are available (Section III). Estimates of PCC-50 and PCC-10 are given in Section III.

The proportion of a population that has renal levels of cadmium above the critical concentration can be calculated based on the two types of distributions mentioned

Table 3

CALCULATED AVERAGE KIDNEY CORTEX CADMIUM CONCENTRATION AT AGE 45 FOR NONSMOKERS WITH CADMIUM INTAKE VIA FOOD ONLY. EUROPEAN BODY WEIGHT (70 KG)

Average daily cadmium intake at age 50		100	200	300	400	500	600	700
Geometric mean cadmium concentration in kidney cortex[a] (mg/kg)		61	102	143	183	224	265	305
Estimated proportion (%) with kidney cortex cadmium above their individual critical concentration	A	2.7	11	22	34	44	53	60
	B	1.8	7.8	17	26	35	44	51

Note: A: PCC-50 = 250 mg Cd/kg and PCC-10 = 180 mg Cd/kg, log-normal distribution of critical concentrations. B: PCC-50 = 300 mg Cd/kg and PCC-10 = 200 mg Cd/kg, log-normal distribution of critical concentrations.

[a]   Assumed to have a log-normal distribution with geometric SD of 2.

above. Such calculations were first reported for cadmium by Sakurai at the 1984 International Conference on Occupational Health in Dublin. It is assumed that the distribution of individual critical concentrations is log-normal, as are the renal cortex cadmium concentrations. This proportion is equal to the probability of an individual in the population having an actual renal cortex concentration (X_1) above his or her critical concentration (X_2) and to the response rate. If the renal cortex concentrations and the critical concentrations are expressed as their logarithms, this probability for a specified PCC-50 and a specified mean renal cortex level can be calculated as:

$$P(X_2 - X_1 < 0) = \Phi \left(\frac{m_1 - m_2}{\sqrt{\sigma_1^2 + \sigma_2^2}} \right)$$

where Φ is the standard normal distribution function; m_1, σ_1 are the mean and standard deviation of the logarithms of the renal cortex levels; and m_2, σ_2 are the mean and standard deviation of the logarithms of the critical concentrations. The difference $X_1 - X_2$ is the difference between two variables that are both normally distributed. According to basic statistical theory, the difference itself follows a normal distribution with the mean ($m_1 - m_2$) and the standard deviation ($\sqrt{\sigma_1^2 + \sigma_2^2}$).

With a geometric standard deviation of 2 in the renal cortex cadmium concentration distributions, σ_1 will be 0.30 (log 2). The different values of m are given in Table 3. PCC-50 was chosen as either 250 mg Cd per kilogram (m = log 250 = 2.4) or 300 mg Cd per kilogram (m = 2.477) (Section III). PCC-10 was chosen as either 180 or 200 mg Cd per kilogram (Section III), which gives σ_2 of 0.11 or 0.14, respectively.

Calculations based on two options are given in Table 3, The higher PCC assumption gives somewhat lower estimated response rates, as expected.

The model can also be used to analyze the situation for workers occupationally exposed to cadmium, using the same methodology described above. We assumed exposure started at age 30 and continued until age 40. The subjects were nonsmokers and ingested, on average, 16 μg Cd per day via food (Chapter 7). As is seen in Table 4, the calculated response rate is about 10% after 10 years exposure to 50 μg Cd per cubic meter.

C. Comparison of Epidemiological Data and Kinetic Model Calculations

As seen in Table 1, epidemiological data have shown that an average daily intake of about 560 μg Cd per day in a population with an average body weight of 70 kg can be

Table 4

CALCULATED AVERAGE CADMIUM CONCENTRATION IN KIDNEY
CORTEX AT AGE 40 FOR NONSMOKING WORKERS WITH CADMIUM
EXPOSURE VIA INHALED DUST AT DIFFERENT CONCENTRATIONS. TEN
YEARS EXPOSURE DURATION FROM AGE 30 TO 40

		100	200	300	400	500	1000	1500
Average daily cadmium intake via dust (μg/day)		100	200	300	400	500	1000	1500
Dose index (mg/day × years) (as used in Figure 2)		1	2	3	4	5	10	15
Average cadmium concentration in air (μg/m³)		16	32	50	65	80	160	240
Geometric mean cadmium concentration in kidney cortex[a] (mg/kg)		48	78	108	139	171	320	471
Estimated proportion (%) with kidney cortex	A	1.2	5.6	13	21	30	63	80
cadmium above their individual critical concentration	B	0.8	3.8	8.9	15	23	54	72

Note: A: PCC-50 = 250 mg Cd/kg and PCC-10 = 180 mg Cd/kg, log-normal distribution of critical concentrations. B: PCC-50 = 300 mg Cd/kg and PCC-10 = 200 mg Cd/kg, log-normal distribution of critical concentrations.

[a] Assumed to have a log-normal distribution with geometric SD of 2.

expected to give rise to a 50% response rate if an excretion of 500 to 700 μg β_2-microglobulin per liter urine is used as an effect criterion. The intakes giving rise to 20, 10, and 5% response rates, respectively, are 240, 160, and 110 μg Cd per day (Table 1). These agree reasonably well with the calculated response rates in Table 3. The agreement seems to be better using alternative A for PCC-50 and PCC-10 than alternative B. If a less sensitive effect parameter (retinol-binding protein about 400 μg/ℓ) is used (Table 2), the necessary intake for the same response rates is higher.

A comparison of Figure 2 and Table 4 also shows a reasonable agreement for occupationally exposed workers. Alternative A seems to be better at the lower response rates and alternative B at the higher response rates. The assumption of a geometric SD of 2 for the renal cortex cadmium concentration distribution was based on populations with exposure to cadmium in food. A different SD may apply for occupational exposures, and this would influence the calculated response rates. In addition, at high dose levels (Figure 2) a large proportion of the dust may have been nonrespirable.[12] As it has been possible to make a comparison with empirical data from only one study, caution must be exercised in any generalization.

By combining the data from epidemiological studies and kinetic model calculations, some estimates of "safe" levels of exposure using different "maximum acceptable risks" can be made. If, e.g., a 1% response rate for tubular proteinuria is considered acceptable, our calculations show that this occurs at about half the dose of the 5% response rate, which is the lowest level at which empirical data are available (Tables 1 and 3). An average cadmium intake via food of 55 μg Cd per day could lead to a 1% response rate if the most sensitive effect parameter is used (Table 1). This is very close to the WHO/FAO provisional tolerable weekly intake of 400 to 500 μg Cd per week[32] or 60 to 70 μg Cd per day. It should be pointed out that the WHO/FAO value refers to the maximum intake for individuals, and thus, in order to stick to the WHO/FAO recommendations, the maximum average intake in a population has to be set at a much lower level.

To the extent that it is possible to generalize from the data from the Swedish battery plant[12] for occupational exposure (Table 4), a 1% response rate may be equivalent to a dose index of about 1 mg Cd per day × years, or 10 years exposure at 16 μg Cd per cubic meter. The 5% response rate is reached at about twice this level (Table 4). No epidemiological data are available for such low response rates. For exposure durations exceeding 10 years, the maximum concentrations have to be reduced even further. A respirable cadmium dust level above about 10 µg Cd per cubic meter is therefore likely to be unacceptable if the maximum response rate is 1%. As the criterion for effect has often been an increase in urinary low molecular weight protein, above the 97.5 percentile, a maximum response rate of 1% would mean that in the exposed population there would be an increase in the prevalence of high urinary excretion of low molecular weight proteins from about 2.5 to 3.5%.

If instead a maximum response rate of 10% is used as a criterion for a "safe" dose, in the above examples, this may be reached by an average daily cadmium intake via food between 150 and 200 µg Cd per day (Tables 1 and 3). A similar approach to estimating "safe" occupational exposure levels will result in an estimate of a dose index of about 3 mg Cd per day × years (Table 4) or about 50 µg Cd per cubic meter for 10 years or 20 µg Cd per cubic meter for 25 years.

V. EXPOSURE LIMITS

Most countries have no legal maximum daily intake of cadmium via food, but the WHO/FAO maximum recommended intake of 400 to 500 µg/week is used as a guideline when setting standards for food. In the U.S. a maximum cadmium concentration of 1 mg/kg in any foodstuff applies. In Japan the same level is used as a maximum for rice.[11]

Occupational standards [threshold limit values (TLV)] have been set in a number of countries.[10] Most of these countries use the TLV set by the American Conference of Governmental Industrial Hygienists at 50 µg/m³. Finland has a TLV of 20 µg/m³ for dust and 10 µg/m³ for fume. Sweden has a TLV of 50 µg/m³ for total dust, 20 µg/m³ for respirable dust and fume in existing workplaces, and 10 µg/m³ for respirable dust and fume in new workplaces.

A working group in the World Health Organization has proposed a "health-based permissible level" of 10 µg/m³.[33]

VI. CONCLUSIONS

The critical organ for long-term cadmium exposure is the kidney. The critical concentration is a fundamental variable in cadmium dose-response analysis. It varies between individuals and it is therefore not possible to give one single value for the critical concentration. On a population level the available data indicate that approximately 50% of individuals have a critical concentration below 250 to 300 mg Cd per kilogram (PCC-50). The corresponding estimate for the 10% level (PCC-10) is 180 to 200 mg Cd per kilogram.

Epidemiological studies indicate that 10% of a population exposed to an average of about 200 µg Cd per day via food is likely to develop tubular proteinuria at age 45. Such proteinuria is here defined as an increased excretion of urinary β_2-microglobulin above the 97.5 percentile in a "normal" population. The same prevalence (10%) of tubular proteinuria is likely to develop after 10 years occupational exposure to 50 µg Cd per cubic meter in workroom air.

Calculations using a kinetic model of cadmium metabolism and distributions of critical concentrations and renal cortex cadmium concentrations in exposed populations produce similar dose-response data as those observed in epidemiological studies. If the calculations are used to extrapolate the dose-response relationship to a 1% response rate for tubular proteinuria, it is seen that these may develop after 45 years exposure to 50 μg Cd per day in food or 10 years exposure to 16 μg Cd per cubic meter in workroom air.

The dose-response relationships for other effects of cadmium are not known in detail. Prevention of the cadmium-induced renal tubular damage is also likely to prevent any other effects.

REFERENCES

1. Elinder, C-G., Kjellström, T., Friberg, L., Lind, B., and Linnman, L., Cadmium in kidney cortex, liver and pancreas from Swedish autopsies, *Arch. Environ. Health,* 31, 292-302, 1976.
2. Ellis, K. J., Morgan, W. D., Zanzi, Il., Yasumara, S., Vartsky, D., and Cohn, S. H., Critical concentrations of cadmium in human renal cortex: dose-effect studies in cadmium smelter workers, *J. Toxicol. Environ. Health,* 7, 691—703, 1981.
3. Ellis, K. J., Yuen, K., Yasumura, S., and Cohn, S. H., Dose-response analysis of cadmium in man: body burden vs. kidney dysfunction, *Environ. Res.,* 33, 216—226, 1984.
4. Finney, D. J., *Probit Analysis,* Cambridge University Press, Cambridge, 1977.
5. Friberg, L., Health hazards in the manufacture of alkaline accumulators with special reference to chronic cadmium poisoning. Doctoral thesis, *Acta Med. Scand.,* 138(Suppl. 240), 1—124, 1950.
6. Friberg, L. and Kjellström, T., Toxic metals — pitfalls in risk estimation, in *Proc. Int. Conf. Heavy Metals in the Environment, Amsterdam, September 1981,* CEP Consultants, Edinburgh, 1981, 1—11.
7. Hassler, E., Exposure to Cadmium and Nickel in an Alkaline Battery Factory, Doctoral thesis, Department of Environmental Hygiene, Karolinska Institute, Stockholm, 1983.
8. Holden, H., Health status of European cadmium workers, in *Occupational Exposure to Cadmium,* Cadmium Association, London, 1980, 330—337.
9. Hutton, M., Cadmium exposure and indicators of kidney function, MARC Rep. No. 29, Monitoring Assessment and Research Centre, Chelsea College, London, 1983.
10. ILO, Occupational exposure limits for airborne toxic substances, Occupational Safety and Health Ser. No. 37, International Labour Office, Geneva, 1977.
11. Japanese Ministry of Health and Welfare, The Essentials of Tentative Countermeasures Against Environmental Pollution by Cadmium, Tokyo, 1969 (in Japanese).
12. Kjellström, T., Accumulation and Renal Effects of Cadmium in Man. A Dose-Response Study, Doctoral thesis, Department of Environmental Hygiene, Karolinska Institute, Stockholm, 1977.
13. Kjellström, T., Problems in the establishment of the critical concentrations for cadmium in kidney cortex, *Jpn. J. Hyg.,* 57, 240, 1982 (in Japanese).
14. Kjellström, T. and Piscator, M., *Quantitative Analysis of β₂-Microglobulin in Urine as an Indicator of Renal Tubular Damage Induced by Cadmium,* Phadedoc No. 1, Diagnostic Communications, Pharmacia Diagnostics AB, Uppsala, Sweden, 1977, 3—21.
15. Kjellström, T., Shiroishi, K., and Evrin, P.-E., Urinary β₂-microglobulin excretion among people exposed to cadmium in the general environment. An epidemiological study in cooperation between Japan and Sweden, *Environ. Res.,* 13, 318—344, 1977a.
16. Kjellström, T., Evrin, P.-E., and Rahnster, B., Dose-response analysis of cadmium-induced tubular proteinuria. A study of urinary β₂-microglobulin excretion among workers in a battery factory, *Environ. Res.,* 13, 303—317, 1977b.
17. Kojima, S., Haga, Y., Kurihara, T., Yamawaki, T., and Kjellström, T., A comparison between fecal cadmium and urinary β₂-microglobulin, total protein, and cadmium among Japanese farmers. An epidemiological study in cooperation between Japan and Sweden, *Environ. Res.,* 14, 436—451, 1977.
18. Kowal, N. E. and Kraemer, D. F., Urinary cadmium and β₂-microglobulin levels of persons aged 20—74 years from a subsample of the national HANES II survey, 1976—1980, in *Cadmium 81, Ed. Proc. 3rd Int. Cadmium Conf. Miami,* Cadmium Association, London, 1982, 119—122.
19. Lauwerys, R. R., Buchet, J. -P., Roels, H. A., Brouwers, J., and Stanescu, C., Epidemiological survey of workers exposed to cadmium, *Arch. Environ. Health,* 28, 145—148, 1974.
20. Lauwerys, R., Roels, H., Bernard, A., and Buchet, J. -P., Renal response to cadmium in a population living in a nonferrous smelter area in Belgium, *Int. Arch. Occup. Environ. Health,* 45, 271—274, 1980.

21. Nogawa, K., Ishizaki, A., and Kawano, S., Statistical observations of the dose-response relationships for cadmium based on epidemiological studies in the Kakehashi River basin, *Environ. Res.*, 15, 185—198, 1978.

22. Nogawa, K., Kobayashi, E., and Honda, R., A study of the relationship between cadmium concentrations in urine and renal effects of cadmium, *Environ. Health Perspect.*, 28, 161—168, 1979.

23. Nordberg, G. F. and Strangert, P., Estimations of a dose-response curve for long-term exposure to methylmercuric compounds in human beings taking into account variability of critical organ concentration and biological half-time: a preliminary communication, in *Effects and Dose-Response Relationships of Toxic Metals,* Nordberg, G. F., Ed., Elsevier, Amsterdam, 1976, 273—282.

24. Nordberg, G. F. and Strangert, P., Fundamental aspects of dose-response relationships and their extrapolation for noncarcinogenic effects of metals, *Environ. Health Perspect.*, 22, 97—102, 1978.

25. Nordberg, G. F. and Strangert, P., to be published.

26. Roels, H. A., Lauwerys, R. R., Buchet, J. -P., Bernard, A., Chettle, D. R., Harvey, T. C., and Al-Haddad, I. K., *In vivo* measurement of liver and kidney cadmium in workers exposed to this metal: its significance with respect to cadmium in blood and urine, *Environ. Res.*, 26, 217—240, 1981.

27. Roels, H., Lauwerys, R., and Dardenne, A. N., The critical level of cadmium in human renal cortex: a reevaluation, *Toxicol. Lett.*, 15, 357—360, 1983.

28. Stewart, M. and Hughes, E. G., Urinary β_2-microglobulin in the biological monitoring of cadmium workers, *Br. J. Ind. Med.*, 38, 170—174, 1981.

29. Strehlow, C. D. and Barltrop, D., Indices of cadmium exposure from contaminated soils in exposed and control populations, in *Proc. Int. Conf. Heavy Metals in the Environment, Amsterdam, September 1981,* CEP Consultants, Edinburgh, 1981, 534—537.

30. Task Group on Metal Interaction, Factors influencing metabolism and toxicity of metals: a Consensus report, *Environ. Health Perspect.*, 25, 3—41, 1978.

31. Task Group on Metal Toxicity, in *Effects and Dose-Response Relationships of Toxic Metals,* Nordberg, G. F., Ed., Elsevier, Amsterdam, 1976, 1-111.

32. WHO, Evaluation of certain food additives and the contaminants mercury, lead and cadmium, Tech. Rep. Ser. 505, World Health Organization, Geneva, 1972.

33. WHO, Recommended health-based limits in occupational exposure to heavy metals. Report of a study group, Tech. Rep. Ser. 647, World Health Organization, Geneva, 1980.

Chapter 14

GENERAL SUMMARY AND CONCLUSIONS AND SOME ASPECTS OF DIAGNOSIS AND TREATMENT OF CHRONIC CADMIUM POISONING

Lars Friberg, Carl-Gustaf Elinder, Tord Kjellström, and Gunnar F. Nordberg

TABLE OF CONTENTS

I. INTRODUCTION

In the preceding chapters a detailed review of the different aspects of cadmium intoxication has been given. In each chapter, whenever motivated, there are special sections for conclusions, which will not generally be repeated here. This chapter will deal with the general aspects of the health problems caused by cadmium. It will also point out some of the more important conclusions and emphasize the need for further information.

The diagnosis and treatment of cadmium poisoning have been touched upon in several of the preceding chapters of this monograph. However, no specific chapter has been devoted to these topics, therefore, some brief comments have been included in this final chapter.

Cadmium can, undoubtedly, constitute a serious health problem. There is much evidence that exposure to this metal both in the industrial and in the general environment has given rise to serious intoxication in human beings. Such effects have been shown from inhaled as well as from ingested cadmium. Most of these intoxications are a result of high exposure in the past but new cases, although as a rule less pronounced, are still appearing.

II. EXPOSURE ROUTES AND METABOLISM

Human beings are exposed to cadmium from a large number of different sources. Absorption of cadmium can take place after ingestion and after inhalation.

Cadmium has rightly been called the dissipated element. It is ubiquitous in both soil and water and is used in a very large number of commercial products which as a rule are not recycled, but enter the environment from industrial and municipal discharges to both air and water. As the soil, when it has been contaminated with cadmium, retains the cadmium for long periods, the possibility for further exposure via, e.g., plants and different grains, like wheat and rice, increases.

Food is generally the most important source of exposure. Certain foods like crustaceans and molluscs as well as kidneys and liver from some mammals contain very high concentrations of cadmium and may constitute a risk if consumed regularly. Generally speaking, however, basic foodstuffs like wheat and rice constitute the major hazard in polluted areas.

After ingestion, the average absorption is approximately 3 to 7% in humans, and somewhat lower in most animal species. When there is an iron deficiency or a low intake of calcium and protein, gastrointestinal absorption may increase considerably and reach values up to 20%.

Absorption after inhalation depends on particle size and solubility and varies between 10 and 60% of the inhaled amount. A major source of cadmium exposure via inhalation is cigarette smoke. From one single cigarette the smoker inhales about 0.1 to 0.2 μg Cd. Taking into consideration the higher absorption after inhalation than after ingestion of cadmium, smoking one pack of cigarettes daily may double the body burden of cadmium. Ambient air, on the other hand, as well as drinking water, do not usually constitute a major problem.

In several occupations industrial exposure constitutes a major risk. The workers are exposed directly via inhalation of cadmium fumes or dust but also via ingestion and/ or inhalation due to contamination of food, cigarettes, and clothing. However, the exposure situations known to us have improved considerably during recent years, and it has been shown that it is possible to keep occupational exposure under control.

In the past there have been difficulties in carrying out accurate analysis of cadmium, particularly in biological material. The method most commonly used for analysis of cadmium is atomic absorption spectrophotometry, but due to interference from the sample matrix considerable errors may occur. Reported concentrations in foodstuffs and in body fluids such as urine and blood have often been too high due to the use of inadequate analytical procedures.

Risks for gross errors in relation to contamination of samples are usually smaller when it comes to analysis of internal organs like the kidneys where concentrations are higher (usually several milligrams per kilogram or more) than in, e.g., blood and urine where concentrations are in the order of less than a microgram or a few micrograms per liter. It is important that reliable quality assurance procedures are included in any monitoring project. Publications without quality assurance data have to be carefully evaluated.

In regard to exposure from the general environment and food as well as in industry, the effects of long-term exposure to low concentrations of cadmium are of particular importance. The critical organ is the kidney. After absorption cadmium is transported to the liver where it stimulates the synthesis of metallothionein. Cadmium is bound to metallothionein and transported via blood to the kidneys where it accumulates with a half-time of decades. There is also a considerable accumulation in the liver and the muscles.

After long-term low-level exposure, approximately one third of the cadmium in the body will be found in the kidneys, and most of the rest of the body burden is found in muscles and liver. Approximately half the body burden is found in the kidneys and liver together. At higher long-term exposure levels, proportionately more cadmium will be found in the liver, and at very high exposures, the liver may contain higher concentrations than the kidneys.

The accumulation in the kidneys is explained by the fact that cadmium metallothionein in plasma is filtered via the glomeruli and then reabsorbed in the tubuli as are other low molecular weight proteins in plasma. Continuous catabolism of the cadmium metallothionein takes place after reabsorption in the tubuli. The "free" cadmium stimulates synthesis of metallothionein in the kidney tubular cells and is bound to this newly formed metallothionein. Kidney damage is prevented by this binding until a stage is reached at which a sufficient proportion of kidney cadmium can no longer be bound to metallothionein. According to a plausible theory, at this stage the nonmetallothionein-bound cadmium will exceed the threshold for toxicity and renal tubular damage occurs. The exact mechanisms which would explain the renal tubular damage after long-term cadmium exposure still remain to be elucidated. Regardless of the details of these mechanisms there is sufficient evidence that after long-term exposure renal damage occurs when the cadmium concentration in renal cortex exceeds a certain critical concentration. This concentration varies among animal species and among individuals of the same species.

III. EFFECTS OF CADMIUM

The manifestation of cadmium intoxication can take several forms, be of an acute or chronic nature, and be a result of local or systemic action. Inhalation of finely dispersed cadmium aerosols may produce acute damage to the lungs in the form of pneumonitis or pulmonary edema. For human beings, even a couple of hours of exposure to a few milligrams per cubic meter is dangerous and may prove fatal. Prolonged exposure to cadmium dust or fumes can give rise to chronic pulmonary disorders, characterized by obstructive changes. As a rule these changes take several years to develop, but sometimes occur after only a few years of exposure.

Dose-response relationships are difficult to establish because time-weighted exposure averages are usually not available and because the pulmonary changes could also be caused by peak exposures which have not been recognized. The lowest levels of cadmium in workroom air associated with respiratory effects have been in the order of 70 μg Cd per cubic meter as total dust and 20 μg/m^3 as respirable dust. A significant increase in deaths caused by obstructive respiratory disease has been observed in several independent studies on workers exposed for long periods of time to high levels of cadmium in air during work.

The kidney damage resulting from chronic cadmium intoxication has been known to exist since the early studies among alkaline battery workers in the late 1940s. Already at that time it was recognized that one important sign of the kidney dysfunction was the excretion in urine of proteins which differed from albumin in regard to certain precipitation reactions and molecular size. In advanced cases there were signs of glomerular as well as tubular damage. It is now widely recognized that the first sign of cadmium-induced kidney dysfunction is the effect on the proximal tubuli, resulting in decreased reabsorption of all plasma proteins that have been filtered through glomeruli. Albumin is poorly filtered and low molecular weight proteins are almost completely filtered. Still, the albumin concentration in plasma is so much higher that the resulting concentration in the tubular fluid will be higher than the concentration of low molecular weight proteins.

The decreased tubular reabsorption leads to an increased urinary excretion of both albumin and low molecular weight proteins. It has been shown in cases of cadmium intoxication that the absolute increase will be highest for albumin, but the relative increase will be highest for low molecular weight proteins.

In particular, there is a very high relative increase of several low molecular weight proteins, e.g., β_2-microglobulin and retinol-binding proteins. Most studies have focused on β_2-microglobulin, but recently attention has been given to studies of retinol-binding proteins. The determination of retinol-binding proteins has certain advantages over that of β_2-microglobulin as there is no interference from the pH of the urine. When tubular damage progresses there will also be an increase in the excretion of glucose, amino acids, and phosphate. In advanced cases both tubular damage and signs of glomerular damage in the form of a decreased glomerular filtration rate have been observed.

Once the cadmium-induced kidney dysfunction is established it seems to be irreversible. This agrees with data on the metabolism of cadmium, which show a continuous build-up of high cadmium levels in the kidneys long after cessation of exposure. If exposure is interrupted at an early stage prognosis in most cases is probably good, but few detailed follow-up studies have been made.

In advanced cases of cadmium intoxication secondary manifestations including severe osteoporosis and osteomalacia have occurred. This form of cadmium intoxication has been seen industrially after inhalation of cadmium and after long-term ingestion of contaminated food. In Japan the fully developed syndrome is known under the name ''Itai-itai byo'' (literally ouch-ouch disease) because of the severe pains accompanying the skeletal disorder. Many thousands of cases of cadmium intoxication with renal dysfunction but without known osteomalacia have also been found in Japan. Contamination of the food, particularly rice, is considered to be the main source of cadmium in these cases of poisoning. The cause of the contamination of rice is high cadmium concentrations in soil caused by both industrial contamination and high background cadmium levels of other origin.

There is firm epidemiological evidence that cadmium has been a necessary factor for the high incidence of osteomalacia seen in some cadmium-contaminated areas. Epidemiological evidence is supported by experimental evidence. Osteomalacia and/or osteoporosis have thus been produced in some animal species after heavy cadmium exposure.

The active metabolite of vitamin D, 1,25-dihydroxycholicalciferol (1,25-DHCC; Kalcitriol®), is formed in the kidney tubular cells. It stimulates calcium absorption from the intestine and is necessary for normal bone mineralization. Cadmium exposure inhibits the formation of 1,25-DHCC both in vitro and in animals in vivo and it is likely that this is one mechanism for the production of cadmium-induced acquired renal osteomalacia.

Cadmium also affects collagen metabolism in animals. The urinary findings of increased proline and hydroxyproline in cadmium-exposed people indicate that the production of collagen fibers may be affected in humans also.

From two countries there are reports of a high incidence of kidney stones in persons with cadmium-induced renal dysfunction. This seems to be another manifestation of disorders of the calcium metabolism caused by cadmium intoxication. The reasons why kidney stones have only been reported from Sweden and the U.K., are not known, but may be dependent on the dietary calcium intake.

There is some evidence, related to the development of osteomalacia in cadmium workers, which may indicate that cadmium alone is not sufficient for the development of osteomalacia. In some countries osteomalacia has not been reported despite heavy exposure in industrial plants, i.e., exposure which resulted in pronounced kidney damage. It is possible that in addition to cadmium exposure, a suboptimal diet was of decisive importance in the causation of osteomalacia in certain industries. There is reason to believe that nutritional factors were also of importance for the development of Itai-itai disease in Japan.

It has been known for a long time that injection of cadmium in animals can cause sarcomata at the site of injection as well as interstitial tumors of the testes. This and some early indications that industrial exposure to cadmium were associated with an increased incidence of cancer of the prostate and pulmonary cancer in humans stimulated further studies on the carcinogenicity of cadmium. The question whether occupational exposure to cadmium may increase the risk for prostatic cancer is still controversial but the most reasonable interpretation, of several epidemiological studies taken together, is that exposure to very high concentrations of cadmium may contribute to the development of prostatic cancer.

A recent experimental study on rats has shown that long-term inhalation of low concentrations of a cadmium chloride aerosol (in doses consistent with those occurring in industry) gives rise to a highly increased and dose-related incidence of lung cancer. In several recent epidemiological studies an excess of lung cancer has been noted among previously heavily exposed cadmium workers. In view of the recent experimental and epidemiological results there are good reasons for suspecting that high exposure to cadmium via inhalation may induce lung cancer in humans. Additional experimental and epidemiological studies are needed. Smokers inhale a very finely dispersed aerosol containing cadmium and one may speculate that cadmium in cigarettes may contribute to the development of lung cancer in smokers.

In some animal experiments it has been possible to produce hypertension after prolonged exposure to cadmium. These effects are somewhat contradictory, however, because in several other experiments no hypertensive effects have been recorded. Experiments on rats, the species in which hypertension has mainly been seen, have shown that apart from species, strain, sex, and diet are of importance. Hypertension has been observed predominantly after exposure to relatively low cadmium levels. At very high exposure levels no increase in blood pressure was seen.

There is no evidence that hypertension has occurred in workers exposed to cadmium. Mortality in cardiovascular disease has not been in excess among cadmium workers. One study of people with cadmium-induced renal damage, exposed to cadmium via food from the general environment, showed an excess death due to cerebrovascular diseases, but another study found no such effects. Generally speaking, there are various uncertainties in connection with the published reports and confounding factors have as a rule not been taken into consideration. A possible hypertensive effect of cadmium should be looked into further. In view of results obtained in animal studies low exposure ranges may be of special interest.

Parenteral administration of high doses of cadmium to male mammals gives rise to acute testicular necrosis within 24 to 48 hr after administration. As the effect has been observed in many mammalian animal species, it seems probable that a similar effect would also occur in human beings if this type of exposure occurred, which is not likely. The necrotizing action of cadmium may be prevented by a previous administration of zinc, selenium, or cobalt. Of special interest is the fact that relatively small doses of cadmium, too small to cause necrosis of the testes, may prevent testicular damage from a higher dose. This preventive action is due to the production of metallothionein as a result of the pretreatment. After long-term exposure to cadmium, the testes are not a critical organ due to the binding of cadmium to metallothionein.

Experimental animal studies have shown that the administration of very high doses of cadmium during pregnancy may give rise to placental necrosis. This type of exposure may also give rise to fetal death or malformations. The mechanism underlying the embryotoxic and teratogenic effects of cadmium is not known in detail but, probably, interactions with fetal zinc uptake are of importance. One report states that pregnant women exposed to high concentrations of cadmium have given birth to children with lower birth weights than normal, but sufficient details were not presented in the report. For the time being, available data do not make possible definite conclusions but the possibility of embryotoxic effects in humans should be borne in mind. Some experimental studies on animals have shown that maternal cadmium exposure may cause decreased placental blood flow which in turn may affect the fetus.

IV. INDICATORS OF DOSE AND DOSE-RESPONSE RELATIONSHIPS

The normal excretion of cadmium via urine is low (about 0.1 to 1 $\mu g/g$ creatinine) and related to kidney and body burden of cadmium. Cadmium levels in urine may therefore serve as a good indicator of the body burden of cadmium in long-term low-level exposure before tubular dysfunction has developed. The gastrointestinal excretion of cadmium is also low, although higher than in urine. Part of it is related to body burden and part to the daily dose. As the major part of cadmium in the feces constitutes unabsorbed cadmium, fecal cadmium cannot be used to estimate body burden but has proven useful to estimate dietary exposure. Cadmium levels in blood increase with exposure. Part of the blood levels reflect body burden, but a substantial part is related to the exposure during the past few months. Blood levels may therefore be used to evaluate body burden as well as recent exposure, depending on circumstances.

When kidney damage occurs, the concentration of cadmium in the kidneys will decrease substantially and urinary excretion will increase. This explains why in advanced cases of cadmium intoxication, e.g., in Itai-itai disease and sometimes in industrial poisoning, kidney levels are low in spite of high liver levels of cadmium. High urinary cadmium levels should always raise the suspicion that kidney dysfunction has already occurred.

The kidney is the critical organ for long-term cadmium exposure. Until recently estimations of critical concentrations in the kidneys have been based on animal data and on cadmium concentrations in kidneys found at autopsies or biopsies. Techniques for in vivo measurement of cadmium in human kidney and liver have now become available. Taking all the evidence together it seems that when kidney cortex concentrations of about 250 mg Cd per kilogram wet weight are reached, approximately 50% of those exposed will have reached their critical concentration and will show signs of kidney dysfunction. At cortex levels of about 180 mg Cd per kilogram, approximately 10% will have reached their critical concentration. Although these estimates are the best possible at present, it should be kept in mind that they are still uncertain.

Calculations with the kinetic model based on the distribution of critical concentrations show that a population with an average cadmium intake via food of about 200 μg/day will reach an average renal cortex concentration of about 100 mg Cd per kilogram. In such a population 10% would have exceeded their own critical concentrations and can therefore be expected to develop renal tubular damage. The 10% risk of renal damage (PCC-10) occurs at a kidney cortex concentration of about 180 mg Cd per kilogram, as mentioned above. The lower level for the population average at which a 10% response occurs is a function of the variability of individual renal levels and individual critical concentrations (Chapter 13, Section IV.B).

This calculation was based on an average gastrointestinal absorption of 5%. In a population where the average absorption is higher, due to low iron or calcium intakes, a correspondingly lower average intake will be sufficient to reach the average renal cortex level mentioned above.

The industrial air exposure needed to reach an average renal cortex cadmium concentration of 100 mg Cd per kilogram is about 50 μg Cd per cubic meter for 10 years or about 20 μg Cd per cubic meter for 25 years. Similar to the situation after exposure to cadmium via food, the dose estimates above can be expected to lead to a 10% prevalence of renal tubular damage.

There is limited empirical evidence from industrial exposure and from exposure via food. In one study, referred to in Chapters 9 and 13, the response rate was 17% after 10 years at 50μg Cd per cubic meter, but the confidence interval was not reported. In regard to effects from the general environment, a 10% increase of tubular proteinuria has been observed after long-term consumption of rice which contained approximately 0.3 to 0.4 mg Cd per kilogram. This corresponds to a daily cadmium intake of about 210 to 280 μg/day, assuming that the average person eats 350 g rice per day and that half of the daily cadmium intake comes from rice. In one area an average daily intake of approximately 150 μg/day, estimated from feces analysis, was associated with a 10% increase in the prevalence of tubular proteinuria. There is thus, reasonable agreement between the calculated estimates of the dose-response relationship and the available epidemiological data on renal tubular damage in cadmium-exposed populations.

V. ASPECTS ON DIAGNOSIS AND TREATMENT OF CHRONIC CADMIUM POISONING

The diagnosis of cadmium poisoning relies on an understanding of the three major disease processes that cadmium can start: obstructive lung disease, renal tubular disease, and a combination of osteoporosis and osteomalacia.

Symptoms of breathlessness and reduction in lung function tests are indicators of respiratory effects. Tobacco smoking results in similar changes in lung function and may therefore cause the deterioration due to cadmium to accelerate. Tobacco smoking will also be an important confounding factor in epidemiological studies of cadmium workers. X-ray changes indicating emphysema are likely to be a later finding and have been seen in severe cases.

After excessive exposure to cadmium in food, renal tubular disease is the first sign of poisoning and this is often likely to be the case after excessive exposure in the workplace. Early diagnosis of renal tubular disease is therefore the best means of detecting cadmium poisoning at an early stage.

Analysis of proteins in urine is the most important method of diagnosis. Renal tubular damage leads to an increase of urinary excretion of all proteins, but particularly high increases of low molecular weight proteins. Sensitive methods to analyze, e.g., β_2-microglobulin and retinol-binding proteins are now available. An increase, beyond the normal range, of these proteins in urine is an early sign of the cadmium-induced damage. At this stage urinary albumin or total protein may well be within the normal range. Urinary amino acids, enzymes, and glucose may also increase, but the relative increase is lower than for low molecular weight proteins. The introduction of screening programs for this type of proteinuria has been discussed.[1]

When the renal damage progresses more severe general proteinuria will be seen, and eventually a reduction of glomerular filtration rate indicating glomerular damage. Blood tests are necessary to diagnose these more severe signs of cadmium poisoning.

Cadmium-induced bone diseases occur mainly at the later stages of cadmium poisoning. The patients are likely to have had kidney disease before the bone damage occurs. The diagnosis of the osteoporosis and/or osteomalacia is based on blood analysis of calcium, phosphate, and alkaline phosphatase and on X-rays of the skeleton, particularly the long bones, pelvis, and ribs.

General aspects on the diagnosis of metal poisoning, including cadmium poisoning, have been reviewed[2] and a guide booklet for physicians on health maintenance of cadmium workers has also been prepared.[3] Some suggestions for monitoring have also been put forward.[4]

There is no specific treatment for chronic cadmium poisoning. The treatment should be symptomatic for the lung, kidney, or bone disease. Cessation of exposure is naturally of great importance to prevent any further poisoning. In addition, as cadmium-induced lung damage and damage from tobacco smoking resemble each other, cessation of smoking is likely to be of benefit for cases with cadmium-induced lung disease.

There is no known treatment that will restore the function of the lung or the kidney. In cases with bone disease, treatment with calcium and vitamin D or its metabolite Kalcitriol® is essential to restore bone function.

Chelation therapy has been tried in animals and cases of human poisoning as a means of reducing the body burden of cadmium. It has not been found effective[2,3] and in fact much of the data indicate that such treatment will increase the kidney damage rather than reduce it.

VI. PREVENTION OF CHRONIC CADMIUM EFFECTS

After long-term relatively low level exposure to cadmium, tubular kidney dysfunction is the critical effect. If such dysfunction is prevented, more serious effects are also avoided. After inhalation, pulmonary effects may occasionally become the critical effects, possibly as a result of high peak exposures due to fluctuations around long-term exposure means. The possibility that exposure to certain cadmium compounds may contribute to the development of prostatic and pulmonary cancer in humans is becoming a question of more and more concern. The use of cadmium should, therefore, be limited by substituting safer materials whenever possible. The workplace environment needs to be improved further to meet the standard proposed by a WHO working group.[5]

On a global scale the major problem is the contamination of the environment and the associated accumulation of cadmium in the human body. In some parts of the world segments of the population have already reached critical concentrations of cadmium in the kidneys or are close to critical levels. Even in countries where exposure is low, the margin of safety is relatively small. Therefore, any changes in exposure levels should be closely monitored and sources of increased exposure should be identified and controlled if necessary.

REFERENCES

1. Kjellström, T. and Piscator, M., *Quantitative Analysis of β_2-Microglobulin in Urine as an Indicator of Renal Tubular Damage Induced by Cadmium,* Phadedoc No. 1, Diagnostic Communications, Pharmacia Diagnostics AB, Uppsala, Sweden, 1977, 3—21.
2. Kazantzis, G., General aspects on individual diagnosis and treatment of metal poisoning, in *Handbook on the Toxicology of Metals,* Friberg, L., Nordberg, G. F., and Vouk, V., Eds., Elsevier, Amsterdam, 1979, 219—236.
3. Lauwerys, R. R., *Health Maintenance of Workers Exposed to Cadmium, a Guide for Physicians,* Cadmium Council, New York, 1980.
4. Piscator, M., Guidelines for occupational exposure to cadmium in Sweden, in *Cadmium 79, Ed. Proc. 2nd Int. Cadmium Conf. Cannes,* Metal Bulletin Ltd., London, 1980, 227—228.
5. WHO, *Recommended Health-Based Limits in Occupational Exposure to Heavy Metals,* Report of a study group, Tech. Rep. Ser. 647, World Health Organization, Geneva, 1980.

Appendix

ITAI-ITAI DISEASE

Tord Kjellström

TABLE OF CONTENTS

I. HISTORY OF ITAI-ITAI DISEASE

In 1946, Dr. Hagino, a general practitioner returning from the war, reopened a private clinic which had operated in Fuchu town, Toyama Prefecture, since his grandfather's time. Dr. Hagino was visited by patients with a painful disease, which he called the "Itai-itai byo", meaning "ouch-ouch disease". According to Hagino,[13] the disease had occurred endemically in the area for several years (Figure 1). Nagasawa and co-workers[52] reported on a "rheumatic disease" in 44 patients living in Fuchu. Most of the patients were women, between 40 and 70 years old, but there were also 13 male patients. Nagasawa and co-workers[52] included only blood cell counts and subjective pain in their description of symptoms, and it is therefore difficult to say to what extent this "rheumatic disease" is identical with the Itai-itai disease.

In a review of the history of Itai-itai disease, Hagino[17] mentioned that in 1947 many of the patients whom he referred to hospitals in Toyama City were diagnosed there as having "kidney trouble or diabetes". In 1947 osteomalacia was suspected and treatment with cod liver oil was tried without effect.[17] It was not until 1955 that Hagino and Kono[18] brought about a more general awareness of the Itai-itai disease, making the first official use of the name at the 17th Meeting of the Japanese Society of Clinical Surgeons. In the same year, treatment with large doses of vitamin D was started. After several months of this treatment the bone symptoms improved. During 1956 and 1957 a number of reports about the disease were written; at that time it was generally believed that Itai-itai disease was a vitamin D-deficient osteomalacia.

Takeuchi in 1973[70] referred to reports from 1906 and concluded that osteomalacia among young adults and rickets among children had been rather common diseases in Toyama Prefecture during the early part of the century. However, the age distribution as well as the geographical distribution were completely different from those of the Itai-itai disease. Many of the patients were children and many of them were living in the northwestern corner of Toyama Prefecture. There was no specific concentration of cases in the Fuchu town area as was the case with Itai-itai disease (Figure 1). During the whole history of Itai-itai disease (1946 to 1984) no report mentions an increase of rickets in children.

During World War II, damage to the rice crop in Fuchu had caused a severe dispute between the farmers and the Kamioka Mining Company.[38] The farmers argued that river water, used for irrigation of the rice fields, had been polluted by waste from the mine. They concluded that metals in the river water had damaged the crops. Kobayashi[37] carried out an investigation, but at that time no analysis of heavy metals in the soil or rice was performed.

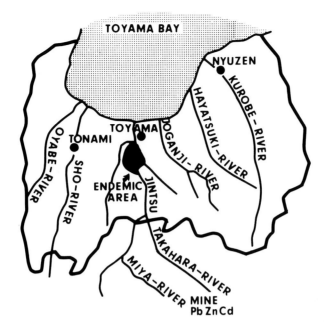

FIGURE 1. Location of population centers within Toyama Prefecture, with the location of the Kamioka mine. (From Ishizaki, A. and Fukushima, M., *Jpn. J. Hyg.*, 23, 271-285, 1968.)

Moritsugi and Kobayashi[44] published data on cadmium analysis of rice performed in 1960. In 20 samples of rice from the endemic area, the average cadmium concentration (wet weight) was more than ten times higher (0.68 mg/kg) than in about 200 samples from other areas of Japan (0.066 mg/kg). In 1961 Hagino and Yoshioka[19] reported on these data at the 34th Meeting of the Japanese Association of Orthopedics, and it was postulated that cadmium played an etiological role in the development of the Itai-itai disease. A detailed report was published by Yoshioka.[83] Epidemiological and clinical studies were started on a larger scale in the 1960s by groups supported by the Japanese government. Based on the results of these studies, the Japanese Ministry of Health and Welfare declared in 1968, "The Itai-Itai disease is caused by chronic cadmium poisoning, on condition of the existence of such inducing factors as pregnancy, lactation, imbalance of internal secretion, aging, deficiency of calcium, etc."[41] The etiology of Itai-itai disease will be discussed in Section VII.

II. CLINICAL FEATURES

A. Symptoms and Signs

The clinical course of the disease has been described by Nakagawa,[53] Takase and coworkers,[68] Hagino,[14-16] Ishizaki,[24] Murata and co-workers,[49,50] and Tsuchiya.[75] The greatest number of the patients have been treated by Hagino and Murata. From Murata and co-workers[50] the clinical manifestations can be summarized as follows. Most of the patients are postmenopausal women with several deliveries (an average of six). The most characteristic features of the disease are lumbar pains and leg myalgia. Pressure on bones, especially the femurs, backbone, and ribs, produces further pain. Another characteristic of the disease is a duck-like gait. These conditions continue for several years until one day the patient experiences a mild trauma and finds herself unable to walk. She is then confined to bed and the clinical condition deteriorates rapidly. Not only the bones of the extremities but also the ribs and other bones are susceptible to multiple fractures after very slight trauma such as coughing. Skeletal deformation takes place with a marked decrease in body height.

Nogawa and Kawano[56] reported that increased prevalence of hypertension was not found in patients with Itai-itai disease nor in persons living near the endemic district (cf. Chapter 11, Section V.B.2.ii).

B. Laboratory Examinations
1. Blood Findings

According to Murata and co-workers,[50] blood examination showed hypochromic anemia in most patients. Serum iron levels were low in many patients, and the erythrocyte sedimentation rate was elevated in some patients. Almost no deviation from the normal has been observed in serum protein and albumin to globulin ratios. The levels of calcium and inorganic phosphorus in serum are often low, while the alkaline phosphatase level is high.

2. Urinary Findings and Renal Function Tests

According to Hagino,[17] proteinuria and glucosuria were common findings in the patients before any treatment had been given. Taga et al.[67] reported bone symptoms in ten patients. Two of these patients had died in 1955 and were autopsied by Kajikawa (Section II.D.1). He reported that the patients had chronic pyelonephritis with renal arteriosclerosis.[33] It can be assumed that there had been severe kidney damage. Hagino[13] reported that he had found proteinuria in 82% and glucosuria in 32% of 71 surviving patients. The methods of determination were not stated.

Nakagawa[53] described 30 patients from Shinbo district (bordering on Fuchu) who were studied in 1955 to 1958. Vitamin D treatment had not been given to the patients prior to examination (Nakagawa, personal communication). All of the patients had proteinuria (sulfosalicylic acid method) and some had glucosuria as well (Nylander method). The tubular function was affected in a number of patients as judged by the PSP excretion test (phenolsulfonphthalein). The lower normal limit of excretion is considered to be 25% after 15 min. In seven out of ten patients the excretion was between 10 and 22.5% and in the other three 25%. Kono and co-workers[40] also reported low ratings (19% after 15 min) on the PSP excretion test in two patients. In four patients studied by Takeuchi and co-workers,[72] the PSP excretion was 6 to 12% after 15 min. The concentration test, dilution test, and phosphorous reabsorption test showed slight kidney damage in all four cases.

Ishizaki and Fukushima[26] reported that among 37 Itai-itai patients "I" (for explanation of symbols used see classification in Section III) studied in the 1967 epidemiological investigation (Section VI.B), only 3 did not have glucosuria (= $^1/_{32}$% or higher) and only 1 did not have proteinuria (= 10 mg/100 mℓ or higher). While urinary calcium levels were normal, urinary phosphate was reduced. The urinary amino acid levels were above normal.[10,24,49,68,72,75]

Quantitative determinations have shown an increase in cadmium concentration and a decrease in zinc concentration in blood and urine, in comparison with control subjects (see Murata et al.[50] and Section IV.A).

In Itai-itai disease, proteinuria, glucosuria, and aminoaciduria are considered signs of renal dysfunction. On separation using paper electrophoresis, a typical tubular protein pattern occurs in cases of Itai-itai disease and also in the preceding stages (proteinuria without bone changes), as shown by Piscator and Tsuchiya.[62] These authors examined 10 female patients with Itai-itai disease and 12 subjects (10 women and 2 men) from the endemic area with only proteinuria and glucosuria. The pattern found was similar to that observed in chronic occupational cadmium poisoning. In Figure 2 some of these patterns are shown (cf. Chapter 9, Figure 11).

mg/g creatinine

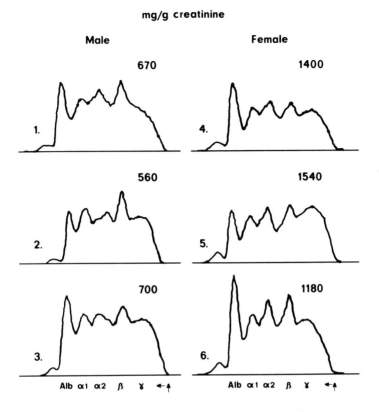

FIGURE 2. Paper electrophoretic patterns in an occupationally exposed male Japanese worker and male and female patients from the endemic district.[62]

Comparisons of disc electrophoretic patterns among cadmium workers, Itai-itai patients, and patients with other kidney diseases have been performed by Tsuchiya[76] and Harada.[20] Cadmium workers and Itai-itai patients showed similar if not quite identical patterns. It must be kept in mind that so far it has not been possible to find comparable sex and age groups of Itai-itai patients and cadmium workers in these comparisons. Still further evidence that the proteinuria is of a tubular type has been obtained by using gel filtration on Sephadex G-100.[62] The main part of the urine proteins from cases and suspected cases of Itai-itai disease had molecular weights less than that of albumin. Among these low molecular weight proteins, immune globulin chains and β_2-microglobulin were found, as has earlier been demonstrated in cadmium workers by Piscator.[59-61]

Electrophoretic examinations of urine proteins have also been performed by Fukushima and Sugita.[9] In eight patients, they found a relatively small percentage of albumin and a dominance of $\alpha 2$-, β-, and γ-proteins, which also indicates a tubular type of proteinuria. Fukuyama[11] and Sano and co-workers[64] compared the patterns of urinary proteins after gel filtration (G-100 and G-75, respectively) in patients with Itai-itai disease, cadmium workers, and patients with various kidney diseases. Cadmium workers, persons from cadmium-exposed areas, and Itai-itai patients displayed similar urinary protein patterns. Ohsawa and Kimura[58] have isolated β_2-microglobulin from the urine of Itai-itai patients by separating on a Sephadex G-100 column and subsequently on a DEAE-cellulose column. The fractions were studied with disc electrophoresis from which the authors concluded that β_2-microglobulin is excreted to a remarkable degree in Itai-itai patients.

Isoelectric focusing using the method described by Vesterberg[78] and Vesterberg and Nise[79] and radioimmunoassay using the method described by Evrin et al.[7] have been employed for analyzing urinary excretion of β_2-microglobulin in ten Itai-itai patients "I" and ten suspected patients "i".[66] The "I" patients had levels between 24 and 55 mg/ℓ and the "i" patients between 6 and 41 mg/ℓ, a 500- to 1000-fold increase over the normal value. The average urinary β_2-microglobulin level in normal Swedish women in the age range 57 to 76 years as analyzed by radioimmunoassay was 78 μg/24 hr (SD 80), roughly corresponding to 0.078 mg/ℓ (SD 0.080).[6] In Japanese and other population groups without renal dysfunction, the average levels are reported to be about 0.1 mg/ℓ (Chapter 9, Section IV.B.5).

Fukushima and co-workers[10] studied urinary amino acid excretion in two Itai-itai patients and in a control group. In each case, amino acid excretion was generally two to five times higher in the patients than in the controls. Arginine, citrulline, proline, and hydroxyproline were 20 to 50 times elevated in the patients. This is in accordance with what is generally found in patients with metabolic bone diseases. Increased amino acid excretion has also been found among cadmium workers without bone disease.[74] In these cases the urinary levels of arginine and citrulline were especially high (Chapter 9, Section IV.B.1.b), but not those of proline and hydroxyproline, which is consistent with the fact that these workers had no bone disease.

3. Other Findings

In many cases tests have shown a tendency towards more or less impaired pancreatic function.[50] Decreased fat absorption has been revealed using ^{131}I triolein and ^{131}I oleic acid methods, but this could have been partly dependent upon a pancreatic disturbance (Murata, personal communication). Microscopic examinations of the gastrointestinal tract showed shortened ciliated epithelia in the small intestine, atrophy of the mucous membrane, and submucosal cell infiltration.[50] In some cases ulceration of the mucous membrane of the small intestine has also been observed.[50] The changes in the gastrointestinal tract have been named "cadmium enteropathy" by Murata.

C. X-Ray Changes

Radiological examinations of the bones of Itai-itai disease patients show evidence of severe osteoporosis and osteomalacia. The most detailed account has been given by Nakagawa.[53] He reported that "high levels of bone atrophy could be demonstrated throughout the entire skeleton". The compact bone became like thin fiber and the spongy bone was weakly translucent. Many bones were "bent" rather than fractured. Nakagawa considered that 30% decalcification is necessary for the bone atrophy to be seen on X-rays. In Itai-itai disease the atrophy was extreme and generalized. This picture of osteoporosis can hardly be considered equivalent to "normal" senile osteoporosis.

Looser's zones or pseudofractures of Milkman (Chapter 10, Figures 11 to 13) are characteristic radiological signs of osteomalacia and these were very common.[53] In the 30 cases studied by Nakagawa, Looser's zones appeared in 342 locations. They were particularly common in femur, pelvic bones, and ribs. They occurred on both sides of the body and were found by X-ray at the sites of local pain and the centers of radiating pain.

Nakagawa[53] classified, from different types of Looser's zones, the fissured transformation zone (Chapter 10, Figure 11), the cuneiform transformation zone, the branchial-transformation zone (similar to Chapter 10, Figure 13), and the banded transformation zone (Chapter 10, Figure 12). It is clear from Nakagawa's detailed study that the appearance of the osteomalacia on X-ray varies between individual patients and that the classical banded Looser's zone should not be expected in every case.

D. Histopathological Changes

The data on histological changes unfortunately are rather incomplete (for cadmium concentration in organs see Section IV.B).

1. Kidneys

In two early autopsies, Kajikawa and co-workers[33] found senile arteriosclerotic contracted kidneys, pyelonephritis, and metastatic calcification. Takeuchi et al.[72] found no marked changes in the glomerules in biopsy specimens from three patients. In the tubules, atrophic changes with flattening of the epithelium were seen. Kajikawa[32] and Takeuchi[70] reported on six other patients. Five had "tubular nephropathy" (tubular atrophy and dilatation with eosinophilic casts and interstitial fibrosis) of varying degrees, and one had suppurative pyelonephritis. In all six patients, kidney soft tissue calcification of slight degree was reported. Since kidney stones have been a common finding among cadmium workers (Chapter 9, Section IV.B.1.c), cadmium-induced disturbances in calcium metabolism may explain the soft tissue calcification. However, the Itai-itai patients were also exposed to large doses of vitamin D as treatment for the illness (Section V).

A review of 11 cases autopsied up to 1974 by Kajikawa and co-workers[34] found varying degrees of renal tubular damage in all except two cases, one with nephrocalcinosis and one with arteriosclerosis (Table 1).

2. Bone

Ishizaki[24] citing Kajikawa and co-workers[33] and his own material, has described histological findings in bone from autopsies. The changes have been consistent with those found in osteomalacia in combination with more or less severe osteoporosis. On the basis of bone biopsies from seven patients, Nakagawa[53] concluded that they had osteomalacia. Kono and co-workers[40] found the Itai-itai disease very similar to common forms of osteomalacia. However, these authors found fewer osteoblasts in Itai-itai disease biopsies than are usually found in osteomalacia. Kajikawa[32] and Takeuchi[70] reported varying degrees of osteomalacia in combination with osteoporosis in eight patients. Further evidence of the occurrence of these bone diseases is given in Table 1.[34] Two cases had osteitis fibrosa and two had osteoporosis without osteomalacia, but all the others showed combined osteoporosis and osteomalacia. The degree of vitamin D and other treatment may of course influence the pathological findings in the bone.

III. DIAGNOSIS OF ITAI-ITAI DISEASE

Ishizaki and Fukushima[26] discussed the clinical diagnosis of Itai-itai disease. They pointed out that the first and second stages could reasonably be considered to be renal dysfunction and bone changes, respectively. However, even if a specific diagnosis of renal impairment were possible, patients so diagnosed who can carry out their daily lives unhampered until bone symptoms develop cannot from a practical standpoint be included in the Itai-itai group.

Because the subjective symptoms and the bone signs clinically observable on X-rays are very unspecific in the early stages, the diagnosis at this stage could reasonably be verified from signs of osteomalacia in blood, including increased alkaline phosphatase and decreased serum inorganic phosphate. These blood findings constitute the basis for the syndrome "latent osteomalacia" introduced by Murata and co-workers,[48] Ishizaki and Fukushima[26] pointed out that a roentgenological diagnosis would not be helpful in the early stages, because detectable X-ray changes would not be evident until the disease had developed to some extent. This was also pointed out by Nakagawa.[53]

Table 1

AUTOPSY CASES DIAGNOSED CLINICALLY AS ITAI-ITAI DISEASE

Name	No. of autopsy	Age	Sex	Time of autopsy	Bone	Kidney	Parathyroid	Other organs	Cause of death	Duration (years)	Location	Urine
T.M.	3415	62	F	1955	OM+++ OP++	Pyelonephritis Arteriosclerosis Metastatic calcification	Normal	Pulmonary congestion and edema	Emaciation	12	Fuchu-machi	P– P–
T.Y.	3429	62	F	1955	OM+++ OP++	Pyelonephritis Arteriosclerosis Metastatic calcification	Slightly hyperplastic	Pulmonary congestion and edema Enterocolitis	Emaciation	5	Fuchu-machi	P– S–
T.K. —a	3615	49	F	1958	Osteitis fibrosa	Nephrocalcinosis Pyelonephritis Hydronephrosis	Hyperplasia (adenoma)	Metastatic calcification	Uremia	10	Izumi-machi, Toyama City	P+ S–
A.M. —a	4207	44	F	1963	Osteitis fibrosa with pseudofracture	Nephrocalcinosis	Hyperplasia (secondary?) adenomatous	Chronic pancreatitis, gastritis and colitis, thrombosis of the portal vein	Portal thrombosis (?)	5	Nyuzen-machi	P+ S–
I.F.	4574	73	F	1965	OM+ OP+ OS+	TN+++	Normal	Chronic pancreatitis Chronic hepatitis	Uremia	12	Yoshikura (Shimbo), Toyama City	P+ S–
S.M.	5467	61	F	1968	OS+ OP+ (Pseudofracture in X-ray film)	Suppurative pyelonephritis	Normal	Verrucous endocarditis (mitral and aortic stenosis with insufficiency)	Cardiac failure	11	Yoshikura (Shimbo), Toyama City	P+ S–
S.S.	5313	79	F	1968	OM+ OP+ OS+	TN+	Normal	Cancer of the stomach	Cancer	8	Tonan-kami-kuriyama, Toyama City	P+ S++

U.T.	—	73	F	1968	OM++ OP± (ribs)	TN++	Normal	Cancer of the stomach	Cancer	8	Fuchu-machi	P++ S++
H.A.	5677	75	F	1969	OM± OP+ OS+	TN++	Normal	Broncho-pneumonia Fibrinous pleuritis (1) Ulcerative colitis	Uremia	2 (?)	Fuchu-machi	P++ S++
E.T.	6118	68	F	1971	OM++ OP++ OS±	TN+++	Normal	Anemia Hypertrophy of the heart Pulmonary edema	Uremia	16	Shimbo, Toyama City	P++ S++
I.S. —[a]	6356	72	F	1971	OP++	Arteriosclerosis Normal	Anemia Malnutrition	Suicide by hanging	21	Fuchu-machi	P+ S++

Note: OM = osteomalacia; OP = osteoporosis; OS = osteosclerosis (including healed pseudofracture); TN = tubular nephropathy (tubular atrophy and dilatation with eosinophilic casts, and interstitial fibrosis); P = proteinuria; S = glucosuria.

[a]   Possibly different disorders from Itai-Itai disease.

From Kajikawa, K., Kitagawa, M., Nakanishi, I., Ueshima, H., Katsuda, S., and Kuroda, K., *J. Juzen Med. Soc.*, 1974 (in Japanese). With permission.

Moreover, as the disease is frequent in menopause and in old age, menopausal or senile osteoporosis could be confused with the Itai-itai disease.[26] To avoid this type of mistake, specific signs of osteomalacia (Milkman's pseudofracture) were made a requirement for the roentgenological diagnosis of Itai-itai disease. It should be pointed out, though, that in animal studies both osteoporosis and osteomalacia have been induced (see Chapter 10) and as mentioned above (Section II.C), the severe cases of Itai-itai disease had a combination of osteoporosis and osteomalacia. This diagnostic requirement, therefore, is quite conservative and will lead to an underestimation of the total number of cases.

As a result of detailed discussions, the Itai-itai disease research group of 1962 to 1965 devised a standard for the diagnosis of Itai-itai disease, to be used in epidemiological research.[36] The following observations should be noted:

1. Subjective symptoms: pain (lumbago, back pain, joint pain); disturbance of gait (duck-gait)
2. Physical examination: pain by pressure; "dwarfism"; kyphosis; restriction of spinal movement
3. X-ray: Milkman's pseudofractures (Looser's zones); fractures (including callus formation); thinned bone cortex; decalcification; deformation; fishbone vertebrae; coxa vara
4. Urine analysis: coinciding positive tests for protein and glucose; protein (+); glucose (+); decreased phosphorus to calcium ratio
5. Serum analysis: increased alkaline phosphatase; decreased serum inorganic phosphate

The subjects were classified into the following five groups: (we have used the original classification codes as used in the earlier reports):

1. I: Definitely Itai-itai patients (typical bone manifestations on X-ray)
2. i: People deeply suspected for Itai-itai disease (bone signs but not typical)
3. (i): People suspected for Itai-itai disease (slight bone signs on X-ray)
4. (O) or O_{ob}: people needing follow-up (urinary and/or blood signs only)
5. O: People with no suspicion of Itai-itai disease

As stated by Kato and Kawano,[36] the methods for diagnosis differed a little from year to year and a sixth group "(I)" is sometimes named: (I) — person cured of Itai-itai disease.

It should be pointed out that because of the selection procedure (see Section V.B), the vast majority in groups "i" and "(i)" had both glucosuria and proteinuria, as did the Itai-itai patients "I". Sometimes only the "I" type of patients is meant when the "number of patients" is discussed, but in other circumstances, other groups are also included in the "patients" referred to. The distinction is not always clear in the published reports.

During recent years more strict criteria have been used to diagnose the disease. If a definite diagnosis of osteomalacia is not made, the case will not be accepted as Itai-itai disease by the Differential Diagnosis Committee on Itai-itai Disease and Cadmium Poisoning established by the Japan Environment Agency or similar committees established in some prefectures. This is clear from the fact that not one single case has been accepted even though several had severe bone findings.[57]

In the early studies of the disease the symptoms and spontaneous fractures would have been sufficient for diagnosis of "typical bone manifestations", but now a non-disputed "pseudofracture" has to be found in order for this particular diagnosis to be

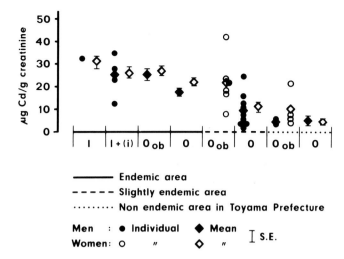

FIGURE 3.   Cadmium concentration in urine of persons from different areas in Toyama Prefecture. (Modified from Ishizaki, A., *J. Jpn. Med. Soc.*, 62, 242-248, 1969b.)

made. Bone biopsy has recently been utilized to diagnose osteomalacia.[45] Abnormal increase of osteoid tissue was found in three cases, which was consistent with osteomalacia. According to the authors these cases had femoral pseudofractures similar to "Looser's zones", but the cases have not been accepted as Itai-itai disease by the Toyama Prefecture Diagnostic Committee for Itai-itai Disease.

In 1982, the same committee reviewed bone biopsy data from eight cases with disputed Itai-itai disease.[80] A control group of 150 autopsy cases who died from other causes than Itai-itai disease was also studied and it was found that criteria for abnormal amount of osteoid tissue could be developed, based on the bone staining method of Yoshiki.[82] If there was more than 25% osteoid tissue in the average bone cross-section and the case had clinical features similar to Itai-itai disease, a diagnosis of the disease could be made. This method is likely to make standardization of the diagnostic procedures possible in the future.

It appears from our discussions with members of the Toyama Prefecture Diagnostic Committee and the equivalent committees in other areas that more rigid criteria for the diagnosis of the disease are used in areas outside the Toyama area. This will make it difficult to compare the occurrence of bone effects in Toyama with the other areas where the incidence rate may be much lower than in Toyama. As shown in Chapter 10 (Table 12), new cases are still being diagnosed in Toyama and a few cases identical to Itai-itai disease (Section VI.H.2) have been reported from other polluted areas.

IV. CONCENTRATIONS OF HEAVY METALS IN URINE AND TISSUES

A. Concentrations of Heavy Metals in Urine

The 1962 to 1965 research group analyzed urine from inhabitants of different parts of Toyama Prefecture. The method of sampling in the various patient categories cannot be obtained from the literature available to us. Values reported, however, showed that cadmium concentrations in urine for persons in the endemic area were higher than for persons in other parts of the Toyama Prefecture.[23,27,75] Lead and zinc values, on the other hand, showed no geographical differences.[27,75] The 1967 research group, as referred to in Ishizaki's report, also analyzed urine concentrations of cadmium (probably by atomic absorption after extraction). Increased concentrations were found in persons who originally came from the endemic area (see Figure 3).

Table 2

CADMIUM CONTENT (µg/g WET WEIGHT) IN ORGANS FROM
FEMALE ITAI-ITAI PATIENTS

	Patient 1	Patient 2	Patient 3	Patient 4	Patient 5
			Age at death		
	79	71	60	73	67
			Cause of death		
Organ	Stomach cancer	Broncho-pneumonia	Endocarditis verrucosa	Stomach cancer	Uremia
Liver	94.1	118.1	63.3	89.0	132
Renal cortex	41.1	31.8	19.8[a]		12
Renal medulla	39.5	26.1	—[a]		10
Lung		2.5	8.0		2.1
Spleen		6.8	6.2		6.0
Pancreas	45.1	64.7			5.2
Stomach				4.8	
Small intestine		3.0	12.5	5.7	9.9
Large intestine		1.7	11.9		
Ribs			2.8		2.6
Bone cortex		1.6			
Bone marrow		1.1			
Skin		4.6	5.1	3.9	
Muscles		14.1			8.3
Brain	0.6				

[a]   Cortex and medulla not separated.

Patients 1 to 4 from Ishizaki et al.;[30] patient 5 from Ishizaki.[25]

B. Concentrations of Cadmium in Organs

Few values of cadmium in organs from Itai-itai patients have been reported. Some early data from one patient described high concentrations in bone. However, in this case the different organs had been stored together in formaldehyde for several years (Kobayashi, personal communication) so that the values must be considered unreliable.

Ishizaki and co-workers[30] and Ishizaki[25] reported organ values (atomic absorption) from autopsies on five Itai-itai patients. The results are given in Table 2. Formaldehyde preservations were not used. At the same time cadmium analyses were performed on organs from 38 controls who died in hospitals in Kanazawa. All of them had been living in nonendemic areas in or near Kanazawa. Results are shown in Figure 4. As can be seen by comparing Table 2 with Figure 4, the cadmium content of the liver was five to ten times higher in Itai-itai patients than in the controls in the same age group. The patients' kidney values, however, were lower than those of the controls. The authors stated that the low kidney values in the Itai-itai patients could be explained by the advanced kidney damage. This decrease of renal cadmium after renal damage has developed has also been seen in animal studies (Chapter 9, Section IV.A.3) and in occupational cadmium poisoning (Chapter 9, Section IV.B.1.b). The tissue cadmium concentrations given should also be compared with "normal values" from Japan and other countries and with data from workers industrially exposed and who have signs of cadmium poisoning (see Chapter 5 and 6). These comparisons show the great accumulation of cadmium in the liver of the Itai-itai patients.

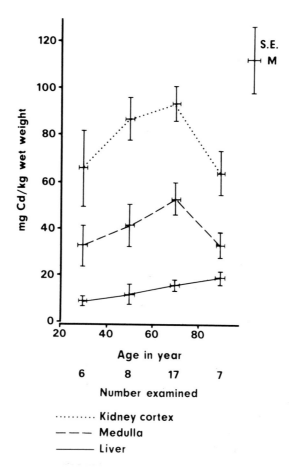

FIGURE 4.   Cadmium content (μg/g wet weight) in controls who died in hospitals in Kanazawa. (From Ishizaki, A. et al. *Jpn. J. Hyg.*, 25, 86, 1970b.)

V. TREATMENT OF ITAI-ITAI DISEASE

No objection-free study on the effects of treatment of Itai-itai disease has been published. There are, however, some data which elucidate to what extent vitamin D has been successful in improving overt osteomalacia symptoms. As such information has some bearing on discussions of the etiology of Itai-itai disease, available data are briefly included here. Takeuchi[70] in a discussion of the etiology of Itai-itai disease, states that the effects of vitamin D were satisfactorily prompt and sufficient, and the patients were never resistant to therapy. The available data referred to below do not support such a conclusion.

When Hagino discovered the first cases, he believed the disease to be vitamin D-deficient osteomalacia, and during 1949 to 1955 tried treatment with cod liver oil (5 to 10 g/day corresponding to about 1000 IU vitamin D) on 30 patients, but this was without effect.[17]

Taga and co-workers[67] reported on the treatment of ten patients with vitamin D. After 3 months of treatment with vitamin D, serum inorganic phosphate had increased and the patients were able to move. There was an improvement in subjective symptoms and X-ray pictures of bones showed improvement. It was not stated how large the doses were or which route of entry was used. Since Hagino was one of the authors of the paper, one may assume that treatment was similar to the method used by him. He later reported[17] that during 1955 to 1957, he administered 50,000 to 100,000 IU vitamin D per day to about 50 patients. The route of entry was not mentioned.

Hagino[13] and Nakagawa and Furumoto[54] reported that large doses of vitamin D were needed to reverse the bone symptoms. Toyoda and co-workers[73] studied three patients who had been treated for 2 months with a diet containing high amounts of calcium and phosphorus; in two cases, the pains were relieved after 2 months. At that time vitamin D treatment (100,000 IU/day, intramuscularly) was started and after 3 months of this treatment the pain had completely disappeared. Two patients displayed the "duck-gait" from the beginning, but after 3 and 5 months of vitamin D treatment these symptoms disappeared. Toyoda and co-workers stated that the X-ray findings revealed a very slow healing process.

Kono and co-workers[40] made balance studies on two Itai-itai patients. Vitamin D treatment brought a change from negative to positive phosphate balance, and the calcium balance treatment became more positive. This agrees with the report by Murata and co-workers[48] that vitamin D was effective for both Itai-itai patients and persons with "latent osteomalacia". As "latent osteomalacia" included patients with only biochemical dysfunction, the improvement in their cases may have been similar to the findings of Kono and co-workers.[40]

In one of the cases reported by Kono and co-workers[40] the vitamin D treatment was ended after a certain time (after how long was not reported) and 3 months later the pains resumed. The patient who had spontaneous pains upon movement could hardly walk 8 months after the end of treatment. Kono and co-workers[40] stated that the disease could not have been caused by malnutrition only, as the symptoms of this patient worsened when the administration of vitamin D and other medications was terminated even though complete nutrients were given.

The most detailed account of treatment and its effect has been given by Nakagawa.[53] He studied a total of 30 patients and divided them into 3 groups according to their movement capability: slight symptoms (duck-gait, some limitation in movement, otherwise few bone symptoms): medium symptoms (can crawl but hardly walk); and extreme symptoms (cannot move at all). Many of the patients with slight symptoms could be treated as out-patients, while most of the other patients were treated as in-patients. Out-patients were given vitamin D tablets (20,000 IU/day = calciferol 0.5 mg) and arrangements were made for better nutrition (more milk, eggs, meat, vegetables) and more sunlight. In-patients received i.m. injections of 100,000 IU every second day plus 10,000 IU/day as tablets. The patients ate regular hospital food and were taken out in the sunlight as much as possible. After leaving the hospital the patients received tablets corresponding to 20,000 IU/day and were given an injection of 200,000 IU every time they came for a check-up (once or twice each month). Check-ups were made of serum calcium level and subjective symptoms. Treatment was continued for 1 to 3 years. Pain decreased after about 3 weeks of treatment for the in-patients and after 2 months of treatment for the out-patients,[53] but the impaired mobility remained for a long time. Persons with slight symptoms could return to everyday life after about 4 months, by which time the duck-gait had disappeared. Patients with medium symptoms were relieved from pains after about 2 months, could stand erect after 5 months, could walk about inside the hospital after 8 months, and returned to everyday life after 13 to 14 months. The effect of treatment was even slower in persons with extreme symptoms: after 5 months they could sit up, after 7 months they could stand up, after 15 months they could walk with a cane, and after 20 months they could walk unaided a couple of meters. When the progress of treatment was studied by X-ray of bones, it was found that pseudofractures persisted after 1 to 3 years of treatment in some cases, but in others they disappeared after 3 to 6 months. There were also slow improvements in the biochemical abnormalities. Serum inorganic phosphate slowly increased to normal and alkaline phosphatase slowly decreased.

Murata and co-workers[51] report on the treatment, with vitamin D (about 100,000 IU/day), of five patients for 15 years. The amounts of vitamin D given varied from year to year. Murata[47] showed that in two patients the pathological fractures had healed after 6 and 24 months, respectively, when judged from X-ray pictures. According to Murata and co-workers[51] in 1956 and 1968, X-ray pictures of one patient showed pathological fractures, but not in 1962. They argued that vitamin D treatment had first cured the patient and then because of overdoses, caused bone porosity. It was not explained why the treatment with 10,000 to 30,000 IU vitamin D per day was continued from 1962 to 1969. Further, an overdose of vitamin D has not been reported to give the same bone findings as a deficiency of vitamin D.

Hagino[17] stated that the patients' symptoms returned when treatment was discontinued, a finding consistent with that of Kono and co-workers.[40] If recidivism did not occur, the very long treatment periods reported by Murata and co-workers[51] would have been unnecessary. Thus, there are clear-cut indications from the studies referred to above that long-term administration of large doses of vitamin D was required in order to bring about improvement of the osteomalacia in Itai-itai disease patients. This would not be the case if the patients had a classical vitamin D-deficient osteomalacia. Thus, all the evidence favors the assumption that Itai-itai disease is a form of vitamin D-resistant osteomalacia. Vitamin D treatment with the large doses that were used in many cases (over 100,000 IU/day) may give rise to side effects, which are mainly the result of hypercalcemia. It can be assumed that the treatment was generally performed in such a way as to avoid acute side effects (nausea, headache, anorexia, etc.). Only incomplete data are available regarding serum calcium concentrations, but Murata and co-workers[51] reported that the patient with the recidivous fracture referred to above had concentrations above 11 mg/100 mℓ in only 2 out of 15 analyses over a 15-year period.

Some persons, diagnosed as being in need of continuous observation regarding Itai-itai disease "O_{ob}", were given "prophylactic treatment" with vitamin D during 1968.[17,51] Average daily dose varied between less than 25,000 and more than 200,000 IU. In 22 out of 48 cases, serum calcium concentration exceeded 11 mg/100 mℓ,[51] but the data presented do not indicate any dose-response relationship between serum calcium and vitamin D dose or between proteinuria or glucosuria and vitamin D dose.

Murata and co-workers[51] presented data on the PSP excretion test, urea clearance, and urea nitrogen from different time periods for five patients treated with large doses of vitamin D. A continuous aggravation of kidney function was obvious. Takeuchi[70] referred to clinical data for three patients studied in 1955 and four patients studied in 1965 which show that the kidney function was damaged in the latter (1965) patients, whereas those of 1955 had normal or only slightly impaired kidney function. Takeuchi did not report to what extent the patients had been treated with vitamin D. As no control patients were used in these studies, it is not clear whether differences in kidney function tests at different times were the result of continued cadmium exposure, vitamin D treatment, or some other cause.

VI. EPIDEMIOLOGICAL STUDIES OF ITAI-ITAI DISEASE

A. Initial Descriptive Studies

Epidemiological studies regarding the Itai-itai disease will be treated in this section, while data on the prevalence of proteinuria and glucosuria in the general population, collected in connection with these studies, are presented in Chapter 9. The limited research on the epidemiology of the disease prior to 1963 must be classified as preliminary or pilot studies. Since no diagnostic standard had been agreed upon, the different studies defined the disease differently. There is no study known to us which shows that symptoms similar to those of the Itai-itai disease occurred in persons living outside the endemic area in the Jintsu River basin, at the time when the disease was discovered there. Murata and co-workers[48] reported that "latent osteomalacia" (Section III) occurred outside the "Shinbo area" in 13 out of 38 persons studied. Whether this meant that "latent osteomalacia" occurred outside the Jintsu River basin is not clear since the exact location of the patients was not defined.

B. Epidemiological Research: 1962 to 1965

An epidemiological survey made between 1962 and 1965 in Fuchu town, Toyama City, Osawano town, Yatsuo town, Nyuzen town, and Tonami City was evaluated according to a diagnostic standard from 1963.[36] All of these places are in Toyama Prefecture. The target group comprised all females over 40 years of age except in Tonami City, where males over 40 years of age were also included in the 1100 subjects examined. Examination of a total of 3645 subjects revealed 28 patients, 14 deeply suspected subjects, and 19 suspected subjects in the endemic districts located along the Jintsu River. On the other hand, in Ota (Toyama City) located along another river in the endemic district, neither patients nor suspected patients were found among 334 females over 40 years of age examined in 1964. The examination was carried out in the same way as for the subjects from the endemic districts. Also in Tonami City and Nyuzen town, located along other rivers in Toyama Prefecture, neither patients nor suspected patients were observed. The participation rate for Ota was reported at about 65%. For the other areas it was not reported, but it may have been equally low. From these studies it was concluded that the Itai-itai disease was located only in specific districts of the Jintsu River basin. Further epidemiological research, therefore, was concentrated along this river, and it confirmed the occurrence of patients only in areas close to the river.

C. Epidemiological Research: 1967

In 1967, a larger, better structured epidemiological study was performed by the Toyama Prefecture Health Authorities in cooperation with Kanazawa University in an attempt to find and treat all Itai-itai patients (the data in this section are from Kato and Kawano[36] and Ishizaki, personal communication, unless otherwise stated).

1. Groups Studied and Selection of Subjects for Final Diagnostic Procedure

A total of 6717 subjects, all inhabitants over 30 years of age and of both sexes, of Fuchu town, certain parts of Toyama City, Osawano town, and Yatsuo town, were included in the first screening which included a questionnaire about symptoms and a semiquantitative determination of protein and glucose in urine. This screening was performed on 6114 persons, 91% of the 6717 subjects originally selected for the study. Based on the results of the first screening, 1911 persons were selected for the second examination. These were the people who had the greatest number of symptoms (more than ten) and/or had both proteinuria and glycosuria (for details see Friberg et al.[8]).

Out of the 1911 people selected for the second examination, 1400 underwent the examination, which included a medical interview, general clinical examination, and X-ray of the right upper arm and shoulder. As a result of the second examination, 451 persons were selected for a final examination. The exact basis for this selection has not been possible to elucidate, but among important criteria were bone changes.

Table 3
RESULTS FROM 1967 FINAL EXAMINATION[26]

District	Total	I	i	(i)	O_{ob}	O
	419	50	17	31	136	185
Fuchu town	227	28	11	14	79	95
Toyama City	105	16	3	8	34	44
Osawano town	53	6	3	6	16	22
Yao town	34			3	7	24
Sex						
M	122	1	1	3	37	80
F	297	49	16	28	93	105
Age						
30—44	39	—	—	—	2	37
45—59	86	6	—	2	28	50
60—74	225	40	10	22	86	67
75+	69	4	7	7	20	31
Mean age		66.6	72.2	68.3	66.1	58.0

Note: I = patient (typical bone signs on X-rays); i = deeply suspected (bone signs but not typical); (i) = suspected (bone signs, slight); O_{ob} = need for continuous observation (urinary and/or blood signs); O = no suspicion.

2. Final Diagnostic Examination (Third Examination)

Out of the 451 persons selected for a final examination, 419 were examined. The examination included a medical interview, X-ray of the pelvis, orthopedic examination, urine analysis for glucose, protein, calcium, phosphorus, creatinine, cadmium, lead, and zinc, as well a serum analysis of alkaline phosphatase and inorganic phosphorus. All of these analyses were not performed for all persons.

3. Number of Cases of Itai-itai Disease up until 1967

It is not possible to state a precise figure for the incidence of Itai-itai disease over the years. The 1967 epidemiological survey was cross-sectional, giving only prevalence data at the time of the study. It is estimated that nearly 100 deaths of Itai-itai disease patients occurred up until the end of 1965.[81] The following numbers of persons with Itai-itai disease were found: group "I" = 50; group "i" = 17; group "(i)" = 31; and group "O_{ob}" = 136 (see Table 3 for the distribution of examined persons according to place of residence, sex, and age). The Toyama Prefectural Health Authorities now continuously register all living patients. Data about subsequent cases are reported in Section VI.D.

4. Age and Sex

According to the 1967 epidemiological research, five male patients or suspected patients were found, as seen in Table 3. Among the observation subjects, about one third were men. Noting the age distribution from the 1967 report in Table 3, one can conclude that osteomalacia has been found only in women of 45 years of age or older. No patient was found in the age group below 44 years, but two observation cases were registered.

5. Incidence

By means of the 1967 epidemiological research and the subsequent registration of all patients by the Toyama Prefectural Health Authorities, a year of onset was estimated for each patient, as seen in Table 4. Great caution must be exercised in interpreting the

Table 4

NUMBER OF PATIENTS WITH ITAI-ITAI DISEASE ACCORDING TO
YEAR OF ONSET OF DISEASE ESTIMATED BY PATIENT'S
COMPLAINTS (PATIENTS REGISTERED BY TOYAMA
PREFECTURE, JUNE 15, 1968)[36]

| | | Starting year | | | | |
		Before 1934	1935—1944	1945—1954	1955—1964	1965—later	Not clear
Total	92	6	10	25	33	12	6
I[a]	51	4	7	16	16	5	3
i + (i)	41	2	3	9	17	7	3

[a]   For explanation of symbols for the Itai-Itai classification, see Table 3.

data. As seen in the table, a majority of the presently registered patients first complained of symptoms in the period between 1945 and 1964, but 10 to 15% did not complain of symptoms until 1965 or later. Many Itai-itai patients who contracted the disease some years earlier had died by 1968.[81]

6. Geographical Distribution of Itai-itai Disease

The locations of the different population centers within the Toyama Prefecture are seen on the maps in Figures 1 and 5.

In Toyama Prefecture cases of the disease have been found only in a limited area around the parts of the Jintsu River near Toyama City, where the river water is used for irrigation of rice fields. The prevalence of the Itai-itai disease in different parts of the endemic area has been best illustrated by the detailed survey of 1967, the results of which can be seen in Figure 6.

The 1967 research group also analyzed the cadmium concentration in the upper stratum of the soil in the paddy fields (see Section VI.F) and found a striking correlation with the prevalence of the Itai-Itai disease even within the endemic area (see Figures 5 and 6).

D. Subsequent Studies in the Toyama Area

Since 1967 annual studies of the cadmium-exposed population have been carried out by the Toyama Institute of Health. Each year different subgroups of the population have been studied. Suspected Itai-itai disease patients or observation cases have generally been studied each year. Other people, mainly in the age group above age 50, have been studied to the extent that resources permitted. The main finding in these studies has been a very high prevalence of renal tubular damage in the cadmium-exposed area (Chapter 9, Section V.B), but some cases of Itai-itai disease have also been diagnosed. Other cases with symptoms of the disease have applied directly to the Toyama Prefecture Itai-itai Disease Diagnostic Committee for recognition as sufferers of this disease.

Since 1967 a total of 132 cases have been recognized; 97 of these have died and 35 are still alive (Chapter 10, Table 12). Most of the deceased cases were never autopsied, but during recent years, greater efforts have been made to confirm the diagnosis by autopsy. Of the 14 recognized cases who died from 1979 to 1982,[35] 8 were autopsied, and of the 15 observation cases who died, 10 were autopsied. Table 12 in Chapter 10 shows that some cases are still being diagnosed. This is not likely to be due to recent cadmium exposure, but to the high exposure in the past.

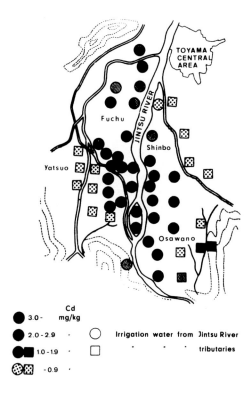

FIGURE 5.   Distribution of cadmium in paddy soil, surface layer.[36,81]

E. Hereditary Factors

Dent and Harris[4] have described a number of renal tubular reabsorption defects based on heredity which have given rise to osteomalacia. Therefore, it is of interest to seek evidence of similar hereditary factors for the Itai-itai disease. Ishizaki[24] reported that in many families in the endemic district, both the farmer's mother and his wife had contracted the disease. However, until 1969, no cases had occurred in women from the endemic district who had married and moved to a family in the nonendemic district. One woman was diagnosed in 1976 after having developed the disease symptoms several years after she left the Fuchu area and went to live in another part of Japan. All evidence indicates that the etiology rests on environmental rather than hereditary factors.

F. Dietary Factors

The classical form of osteomalacia is based on a deficiency of vitamin D. In the beginning of the research on the Itai-itai disease, dietary and climatological factors were assumed to play an important role in its etiology.

The Toyama Prefectural Health Authorities performed a survey on the dietary conditions of 200 households in the endemic area during 1955 to 1956. The results are given in Table 5. For comparison, averages for Sweden in 1960 are also given. It is evident that the dietary standard in the Itai-itai patients' homes and in the endemic area as a whole was not much different from the mean in the Toyama Prefecture or from the Japanese average. It is not known to what extent the figures for the different areas in Japan are directly comparable to the figures for Sweden. However, for certain dietary components the large differences noted are not likely to disappear, even if the

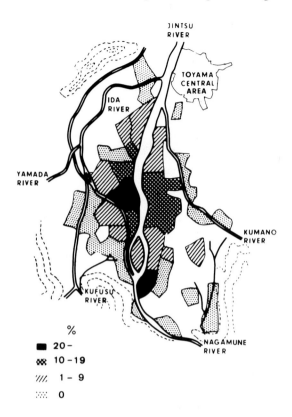

FIGURE 6.   Percentage of women over 50 years of age with Itai-itai disease [I, i, or (i)] at the examination of 1967.[26,36,81]

Table 5
NUTRITIONAL DATA FROM JAPAN AND SWEDEN

Foodstuff	Itai-Itai patients homes[a]	Kumano district[a]	Mean in rural district (Toyama Prefecture)	Mean in Japan (1955)	Mean in Sweden (1960)
Calorie (cal)	2139	2209	2237	2074	2868
Protein (g)	73	74	62	69	74
Fat (g)	14	17	17	21	124
Ca (mg)	408	374	331	364	938
P (mg)	1556	1432	1433	1822	1378
Fe (mg)	34	15	23	6	14
Vitamin A (IU)	1325	1754	1376	2814	3267
Vitamin B$_1$ (mg)	0.8	1.1	1.0	1.1	1.8
Vitamin B$_2$ (mg)	0.6	0.6	0.7	0.7	1.8
Vitamin C (mg)	69	93	64	66	75
Vitamin D (mg)					5.1

[a]   Total: 200 households.

Japanese data for 1955 and 1956 from Takase et al.[68] or Tsuchiya.[75] Swedish data for 1960 from Blix et al.[2]

figures are corrected for the different methods of calculation. As vitamin D is a known essential dietary factor in relation to osteomalacia, it is unfortunate that no Japanese figures for this vitamin were available. The vitamin A and fat intakes in the Itai-itai patients were lower than the Japanese average and considerably lower than the Swedish average. Therefore, it is probable that the vitamin D intake was also lower. If this is the case, extended exposure to sunlight could compensate for the low intake of vitamin D. However, since the endemic area has a rainy climate, and since the women dress in such a way that they screen off a great part of the sunshine, this type of compensation is not likely. A higher calcium intake could also partly compensate for a low level of vitamin D, but the calcium intake in Japan was very low.

The figures from the investigation of 1955 to 1956 in the homes of the Itai-itai patients represented a mean intake for all family members. Several investigations in various countries have shown that the calcium intake is usually much lower in women; therefore, it is likely that the women in the endemic area consume significantly less calcium than indicated by the figures. It must also be remembered that the nutritional situation in Japan during and immediately after the Second World War was poor. Insull and co-workers[21] documented acute generalized malnutrition problems arising in 1936 and 1945. These authors cited evidence showing that the heights of children during and directly after the war were significantly retarded.

G. Concentrations of Heavy Metals in the Environment: Daily Intake of Cadmium

As has been discussed, most of the Itai-itai patients displayed their symptoms many years ago. If the disease was caused by heavy metals, even exposure several decades earlier must have been of importance. However, no quantitative information on this type of exposure is available.

After the proposal of the theory that cadmium was a cause of the Itai-itai disease, the Jintsu River was suspected to be the carrier of the cadmium from the Kamioka mine to the endemic area. Kobayashi[38] stated that the increased production at the mine, together with faulty treatment of waste water, heavily polluted the Jintsu River during World War II. He explained that particles carried by the river were deposited in the rice fields. Damage to the rice crop prompted a severe dispute between the farmers and the owners of the mine. Studies on cadmium content in year rings from cedar trees also indicate a higher concentration of cadmium in the Jintsu River area several years ago.[29] On the other hand, Nitta[55] reported from geological studies a "remarkably high" natural cadmium content in the soil of the Jintsu River basin. He stated that the distribution of cadmium in soil fractions indicated that it had been sedimentous in this specific area for hundreds of years.

In 1959, Kobayashi[39] and Yoshioka[84] as stated in a report by Yamagata and Shigematsu,[81] analyzed samples of rice from the endemic area. Values (milligrams per kilogram in ash) in polished rice were given as 120 to 350 mg/kg in exposed areas, compared with 21 mg/kg in a control area. In the root of rice plants, corresponding figures were 690 to 1300 and 35 mg/kg, respectively. No information about the representativeness of the samples was reported. Moritsugi and Kobayashi[44] reported cadmium concentrations in over 200 samples of polished rice from throughout Japan. Analysis was performed with the dithizone colorimetric method described by Saltzman.[63] The average of 20 samples from the Itai-itai disease endemic area was 0.68 mg/kg in rice "kept in an air-dried condition at room temperature", while the average of samples from other parts of Japan was 0.066 mg/kg (wet weight).

In the samples of water taken upstream from the Kamioka mine and tributaries of the Jintsu River, cadmium was detected in trace amounts or not at all.[81] In water downstream from the mine a maximum value of 9 $\mu g/\ell$ was found. Out of four samples from the drainage from the mine, three contained 5 to 60 $\mu g/\ell$ (pH about 7 to 8)

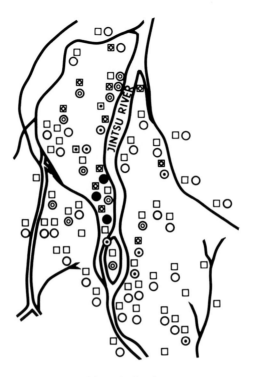

Polished rice	Unpolished rice	Cd
	● 2.00 –	mg/kg
▣	◉ 1.00 – 1.99	"
⊠	◎ 0.50 – 0.99	"
□	○ 0.49	"

FIGURE 7.   Concentration of cadmium in rice in the Fuchu area in 1967. (From Ishizaki, A. et al., *Annual Meeting of the Japanese Public Health Association,* Japanese Public Health Association, Kyoto, 1968.)

and one contained 4000 μg/ℓ (pH 2.8) of cadmium. Near the drainage area of the mine, cadmium concentrations of 363 and 382 mg/kg were found in suspended materials.

Samples of paddy soils irrigated with water from the Jintsu River and from its tributaries have been analyzed by Kato and Kawano[36] and Yamagata and Shigematsu.[81] The results, which revealed much higher concentrations in river-irrigated soil, are shown in Figure 5. Figure 6 shows the prevalence of positive and suspected cases of Itai-itai disease in different areas around the Jintsu River and its tributaries. Unfortunately, the maps cover only a limited area; comparable data from more distant districts would have been of great value. However, when Figures 5 and 6 are considered together, the association between high soil concentrations of cadmium and the Itai-itai disease is obvious. Ishizaki and co-workers[28] recorded values of over 2 mg/kg for cadmium concentration in rice from polluted districts, very seldom finding values under 0.5 mg/kg in such districts. The variation in cadmium concentration of individual rice samples (Figure 7) taken in close proximity to each other is larger than that of soil samples.

Based on the National Nutritional Survey of 1966 and available data on cadmium content in food, a representative value for the daily intake of cadmium in the Japanese diet at that time has been estimated as 60 µg.[42] A breakdown of the 60 µg indicated that 23 µg of cadmium would come from an intake of 335 g of polished rice with a concentration of 0.07 mg/kg (estimated from data by Moritsugi and Kobayashi[44].)The daily intake of cadmium via food in the endemic area was calculated at 600 µg by assuming an average cadmium concentration in rice of 1 mg/kg and a concentration in other foodstuffs produced and consumed locally of about ten times the value for Japan as a whole. An additional intake from river water used as drinking water has also been assumed. The cadmium content of river water as reported by Yamagata and Shigematsu[81] is too low to contribute significantly to the daily intake. In a calculation of daily intake in the endemic area, the Ministry of Health and Welfare[42] used an assumed water concentration during World War II of about 500 µg/ℓ and arrived at a cadmium intake from water of about 1 mg. No data have been reported in support of this high estimate, so the actual average long-term daily intake in the area is not known. It has been mentioned in other studies, though, and this is discussed in Chapter 10, Section VII.

The concentrations of lead in the soil are about 100 times higher than the concentrations of cadmium. On the other hand, Tsuchiya[75] reported average values for lead of about 0.2 mg/kg and for cadmium of about 0.8 mg/kg in rice samples from the Itai-itai disease endemic area. In the same report, he noted that the average urinary lead and zinc excretion was the same for Itai-itai patients, other persons in the endemic area, and persons in the control area. On the other hand, cadmium excretion was lower in persons in the control area than in those in the endemic area. No studies on lead exposure nor on the possible effects of interaction of lead and cadmium have been performed.

H. Later Screening Studies

After the declaration by the Japanese Ministry of Health and Welfare[41] that Itai-itai disease was a "pollution disease" caused by cadmium, a number of studies were initiated with the aim to locate other cadmium-polluted areas and to discover if possible whether Itai-itai disease occurred in any of these areas.

1. Methodology
a. Method for Selection of Areas Studied for Cadmium Pollution

The Ministry of Health and Welfare[42] published a standard method for selection of "areas requiring observation of cadmium pollution". If values of cadmium concentration in drinking or irrigation water higher than 10 µg/ℓ or values in unpolished rice higher than 0.4 mg/kg were found within a village, a closer investigation of environmental cadmium should be conducted. If the average daily cadmium intake in the area is calculated to be more than 300 µg, the village should be designated as an"area requiring observation". In most polluted districts a large number of villages are included in the "area requiring observation" (observation area) category. Decisions regarding the extent to which studies on health effects of cadmium should be carried out are made at the prefectural level. In one study (Ikuno, Chapter 9, Section V.B), the selection of target areas for health screening was performed by means of preliminary urine analysis on pooled samples, using 9 µg/ℓ as an upper limit. It should be pointed out that in some of the areas discussed in Chapter 9, Section V.B (e.g., Annaka), cadmium concentrations in rice above 0.4 mg/kg have been recorded outside the observation area. On the other hand, in large parts of the observation area in Bandai, cadmium concentrations in rice did not exceed 0.4 mg/kg, according to the published reports.

Control areas "regarded as free from any man-made cadmium contamination"[42] are chosen on the basis of their similarity to the observation area with regard to population structure, living conditions, climate, etc. As "man-made" cadmium does not include natural cadmium some areas have been used as "controls" in spite of reported average daily cadmium intakes far above the Japanese average.

b. Standard Methods for Screening of Cadmium-Related Disease

Two types of health screening were used between 1969 and 1972. The old type of health screening, used until 1971, was divided into two principal stages, the first of which was a general screening of the population over 30 years of age in defined areas. Included in the first screening were interview, proteinuria analysis (trichloracetic acid), glucosuria analysis (test tape), and blood pressure measurement.[42] All persons confirmed as having proteinuria in the first screening should have continued to the second. This rule was not strictly observed in all areas, which is obvious when going through the listings of the individual results.[31] For example, in Bandai, persons with glucosuria but not proteinuria were included in the target group for the second screening. Moreover, the participation rate in the second screening did not reach 100%, which means that there was a difference between the number with proteinuria and the number studied in the second screening. However, this difference is less than 10% in all the cases we were able to check. In the second screening a more detailed study on blood and urine was stipulated. The results of this screening would be used for evaluating whether a person had any effects due to cadmium exposure.

In May 1971, the Ministry of Health and Welfare published a new standard method for screening cadmium-related disease.[43] The target population was persons over 30 years of age. For qualitative proteinuria measurements, both the trichloracetic acid method and the sulfosalicylic acid method were recommended. The new standard method called for measurement of cadmium concentration in urine at the first screening, as well as qualitative proteinuria analysis and blood pressure measurement. Glucosuria measurement was moved to the second screening, where quantitative measurement of proteinuria was also to be made, by any quantitative method as long as it was clearly described. Disc electrophoresis of urinary proteins was to be part of the second screening. If the disc electrophoresis showed a tubular pattern, a third and final screening was performed. It included a number of blood and urine tests as well as renal function tests and X-ray of certain bones.

The study group for differential diagnosis of cadmium poisoning and Itai-itai disease, set up by the Japan Environment Agency, discusses the results of all persons undergoing the third screening and draws a final conclusion. This type of study group is also organized in each prefecture having an "area requiring observation". The conclusions of the national study group are mainly made on an individual basis with possible diagnoses being "Itai-itai disease", "cadmium poisoning", or "neither cadmium poisoning nor Itai-itai disease".

The diagnostic criteria for Itai-Itai disease used in Toyama Prefecture (Section III) have not been used in other areas. No definite criteria for Itai-itai disease or "cadmium poisoning" have been set, but the study group consults various clinical specialists and discusses the clinical and laboratory findings from case to case (Tsuchiya, personal communication in 1974). As no case of Itai-itai disease or "cadmium poisoning" has ever been diagnosed by this committee, as a result of the third screening (Section III), it is clear that neither cadmium exposure in combination with proteinuria, glucosuria, renal tubular reabsorption defects, severe osteomalacia, and osteoporosis nor multiple fractures, has been considered as "cadmium poisoning".

As noted previously, the screening procedures were designed to detect suspected cases of Itai-itai disease. As the screenings do not, primarily, aim at detecting the prevalence of proteinura, the methods of analyzing and classifying proteinura are not uniform from district to district and in some cases not even within districts. Comparisons of the different areas must be made with caution.

Standard methods for proteinuria analysis and daily intake measurement were initially developed (for details see Friberg et al.[8]), but these were later superseded by, for instance, measurement of low molecular weight proteinuria. The three-step approach for finding cases of Itai-itai disease has been maintained in the official studies. This makes it difficult to interpret prevalence data from any other than the first screenings. Some of these are of interest in analyzing the occurrence of renal effects and have been reviewed in some detail in Chapter 9. A revised method for health screenings of the populations of polluted areas has been used since 1976.[65] It is again based on a three-stage procedure. The findings of renal effects are discussed in Chapter 9, Section V.B.

2. Number of People Screened and Cases Found

The total number of people screened in the different studies is about 10,000 (Chapter 9, Table 12) and the total exposed population is estimated to be at least 100,000 (Chapter 3, Table 6). The official Diagnostic Committee only tentatively accepted one case from Tsushima (in 1979) as true Itai-itai disease, in spite of the fact that a number of cases have been reviewed which showed very severe renal tubular damage, increased serum alkaline phosphatase, severe bone changes, as well as high cadmium concentrations in the environment, urine, or autopsy tissues.

The details of the case from Tsushima were reported by Takebayashi.[69] The case was a woman, aged 75 at the time of autopsy, who had been screened for cadmium exposure since 1962. Since that time her urine cadmium concentration decreased progressively from 11.5 to about 3 to 4 $\mu g/\ell$ in 1970 to 1974 (Table 6). She had persistent proteinuria and glycosuria and also very high excretion of low molecular weight proteins (Table 6). The plasma alkaline phosphatase level increased with time (from 7 to 14 KA units in 1967 to 1971 up to 26.5 KA units in 1974), while at the same time plasma calcium was relatively low (8 to 9.5 mg/100 mℓ). While there was a decreased concentration of most amino acids in urine between 1970 and 1973, the concentration of hydroxyproline and proline increased (Table 7). Histopathological study at the autopsy showed severe renal tubular damage and a mixture of osteoporosis and osteomalacia in bone. After first considering this case as being the first accepted case of Itai-itai disease exposed outside Toyama Prefecture, the Committee changed its mind on the basis of a bone biopsy carried out 2 years before the case died (Tsuchiya, personal communication). This biopsy did not show osteomalacia according to the Committee and therefore it was decided that the case was not a "true" Itai-itai disease case. It is interesting to note that the data indicate that the woman's bone disease apparently deteriorated in spite of treatment.

Five cases with symptoms and findings very similar to those seen in Itai-itai disease were reported from the polluted Ikuno area of Hyogo Prefecture.[57] They were all living along a cadmium-polluted river (Chapter 10, Section V.E).

The first case was found on "the day of the aged" in 1972. She did not come to a public meeting for the aged, so the mayor of Kohdera town where she lived went to see her at home (Shibata, personal communication). She was bedridden and in severe pain. Her arms and legs were bent in the most extraordinary way. The mayor had taken an interest in cadmium poisoning and Itai-itai disease because the strawberry production in his town had been affected by the metal-polluted river water used for irrigation. He therefore called the local doctor (Dr. I. Shibata) and the investigations were started. It should be pointed out that screening studies for cadmium poisoning and Itai-itai disease carried out by the local health authorities had started in the Ikuno area in 1971. The patient, Mrs. A, was living outside the area officially designated as polluted and therefore not found in that study.

Table 6
LABORATORY DATA IN URINE FOR ONE CASE OF SUSPECTED ITAI-ITAI DISEASE
FROM TSUSHIMA AREA

	Age													
	62	63	64	65	66	67	68	69	70	71	72	73	74	75
Heavy metal in urine														
Cadmium (μg/ℓ)	11.5		7.1	7.2		4.2	3.0	2.2	3.8	1.9	3.0	3.0	4.0	
Zinc (μg/ℓ)						38.0	108.0	58.8	200	114	57	160	80	
Lead (μg/ℓ)											36.0	21.0	5.0	
Copper (μg/ℓ)										66.7	33.0	40.0	62.0	
Calcium (mg/ℓ)						75.0	60.0	36.3	40.0					
Proteinuria (mg/dℓ)			+	+	+	+	+	+	+	40	35	50	100	150
Glycosuria (mg/dℓ)			+	+	+	+	+	+	+	0.2	0.3	0.3	0.3	+
PSP (%)														
15 min				5			5	5—10	5	5		3.5		
Total				25			25	20—40	20			17.5		
Low molecular protein and amino acid in urine														
β_2-m-g (mg/dℓ)									7.0	3.5	5.5	3.8	3.1	
Lysozyme (mg/dℓ)										1.2	5.2	10.0	5.0	
RBP (mg/dℓ)									4.6	5.2	4.1	3.3	5.8	
Total amino acid (mM/dℓ)									13.3	8.9	16.9	11.2	8.8	
I TRP									46.6	36.6	23.5	12.6	25.1	
Fishberg condition				1010			1019	1018	1017			1015		

From Takebayashi, S., *Cadmium-Induced Osteopathy*, Shigematsu, I. and Nomiyama, K., Eds., Japan Public Health Association, Tokyo, 1980, 124—138. With permission.

Table 7

AMINO ACIDS IN URINE OF ONE CASE OF
SUSPECTED ITAI-ITAI DISEASE FROM
TSUSHIMA AREA

	Age			
	70	71	72	73
Arginine (nmol/mℓ)	1261	102	210	77
Taurine	1050	332	375	180
Histidine	827	171	340	35
Hydroxyproline	65	157	35	294
Lysine	1343	176	455	223
Threonine	972	453	770	466
Serine	635	315	480	221
Asparagine	565	229	505	124
Glutamine	5572	861	1785	Trace
Proline	429	363	375	666
Glycine	2161	694	1835	1025
Alanine	1996	630	1690	1139
Citrulline	430	121	—	134
Cystine	308	154	180	386
Total amino acid (nM/ℓ)	20,095	5,829	10,575	7,755

From Takebayashi, S., *Cadmium-Induced Osteopathy*, Shigematsu,
I. and Nomiyama, K., Eds., Japan Public Health Association, Tokyo, 1980, 124—138. With permission.

An autopsy was carried out on one of the cases and it was shown that the tissue
cadmium levels were as high as those found in Itai-itai disease patients in Toyama
(Chapter 10, Section V.E). None of the cases was recognized as Itai-itai disease by the
Diagnostic Committee on the basis of lack of unanimous diagnosis of osteomalacia.

Two highly suspected cases have been found in the Kakehashi river area.[22] These two
cases had all the signs of renal tubular damage and "Looser's zones" as seen on bone
X-rays (Chapter 10, Section V.E).

It is clear that even if a few cases are disputable, Itai-itai disease cannot be common
in the other polluted areas as so many people were screened and so few cases with
suspected bone changes were found.

VII. ETIOLOGY OF ITAI-ITAI DISEASE

Clinical data have shown that Itai-itai disease can be classified as a combined form
of osteomalacia and osteoporosis. Generally, osteomalacia can be divided into the following categories:

1. Vitamin D deficiency
2. Malabsorption of vitamin D and bone minerals
3. So-called vitamin D-resistant osteomalacia

Primary hyperparathyroidism may sometimes lead to osteomalacia. Underlying
causes of osteomalacia may also lead to secondary hyperparathyroidism. Differential
diagnosis may therefore be difficult, but primary hyperparathyroidism is a very rare
disease.

Occasionally, it has been difficult to differentiate between Itai-itai disease and hyperparathyroidism; this was discussed by Takase and co-workers.[68] In primary hyperparathyroidism the patients have an elevated blood calcium level not seen in cases of Itai-itai disease. However, a more or less decreased calcium level, seen in cases of osteomalacia, can stimulate increased parathyroid activity, so-called secondary hyperparathyroidism. The etiology of Itai-itai disease has been discussed by Takeuchi and Naito,[71] Takeuchi,[70] and Tsuchiya.[77]

A. Vitamin D Deficiency

This type of osteomalacia is dependent on the combination of lack or low intake of vitamin D in the diet and deprivation of ultraviolet irradiation. The combination together with a low intake of calcium and simultaneous high demand for calcium and vitamin D during pregnancy and lactation in some cases have given rise to osteomalacia even in countries with much sunshine, as documented by Groen and co-workers[12] (see also the review by Arnstein and co-workers[1]). In the areas where Itai-itai disease is found, the consumption of foodstuffs rich in calcium and vitamin D such as milk and milk products is very low. The calcium intake is considerably lower than in a country like Sweden, as seen in Table 5. However, the consumption is not lower than in other parts of Japan where Itai-itai disease is not found. As we pointed out before, the weather in this part of Japan is very gloomy with lots of rain and snow. The women also wear their clothes in such a manner that the main part of the sunlight is screened out.

Osteomalacia and rickets have also been seen more frequently in the Toyama Prefecture than in other parts of Japan.[33] However, available data (Sections I and VI) do not show agglomeration in any specific areas, e.g., Fuchu. Dietary conditions cannot be the sole etiological factors underlying the disease, because dietary conditions are similar in nearby villages and towns where the disease has not been found. However, the dietary habits (intake of calcium, vitamin D, and protein) may well be a contributing factor for causation of the disease.

Takeuchi[70] suggested that religious practices existing in the Fuchu area only, at the time of the occurrence of Itai-itai disease, would have made the women more susceptible to vitamin D deficiency. According to Takeuchi, at the beginning of this century among the people of Toyama Prefecture, rickets and osteomalacia were regarded as diseases induced by the wrath of God. The family confined the patient strictly within the house. Dietary deficiency and lack of sunlight exacerbated osteomalacia. Takeuchi claims that Fuchu area was the only part of Toyama Prefecture where this religious belief has persisted until quite recently. However, no data have been presented to support a close association between this practice and cadmium exposure.

B. Malabsorption Syndrome

Osteomalacia can also be caused by a deficient absorption of vitamin D and bone minerals. Chronic pancreatic disease, hepatobiliary disease, and resection of parts of the gastrointestinal tract have given rise to this type of osteomalacia, as reviewed by Arnstein and co-workers,[1] Boström,[3] and Muldowney.[46] In this context the clinical observations reported by Murata and co-workers concerning the function of the gastrointestinal tract and pancreas are of interest. They reported a tendency toward a declined function of the pancreas in many cases. Furthermore, examinations of the gastrointestinal tract showed shortened ciliated epithelia as well as atrophy of the mucous membrane and submucosal cell infiltration in the small intestine. Fat absorption tests showed decreased fat absorption in many cases. These changes in the gastrointestinal tract have been called "cadmium enteropathy" by Murata and co-workers.[50]

It has been impossible to judge the prevalence of these changes. However, in cases in which changes are present, they could act as a contributory etiological factor towards Itai-itai disease.

C. So-Called Vitamin D-Resistant Osteomalacia (Renal Osteomalacia)

This form of osteomalacia is caused by a renal tubular dysfunction which gives rise to damage to vitamin D metabolism in the kidney (Chapter 10, Section II.B) and to losses of bone minerals, primarily phosphate, through the kidneys. There is also an increased excretion of amino acids and glucose. Both hereditary[4] and acquired forms[5] have been described. There does not seem to be any clear difference between what is termed "acquired" and "hereditary", and they can be spoken of collectively as a form of osteomalacia, with which the clinical findings in Itai-itai disease are in accord. The mechanism for this type of osteomalacia in cadmium poisoning has become clarified in the last 10 years (Chapter 10). The renal damage caused by cadmium leads eventually to a decreased production of active vitamin D metabolites (Chapter 10, Section III.A). Available data on treatment (Section V) also support the conclusion that Itai-itai disease can be classified as a vitamin D-resistant form of osteomalacia.

D. Osteoporosis

Osteoporosis is a disease where the hard bone tissue is replaced by fat and fibrous tissue. The mass of trabecular bone is decreased, but the mineralization of the remaining bone mass is not affected. It leads to weaker bone structure but if fractures do not occur, there are no specific symptoms of pain from the bones. Osteoporosis is facilitated by a low dietary calcium and protein intake, by the decrease of blood estrogen levels at menopause in women, and by frequent pregnancies. It is therefore likely that osteoporosis was a fairly common disease among women of older age in Toyama Prefecture in the 1940s and 1950s. However, the very high prevalence of osteoporosis among the Itai-itai disease patients (see Section II.C and II.D) is likely to be higher than in the whole female population of the area. Osteoporosis in Itai-itai disease can either be an underlying disease, which increases the risk that the osteomalacia-causing factors lead to clinical osteomalacia, or it can be a disease caused by the same factors that cause osteomalacia, including cadmium (Chapter 10, Section III.B).

The nutritional conditions of the women in the afflicted area certainly increased the risk of both osteoporosis and osteomalacia. Over the period of time of the outbreak of Itai-itai disease, there were no reports of rickets being common among children, so a low vitamin D level in the diet is not likely to have been a sufficient factor to cause osteomalacia. In view of the high cadmium exposure and the animal experiments showing both osteoporosis and osteomalacia caused by cadmium (see Chapter 10, Section III), the cadmium exposure in the area may also contribute to a combined effect of osteoporosis and osteomalacia.

E. Osteolathyrism

This disease, which has been induced in animals (see Chapter 10, Section III.C) by lysyl oxidase inhibitors, has never been reported to occur in human beings. The exact clinical picture is, therefore, unknown. The main pathological findings would occur in cartilage tissue rather than in bone tissue, and it appears that no pathological studies of Itai-itai disease patients have included detailed studies of cartilage. The complex mechanism for causation of Itai-itai disease may very well include lysyl oxidase inhibition (Chapter 10, Section VI).

F. Cadmium as an Etiological Factor

It has been firmly established (Chapter 9, Section IV.B.2) that exposure to cadmium can give rise to tubular damage of the kidneys with proteinuria and glucosuria. The kidney damage seen in Itai-itai disease has been very similar to that seen in classical industrial chronic cadmium poisoning. Bone changes have not been a common finding in either condition. There are, however, reports from France, the U.K., and the Soviet Union which taken together show that both in male and female workers occupationally exposed to cadmium, bone changes similar to those seen in Itai-itai disease have occurred (Chapter 10, Section IV).

Data from industrial exposure refer mostly to males, while almost all of the Itai-itai patients have been females. It should be stressed, however, that in the endemic area the prevalence of proteinuria and glucosuria has been extremely high in males also and only slightly lower than in females. Proteinuria and glucosuria were already common findings among patients before vitamin D treatment was started (Section II.B.2). When tubular damage, directly or indirectly, disturbs the metabolism of calcium and phosphorus, women will be affected more than men. This was particularly the case in the women involved who had given birth to several children and who lived in an area with a low intake of calcium and probably also of vitamin D.

Although no data exist concerning cadmium exposure several years ago, when it was supposedly highest, recent information on cadmium concentrations in rice and paddy fields has shown a close correlation between high cadmium concentrations and the occurrence of Itai-itai disease. Furthermore, a causal relationship between cadmium and renal tubular damage in polluted areas of Japan has been shown conclusively (Chapter 9, Section IV.B.2). Questions have been raised as to whether or not cadmium exposure has been high enough to cause the bone changes in Itai-itai disease and as to why such effects have been so rare in the polluted areas other than Fuchu. First of all, approximately ten cases with symptoms and signs very similar to Itai-itai disease have been seen (Chapter 10, Section V.E) which have not received an official diagnosis. The diagnostic criteria for Itai-itai disease used in the Fuchu area have not been employed in other areas (Section VI.H). Further, all data favor a higher exposure in Fuchu than in other polluted areas, particularly if the possibility of a still higher exposure in the past is taken into consideration. Looking at the few autopsy data on hand, it seems that exposure has been as high as that of workers with cadmium intoxication. In Fuchu, liver cadmium values have been found to be high in persons who have low kidney values at the same time, indicating pronounced kidney damage (Chapter 9, Section VI.B.2).

Added to what has been said so far, Toyama Prefecture has apparently long been considered an area in which rickets may occur, possibly due to lack of sunlight brought about by the meteorological and other conditions discussed in Section I. No doubt the population in the Fuchu area was particularly susceptible to the development of osteoporosis and osteomalacia, but considering the close epidemiological association between cadmium-exposure and Itai-itai disease, it is very unlikely that such a concentration of this bone disease would have occurred in the absence of a high cadmium exposure.

VIII. CONCLUSIONS

It has been shown conclusively that cadmium exposure in polluted areas of Japan has caused widespread renal tubular dysfunction which is manifested above all in a high prevalence of low molecular weight proteinuria.

The Itai-itai disease is a bone disease, characterized by the finding of a combination of osteoporosis, osteomalacia, and renal tubular dysfunction, which is also an expression of chronic cadmium poisoning. It was particularly prevalent in the Japanese area with the highest level of cadmium exposure, but it has also been sporadically found in other areas, and some new cases are still being discovered. We have concluded that a high cadmium dose is a necessary factor in the causation of the disease. There is reason to believe, however, that inadequate consumption of certain essential food elements and vitamins has been a contributing factor towards the occurrence of Itai-itai disease.

REFERENCES

1. Arnstein, A. R., Frame, B., and Frost, H. M., Recent progress in osteomalacia and rickets, *Ann. Intern. Med.*, 67, 1296—1330, 1967.
2. Blix, G., Wretlind, A., Bergström, S., and Westin, S. I., The food intake of the Swedish people, *Vår. Föda*, 17, 1—23, 1965 (in Swedish).
3. Boström, H., Osteomalacia, *Läkartidningen*, 64, 4679, 1967 (in Swedish).
4. Dent, C. E. and Harris, H., Hereditary forms of rickets and osteomalacia, *J. Bone Jt. Surg. (Br.)*, 38, 204—226, 1956.
5. deSeze, S., Lichtwitz, A., Hioco, D., Bordier, P., and Miravet, L., Hypophosphataemic osteomalacia in the adult with defective renal tubular function, *Ann. Rheum. Dis.*, 23, 33—44, 1964.
6. Evrin, P. -E. and Wibell, L., The serum levels and urinary excretion of β_2-microglobulin in apparently healthy subjects, *Scand. J. Clin. Lab. Invest.*, 29, 69—74, 1972.
7. Evrin, P. -E., Peterson, P. A., Wide, L., and Berggård, I., Radioimmunoassay of β_2-microglobulin in human biological fluids, *Scand. J. Clin. Lab. Invest.*, 28, 440—443, 1971.
8. Friberg, L., Piscator, M., Nordberg, G. F., and Kjellström, T., *Cadmium in the Environment*, 2nd ed., CRC Press, Boca Raton, Fla., 1974.
9. Fukushima, M. and Sugita, Y., Some urinary findings on Itai-itai diseased patients (2nd report). On proteins excreted in the urine from the patients, *Jpn. J. Public Health*, 17, 175, 1970 (in Japanese).
10. Fukushima, M., Kobayashi, S., and Sakamoto, R., Urinary free amino acids in Itai-itai patients and among inhabitants in a cadmium polluted area, in *Kankyo Hoken Report No. 24*, Japanese Public Health Association, Tokyo, September 1973, 53—57 (in Japanese).
11. Fukuyama, Y., Proteinuria in Itai-itai disease and cadmium poisoning, *Med. Biol.*, 84, 41—46, 1972 (in Japanese).
12. Groen, J. J., Eshchar, J., Ben-Ishay, D., Alkan, W. J., and Ben Assa, B. I., Osteomalacia among the Bedouin of the Negev Desert, *Arch. Intern. Med.*, 116, 195—204, 1965.
13. Hagino, N., About investigations on Itai-itai disease, *J. Toyama Med. Assoc.*, Dec. 21, 7, 1957 (in Japanese).
14. Hagino, N., Itai-itai disease, *Medicina*, 5, 99—102, 1968a (in Japanese).
15. Hagino, N., Itai-itai disease, *Accident Med.*, 11, 1390—1393, 1968b (in Japanese).
16. Hagino, N., Cd poisoning symptoms, *Gen. Clin.*, 18, 1366—1370, 1969 (in Japanese).
17. Hagino, N., Itai-itai disease and vitamin D, *Dig. Sci. Labour*, 28, 32—46, 1973 (in Japanese).
18. Hagino, N. and Kono, M., Itai-itai disease, Proc. 17th Mtg. Japanese Soc. of Clinical Surgeons, 1955.
19. Hagino, N. and Yoshioka, K., A study on the cause of Itai-itai disease, *J. Jpn. Orthoped. Assoc.*, 35, 812—815, 1961 (in Japanese).
20. Harada, A., Findings on urinary protein electrophoresis, in *Kankyo Hoken Report No. 11*, Japanese Public Health Association, April 1972, 87—98 (in Japanese).
21. Insull, W., Jr., Oiso, T., and Tsuchiya, K., Diet and nutritional status of Japanese, *Am. J. Clin. Nutr.*, 21, 753—777, 1968.
22. Ishikawa Prefecture, *White paper on the Environment of Ishikawa Prefecture*, Annual Report 1981, Ishikawa Prefecture Pollution Bureau, Kanazawa, 182, 1981, (in Japanese).
23. Ishizaki, A., Observations on the influence on the body of cadmium in the food, *Clin. Nutr.*, 35, 28, 1969a (in Japanese).
24. Ishizaki, A., On the so-called Itai-itai disease, *J. Jpn. Med. Soc., (Nihon Ishikai Zasshi)*, 62, 242—248, 1969b (in Japanese).
25. Ishizaki, A., About the Cd and Zn concentrations in organs of Itai-itai disease patients and in inhabitants of Hokuriku area, in *Kankyo Hoken Report No. 11*, Japanese Public Health Association, April 1972, 154 (in Japanese).
26. Ishizaki, A. and Fukushima, M., Studies on "Itai-itai" disease (review), *Jpn. J. Hyg.*, 23, 271—285, 1968 (in Japanese).
27. Ishizaki, A., Nomura, K., Tanabe, S., and Sakamoto, M., Observations on urinary and fecal excretion of heavy metals (Cd, Pb and Zn) in the patients of the so-called "Itai-itai" disease, *Jpn. J. Hyg.*, 20, 261—267, 1965 (in Japanese).

28. Ishizaki, A., Fukushima, M., and Sakaomoto, M., Cadmium content of rice eaten in the Itai-itai disease area, in *Annual Meeting of the Japanese Public Health Association,* Japanese Public Health Association, Kyoto, 1968 (in Japanese).

29. Ishizaki, A., Fukushima, M., Kurachi, T., Sakamoto, M., and Hayashi, E., The relationship between Cd and Zn contents and the year rings of the Sugi tree in the basin of the Jintzu river, *Jpn. J. Hyg.,* 25, 376, 1970a (in Japanese).

30. Ishizaki, A., Fukushima, M., and Sakamoto, M., On the accumulation of cadmium in the bodies of Itai-itai disease patients, *Jpn. J. Hyg.,* 25, 86, 1970b (in Japanese).

31. Japanese Public Health Association, *Research about Differential Diagnosis of Itai-Itai Disease and Cadmium Poisoning,* Japanese Public Health Association, Tokyo, 1970 (in Japanese).

32. Kajikawa, K., Pathological studies on renal lesions of Itai-itai patients, in *Kankyo Hoken Report No. 24,* Japanese Public Health Association, September 1973, 29—32 (in Japanese).

33. Kajikawa, K., Okuno, S., Igawa, K., and Hirono, R., A bone disease which occurred in the Toyama Prefecture, so-called "Itai-itai byo" (painful disease), *Trans. Soc. Pathol. Jpn.,* 46, 655—662, 1957 (in Japanese with English summary).

34. Kajikawa, K., Kitagawa, M., Nakanishi, I., Ueshima, H., Katsuda, S., and Kuroda, K., A pathological study of Itai-itai disease, *J. Juzen Med. Soc.,* 1974 (one issue per year; in Japanese with English summary).

35. Kato, T., A review of autopsy data on Itai-itai disease patients, in *Kankyo Hoken Report No. 49,* Japanese Public Health Association, Tokyo, 1983, 121—124 (in Japanese).

36. Kato, T. and Kawano, S., Review of past and present of Itai-itai disease. On the process of research development, *Curr. Med.,* 16, 29, 1968 (in Japanese).

37. Kobayashi, J., Agricultural damage in the Jintsu river basin caused by the Kamioka Mine, a report to the Ministry of Agriculture, July 1943 (in Japanese).

38. Kobayashi, J., *Mizu No Kennoo Shindan* (Health Examination of Water), Iwanami Press Ltd., Tokyo, 1971 (in Japanese).

39. Kobayashi, J., see Yamagata, N. and Shigematsu, I., Cadmium pollution in perspective, *Bull. Inst. Publ. Health (Tokyo),* 19, 1970.

40. Kono, M., Fukuzawa, K., Yamagiri, K., and Fujii, A., On progress in the treatment of the so-called Itai-itai disease, *J. Orthop. Soc. Jpn.,* 32, 61—63, 1958, (in Japanese; translation by Seizaburo Aoki, Japanese Language Translation Service, Fujisawa, Japan).

41. Ministry of Health and Welfare, Opinion of the Welfare Ministry with Regard to "Ouch-Ouch" Disease and Its Causes, May 8, 1968 (in Japanese).

42. Ministry of Health and Welfare, The Opinion of the Ministry of Health and Welfare as Regards Environmental Pollution by Cadmium and Countermeasures in the Future, March 27, 1969 (in Japanese).

43. Ministry of Health and Welfare, Method of Health Examination, A Part of Provisional Countermeasures Against Environmental Pollution by Cadmium, May 19, 1971 (in Japanese).

44. Moritsugi, M. and Kobayashi, J., Study on trace metals in biomaterials. II. Cadmium content in polished rice, *Ber. Ohara Inst. Landwirtsch. Biol. Okayama Univer.,* 12, 145—158, 1964 (in Japanese).

45. Mukawa, A., Nogawa, K., and Hagino, N., Bone biopsy performed on women living on the cadmium-polluted Jinzu river basin, *Jpn. J. Hyg.,* 35, 761—773, 1980.

46. Muldowney, F. P., Metabolic bone disease secondary to renal and intestinal disorders, *Calif. Med.,* 110, 397—409, 1969.

47. Murata, I., Chronic entero-osteo-nephropathy cadmium, *J. Jpn. Med. Assoc.,* 65, 15—42, 1971 (in Japanese).

48. Murata, I., Nakagawa, S., and Yoshimoto, A., A study of 50 patients with osteomalacia and tubular absorption damage, *J. Jpn. Orthoped. Surg. Soc.,* 32, 726—727, 1958 (in Japanese).

49. Murata, I., Hirono, T., Saeki, Y., and Nakagawa, S., Cadmium enteropathy, renal osteomalacia ("Itai-itai" disease in Japan), *Proc. 12th Int. Congr. Radiol.,* Tokyo, October 1969, 34—42.

50. Murata, I., Hirono, T., Saeki, Y., and Nakagawa, S., Cadmium enteropathy, renal osteomalacia ("Itai-itai" disease in Japan), *Bull. Soc. Int. Chir.,* 1, 34—41, 1970.

51. Murata, I., Nakagawa, S., and Hirono, T., Clinical progress of Itai-itai disease, in *Kankyo Hoken Report No. 11,* Japanese Public Health Association, April 1972, 132—139 (in Japanese).

52. Nagasawa, T., Nagasawa, S., Kawada, Y., Horiguchi, Y., Nomura, T., and Otsuka, R., The extensive occurrence of the rheumatic disease in some villages along the Jintsu River in the Toyama Prefecture, *J. Juzen Med. Soc.,* 50, 232—236, 1947; cited in Kato, T. and Kawano, S., *Curr. Med.,* 16, 29, 1968 (in Japanese).

53. Nakagawa, S., A study of osteomalacia in Toyama Prefecture (so-called Itai-itai disease), *J. Radiol. Phys. Ther. Univ. Kanizawa,* 56, 1—51, 1960 (in Japanese with English summary).

54. Nakagawa, S. and Furumoto, S., About osteomalacia poisoning occurring in Toyama Prefecture, *J. Toyama Med. Assoc.,* Dec. 21, 7, 1957 (in Japanese).

55. Nitta, T., Geochemical study on heavy metals in the Jintsu River area; especially cadmium, *J. Soc. Min. Geol. Jpn.,* 22, 191—204, 1972 (in Japanese).

56. Nogawa, K. and Kawano, S., A survey of the blood pressure of women suspected of Itai-Itai disease, *J. Juzen Med. Soc.,* 77, 357, 1969 (in Japanese).

57. Nogawa, K., Ishizaki, A., Fukushima, M., Shibata, I., and Hagino, N., Studies on the women with acquired Fanconi syndrome observed in the Ichi River basin polluted by cadmium, *Environ. Res.,* 10, 280—307, 1975.

58. Ohsawa, M. and Kimura, M., Isolation of β_2-microglobulin from the urine of patients with Itai-itai disease, *Experientia,* 29, 556—558, 1973.

59. Piscator, M., Proteinuria in chronic cadmium poisoning. III. Electrophoretic and immunoelectrophoretic studies on urinary proteins from cadmium workers, with special reference to the excretion of low molecular weight proteins. *Arch. Environ. Health,* 12, 335—344, 1966a.

60. Piscator, M., Proteinuria in chronic cadmium poisoning. IV. Gel filtration and ion-exchange chromatography of urinary proteins from cadmium workers, *Arch. Environ. Health,* 12, 345—356, 1966b.

61. Piscator, M., *Proteinuria in Chronic Cadmium Poisoning,* Beckman's, Stockholm, 1966c.

62. Piscator, M. and Tsuchiya, K., Eds., in *Cadmium in the Environment,* Friberg, L., Piscator, M., and Nordberg, G. F., Eds., Chemical Rubber Co., CRC Press, Cleveland, 1971, 113.

63. Saltzman, B. E., Colorimetric microdetermination of cadmium with dithizone, *Anal. Chem.,* 25, 493—496, 1953.

64. Sano, S., Iguchi, H., and Kawanishi, S., Urinary protein of Itai-itai disease patients, in *Kankyo Hoken Report No. 24,* Japanese Public Health Association, September 1973, 91 (in Japanese).

65. Shigematsu, I., Minowa, M., Yoshida, T., and Miyamoto, K., Recent results of health examinations on the general population in cadmium polluted and control areas of Japan, *Environ. Health Perspect.,* 28, 205—210, 1979.

66. Shiroishi, K., Kjellström, T., Kubota, K., Evrin P.-E., Anayama, M., Vesterberg, O., Shimada, T., Piscator, M., Iwata, I., and Nishino, H., Urine analysis for detection of cadmium-induced renal changes, with special reference to β_2-microglobulin, *Environ. Res.,* 13, 407—424, 1977.

67. Taga, I., Murata, I., Nakagawa, S., Furumoto, S., and Hagino, N., About a common disease in Kumano village of Toyama Prefecture, *J. Jpn. Orthoped. Assoc.,* 30, 33—35, 1956 (in Japanese).

68. Takase, T., Shigematsu, I., Mizumoto, K., Akasu, F., Ishizaki, A., Nomura, S., Sawada, T., Hisada, K., Kajikawa, K., Kawano, S., Waseda, M., Kobayashi, I., and Hiramatsu, H., On the pathogenesis of so-called "Itai-itai" disease patients in Toyama Prefecture, *Jpn. J. Clin. Med.,* 25, 200—219, 1967.

69. Takebayashi, S., First autopsy case, suspicious of cadmium intoxication, from the cadmium-polluted area in Tsushima, Nagasaki Prefecture, in *Cadmium-Induced Osteopathy,* Shigematsu, I. and Nomiyama, K., Eds., Japan Public Health Association, Tokyo, 1980, 124—138.

70. Takeuchi, J., The etiology of Itai-itai disease, *Jpn. J. Clin. Med.,* 31, 132—141, 1973 (in Japanese).

71. Takeuchi, J. and Naito, P., About the etiology of Itai-itai disease, *Strides Med.,* 80, 609—616, 1972 (in Japanese).

72. Takeuchi, J., Shinoda, A., Kobayashi, K., Nakamoto, Y., Takazawa, I., and Kurosake, M., Renal involvement in Itai-itai disease, *Intern. Med.,* 21, 876, 1968 (in Japanese).

73. Toyoda, B., Wada, K., Takahashi, M., Igawa, K., and Saino, S., On the so-called "Itai-itai byo", a kind of osteomalacia, *J. Jpn. Assoc. Rural Med.,* 6, 55, 1957.

74. Toyoshima, I., Seino, A., and Tsuchiya, K., Urinary amino acids in cadmium workers, in inhabitants of a cadmium-polluted area, and in Itai-itai disease patients, in *Kankyo Hoken Report No. 24,* Japanese Public Health Association, September 1973, 65—71 (in Japanese).

75. Tsuchiya, K., Causation of ouch-ouch disease, an introductory review. I. Nature of the disease, *Keio J. Med.,* 18, 181—194, 1969.

76. Tsuchiya, K., Results of long-time observation of cadmium workers, in *Kankyo Hoken Report No. 11,* Japanese Public Health Association, April 1972, 72—76 (in Japanese).

77. Tsuchiya, K., Ed., *Cadmium Studies in Japan — A Review,* Elsevier, Amsterdam, 1978.

78. Vesterberg, O., Isoelectric focusing of proteins in polyacrylamide gels, *Biochim. Biophys. Acta.,* 257, 11—19, 1972.

79. Vesterberg, O. and Nise, G., Urinary proteins studied by use of isoelectric focusing. I. Tubular malfunction in association with exposure to cadmium, *Clin. Chem.,* 19, 1179—1183, 1973.

80. Watanabe, H., in print, 1985.

81. Yamagata, N. and Shigematsu, I., Cadmium pollution in perspective, *Bull. Inst. Public Health (Tokyo),* 19, 1—27, 1970.

82. Yoshiki, S., A simple histological method for identification of osteoid matrix in decalcified bone, *Stain Technol.,* 48, 233—238, 1973.

83. Yoshioka, K., *Study of Itai-Itai Disease, from Mineral Hazard on Agricultural to Mineral Disease of Human Being (Itai-Itai Disease)*, Tatara Shobun Publ. Co., Yonago, Japan, 1970 (in Japanese with English summary).

84. Yoshioka, K., see Yamagata, N. and Shigematsu, I., Cadmium pollution in Perspective, *Bull. Inst. Publ. Health (Toyko)*, 19, 1970.

INDEX

A

Abnormal β_2-microglobulin excretion, defined, 80
Absorption
    after ingestion and inhalation, differences in, 248
    gastrointestinal, cadmium, see Gastrointestinal
        cadmium absorption
    intestinal calcium, see Intestinal calcium
        absorption
Absorption rates, 234, 248
Acceptable daily intake, 241
N-Acetyl-β-$_D$-glucosaminidase, 55
Acid phosphatase, 27, 139
Acquired renal osteomalacia, 285
Acute oral LD_{50}, cadmium compounds, 161
Acute toxicity of cadmium
    blood pressure and cardiovascular effects studies,
        169—170
    carcinogenic and mutagenic effects studies, 206—
        208, 222—226
    dose-effect relationship, 233
    embryotoxic and teratogenic effects studies,
        185—190, 252
    endocrine organ effects studies, 180—181, 192—
        193
    gastrointestinal effects studies, 161
    hematopoietic effects studies, 167
    immunological system effects studies, 192
    liver effects studies, 161—163
    nervous system effects studies, 170, 190—192
    renal effects studies, 25—29, 98, 233—234
    reproductive organ effects studies, 160, 170,
        179—184, 252
    respiratory effects studies, 2—5, 16
Adenocarcinomas, 209—210
Adenomas, 209—211
Adenyl cyclase, 114, 117—119, 152, 164, 166, 182
ADI, see Acceptable daily intake
Age-adjusted incidence, cancer, 219
Age-adjusted mortality rates, cancer, 220—221
Age-related effects
    bone effects studies, 128, 145, 147, 153
    embryotoxic and teratogenic effects studies, 185
    Itai-itai disease studies, 258—259, 261, 266,
        273, 276
    nervous system effects studies, 191—192
    proteinuria, epidemiological studies, 80—86
    renal effects studies, 22, 42, 51—52, 55—59,
        70, 80—87, 89, 235, 238, 241
    reproductive organ effects studies, 185
Age-standardized mortality rate ratios, 220—222
Air concentrations
    general environment, 78—79, 178, 248
    workplace, see Workplace air concentrations
Airflow obstructive syndrome, 8, 10—11, 14—16
Alanine aminotransferase, 163, 165—166
ALAT, see Alanine aminotransferase
Albumin

absolute excretable amounts, 23—25, 60—63
    Itai-itai disease studies, 260—261
    β_2-microglobulin related to, 23—25, 60—64
    renal effects studies, 23—25, 29, 33, 48, 54,
        59—66, 68, 75, 250
    retinol-binding protein and, 62
    tubular reabsorption of, 23—25
    urinary excretion of, 23—25, 60—64, 250
Albuminuria, 48, 97
Alkaline phosphatase
    bone effects studies, 112—113, 120, 126, 136,
        139, 142, 144, 146—147, 149—151, 254
    decreased activity in femur, 139
    embryotoxic and teratogenic effects, 189
    Itai-itai disease studies, 146—147, 149—151,
        260, 263, 266, 271, 273, 281
    liver effects studies, 163—166
    renal effects studies, 38
    respiratory effects studies, 3
    serum concentrations, 112—113, 120, 126, 136,
        142, 146, 149, 151, 260, 263, 266, 271,
        273, 281
Alkaline phosphataseuria, 37
Alkaline phosphate, 37
Allergy, contact, 193—194
Alpha$_1$-antitrypsin, 15—16
1-Alpha-OH vitamin D3, 121, 135—136
Alveolar cells, damage to, 3, 6, 16
Amino acids
    tubular reabsorption of, 26—27
    urinary concentrations, Itai-itai disease, 281—283
Aminoaciduria
    bone effects studies, 142, 146
    Itai-itai disease studies, 146, 260
    renal effects studies, 35—37, 45, 49, 51—52,
        55, 58, 72—73, 90—92, 98
β-Amino-propio-nitrile, 115
cAMP, see Cyclic AMP
AMRR, see Age-standardized mortality rate ratios
Analysis, cadmium, methods, accuracy of, 248—
    249
Anemia, 167—169
Aniline hydroxylase, 162
Animal studies, see also specific animals by name
    blood pressure and cardiovascular effects, 170—
        176, 179, 251—252
    bone effects, 112, 115—139, 151—153, 250—
        251
    carcinogenic and mutagenic effects, 206—211,
        222—226, 251
    embryotoxic and teratogenic effects, 185—190,
        252
    endocrine organ effects, 180—181, 192—193
    gastrointestinal effects, 161
    hematopoietic effects, 167—169
    immunological system effects, 192
    liver effects, 161—166
    lung cancer, 209—211

S